WAVES AND OSCILLATIONS

BY THE SAME AUTHORS :

A TEXT BOOK OF SOUND

WAVES AND OSCILLATIONS

[For B.Sc. (Pass, Honours and Subsidiary), Pre-Medical, Engineering Students of Indian Universities and Students appearing for IAS and other Competitive Examinations]

N. SUBRAHMANYAM
Department of Physics,
Kirori Mal College,
University of Delhi

BRIJ LAL
Reader in Physics,
Hindu College,
University of Delhi

SECOND REVISED EDITION

VIKAS PUBLISHING HOUSE PVT LTD

VIKAS® PUBLISHING HOUSE PRIVATE LIMITED
E-28, Sector-8, Noida-201301 (UP), India
Phone: +91-120-4078900 • Fax: +91-120-4078999
Registered Office: A-27, 2nd Floor, Mohan Co-operative Industrial Estate, New Delhi-110044
E-mail: helpline@vikaspublishing.com • Website: www.vikaspublishing.com

SALES AND BRANCH OFFICES OF VIKAS® AND S CHAND AND COMPANY LIMITED

Chennai : Ph: 044-2363 2120, chennai@schandpublishing.com
Guwahati : Ph: 0361-2738 811, 2735 640, guwahati@schandpublishing.com
Hyderabad : Ph: 040-4018 6018, hyderabad@schandpublishing.com
Jalandhar : Ph: 0181-4645 630, jalandhar@schandpublishing.com
Kolkata : Ph: 033-2335 7458, 2335 3914, kolkata@schandpublishing.com
Lucknow : Ph: 0522-4003 633, lucknow@schandpublishing.com
Mumbai : Ph: 022-2500 0297, mumbai@schandpublishing.com
Patna : Ph: 0612-2260 011, patna@schandpublishing.com

All rights reserved. No part of this publication may be reproduced or copied in any material form (including photo copying or storing it in any medium in form of graphics, electronic or mechanical means and whether or not transient or incidental to some other use of this publication) without written permission of the copyright owner. Any breach of this will entail legal action and prosecution without further notice.

***Jurisdiction**: All disputes with respect to this publication shall be subject to the jurisdiction of the Courts, tribunals and forums of New Delhi, India only.*

Waves and Oscillations
ISBN: 978-07069-8543-6

Product Code: V6WOS68PHYS10ENAB94O

Second Revised Edition 1994
Reprinted several times between 1994 and 2023, 2024 (Twice)
Reprint 2025 (Twice)

VIKAS® is the registered trademark of Vikas Publishing House Private Limited
Copyright © Vikas Publishing House Private Limited, 1974, 1994

PRINTED IN INDIA
By Vikas Publishing House Private Limited, Plot 20/4, Site-IV, Industrial Area Sahibabad, Ghaziabad-201010 and Published by Vikas Publishing House Private Limited, E-28, Sector-8, Noida-201301

Disclaimer: The publishers have taken all care to ensure highest standard of quality as regards typesetting, proofreading, accuracy of textual material, printing and binding. However, neither they nor the author accept responsibility for any loss occasioned as a result of any misprint or mistake found in this publication.

Preface to the Revised Edition

The present edition has been thoroughly revised and enlarged. Keeping in view the suggestions received from the various teachers, students and the National Book Trust many chapters have been improved by adding new topics and a new chapter on Fourier Analysis has been introduced. The subject matter has also been rearranged. Some of the important topics that have been included in this edition are—Free Oscillations with two degrees of freedom, Properties of a mode, Two Coupled LC Circuits, Damped SHM in an Electrical Circuit. What propagates in wave motion? Theory of resonator, Vibrations in rods and plates, Speech, human voice and human ear, Limits of audibility and Tracking of artificial satellites.

A large number of new solved numerical examples on important topics have been included so as to provide better understanding and practice to the students. At the end of each chapter, Exercises are also brought up-to-date, by including questions from recent University and other competitive examinations.

We are grateful to the teachers and students for the favourable response given to the book. We also welcome suggestions for further improvement of the book.

N. SUBRAHMANYAM
BRIJ LAL

Preface to the First Edition

The book intended to meet the needs of B.Sc.(Pass, Honours and Subsidiary) and engineering students, covers the topics included in the latest syllabi of various Indian Universities.

The subject-matter is divided into eleven chapters. Each chapter is self-contained and is treated in a comprehensive way, using the S.I. system of units. Harmonic oscillators, linearity and superposition principle, oscillations with one degree of freedom, resonance and sharpness of resonance, quality factor, Doppler effect in sound and light, medical applications of ultrasonics, acoustic intensity, acoustic measurements, wave velocity and group velocity, Maxwell's equations, Propagation of electromagnetic waves in isotropic media, De Broglie waves, Hersenburg's uncertainty principle and special theory of relativity are some of the important topics which have been given special attention. Solved numerical problems, (in S.I. units), wherever necessary, are given in the text and exercises at the end of each chapter, contain questions which are largely drawn from the university question papers of recent years. The book contains a large number of diagrams.

We hope that the book will be found useful both by students and teachers.

July 11, 1974

N. SUBRAHMANYAM
BRIJ LAL

CONTENTS

CHAPTER 1

Harmonic Oscillators

1.1. Introduction

In every day life we come across numerous things that move. These motions are of two types, *viz.* *(i)* the motion in which the body moves about a mean position *i.e.* a fixed point and *(ii)* the motion in which the body moves from one place to the other with respect of time. The first type of motion of a body about a mean position is called oscillatory motion. A moving train, flying aeroplane, moving ball etc., correspond to the second type of motion. Examples of oscillatory motion are : an oscillating pendulum, vibrations of a stretched string movement of water in a cup, vibration of electrons, movement of light in a laser beam etc.

Sometimes both the types of motion are exhibited in the same phenomenon depending on our point of view. The sea waves appear to move towards the beach but the water moves up and down about the mean position. When a stretched rope is displaced, the displacement pulse travels from one end to the other but the material of the rope vibrates about the mean position without travelling forward.

1.2. Simple Harmonic Motion

Let P be a particle moving on the circumference of a circle of radius a with a uniform velocity v (Fig.1.1). Let ω be the uniform angular velocity of the particle $(v=a\omega)$. The circle along which P moves is called the **circle of reference.** As the particle P moves round the circle continuously with uniform velocity, the foot of the perpendicular M, vibrates along the diameter YY'. If the motion of P is uniform, then the motion of M is periodic *i.e.*, it takes the same time to vibrate once between the

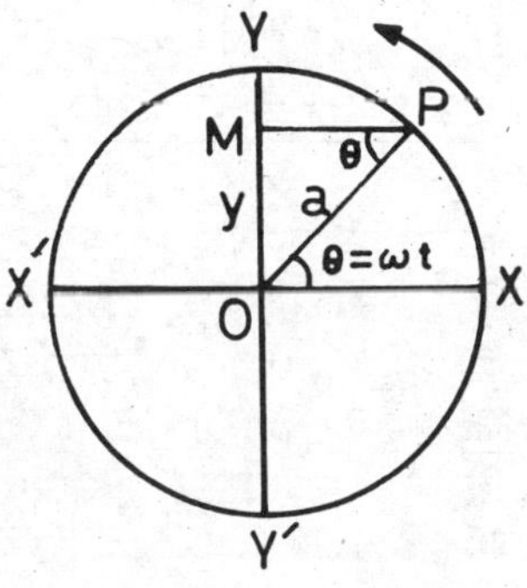

Fig. 1.1.

points Y and Y'. At any instant the distant of M from the centre O of the circle is called the **displacement**. If the particle moved from X to P in time t, then $\angle POX = \angle MPO = \theta = \omega t$.

From the $\Delta\, MPO$,

$$\sin\theta = \sin\omega t = \frac{OM}{a}$$

or

$$OM = y = a\sin\omega t$$

OM is called the displacement of the vibrating particle. The displacement of a vibrating particle at any instant can be defined as its distance from the mean position of rest. The maximum displacement of a vibrating particle is called its **amplitude.**

$$\text{Displacement} = y = a\sin\omega t \qquad \ldots(1)$$

The rate of change of displacement is called the **velocity** of the vibrating particle.

$$\therefore \quad \text{Velocity} = \frac{dy}{dt} = +a\,\omega\cos\omega t \qquad \ldots(2)$$

The rate of change of velocity of a vibrating particle is called its acceleration.

$$\therefore \quad \text{Acceleration} = \text{Rate of change of velocity}$$

$$= \frac{d}{dt}\left(\frac{dy}{dt}\right)$$

$$= \frac{d^2 y}{dt^2} = -a\omega^2\sin\omega t$$

$$= -\omega^2 .\, a\sin\omega t = -\omega^2 y \qquad \ldots(3)$$

Angle ωt	Position of the vibrating particle M	Displacement $y = a\sin\omega t$	Velocity $\frac{dy}{dt} = a\omega\cos\omega t$	Acceleration $\frac{d^2 y}{dt^2} = -a\omega^2\sin\omega t$
O	O	Zero	$+a\,\omega$	Zero
$\frac{\pi}{2}$	Y	$+a$	Zero	$-a\omega^2$
π	O	Zero	$-a\,\omega$	Zero
$\frac{3\pi}{2}$	Y'	$-a$	Zero	$+a\omega^2$
2π	O	Zero	$+a\omega$	Zero

The changes in the displacement, velocity and acceleration of a vibrating particle in one complete vibration are given in the table.

Oscillatory behaviour. At the extreme positions, when y is maximum, dy/dt is zero. The acceleration d^2y/dt^2 is maximum and directed towards the mean position. This return force induces a negative velocity. When the displacement y is zero, the velocity dy/dt is maximum and is –ve. When the displacement is negative maximum, the velocity dy/dt is zero and the acceleration is maximum in the positive direction. This return force again induces a velocity in the positive direction which becomes positive maximum when the displacement is zero. The particle overshoots the mean position due to its velocity. The process repeats itself periodically. Thus the system oscillates. In this process, displacement y, velocity dy/dt and acceleration d^2y/dt^2 continuously change with respect to time.

Thus, the velocity of the vibrating particle is maximum (in the direction OY or OY') at the mean position of rest and zero at the maximum positions of vibration. The acceleration of the vibrating particle is zero at the mean position of rest and maximum at the maximum positions of vibration. The acceleration is always directed towards the mean position of rest and is directly proportional to the displacement of the vibrating particle. This type of motion where the acceleration is directed towards a fixed point (the mean position of rest) and is proportional to the displacement of the vibrating particle is called **simple harmonic motion.**

Further,

$$\text{Acceleration} = \frac{d^2y}{dt^2} = -\omega^2 y$$

$$= -\omega^2 \times \text{displacement}$$

Numerically $$\omega^2 = \frac{\text{Acceleration}}{\text{Displacement}}$$

or $$\omega = 2\pi n = \sqrt{\frac{\text{Acceleration}}{\text{Displacement}}}$$

or $$\frac{2\pi}{T} = \sqrt{\frac{\text{Acceleration}}{\text{Displacement}}}$$

or $$T = 2\pi\sqrt{\frac{\text{Displacement}}{\text{Acceleration}}}$$

$$= 2\pi\sqrt{K}$$

Thus, in general, the time period of a particle vibrating simple harmonically is given by $T = 2\pi\sqrt{K}$ where K is the displacement per unit acceleration.

If the particle P revolves round the circle, n times per second, then the angular velocity ω is given by

$$\omega = 2\pi n = \frac{2\pi}{T}$$

$$\left(\because \quad n = \frac{1}{T} \text{ where } T \text{ is the time period} \right)$$

or $$y = a \sin 2\pi\, nt = a \sin 2\pi\, \frac{t}{T}.$$

On the other hand, if the time is counted [(Fig. 1.2 (*i*)] from the instant *P* is at *S* ($\angle SOX = \alpha$) then the displacement

$$y == a \sin (\omega t + \alpha)$$
$$= a \sin \left(\frac{2\pi\, t}{T} + \alpha \right)$$

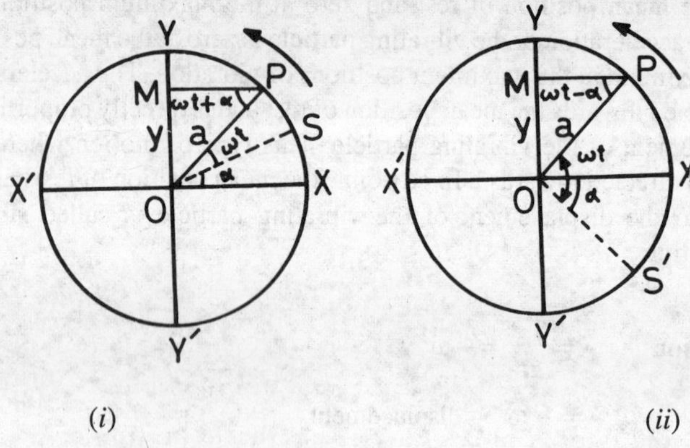

Fig. 1.2.

If the time is counted from the instant *P* is at *S*′ [Fig. 1.2 (*ii*)], then

$$y = a \sin (\omega t - \alpha)$$
$$= a \sin \left(\frac{2\pi\, t}{T} - \alpha \right)$$

Phase of the vibrating particle. (*i*) The phase of a vibrating particle is defined as the ratio of the displacement of the vibrating article at any instant to the amplitude of the virbrating particle *(y/a)* or (*ii*) it is also defined as the fraction of the time interval that has lapsed since the particle crossed the mean position of rest in the positive direction or *(iii)* it is also equal to the angle swept by the radius vector since the virbrating particle last crossed its mean position of rest *e.g.*, in the above equations ωt, $(\omega t + \alpha)$ or $(\omega t - a)$ are called phase angles. The initial phase angle when $t = 0$, is called the epoch. Thus α is called the **epoch** in the above expressions.

1.3. Differential Equation of SHM

For a particle vibrating simple harmonically, the general equation of displacement is,

$$y = a \sin(\omega t + \alpha) \quad \ldots (1)$$

Here y is displacement and a is the amplitude and α is epoch of the vibrating particle.

Differentiating equation (1) with respect to time

$$\frac{dy}{dt} = a\,\omega \cos(\omega t + \alpha) \quad \ldots (2)$$

Here dy/dt represents the velocity of the vibrating particle.

Differentiating equation (2) with respect to time

$$\frac{d^2 y}{dt^2} = -a\,\omega^2 \sin(\omega t + \alpha)$$

But $\quad a \sin(\omega t + \alpha) = y$

$$\therefore \quad \frac{d^2 y}{dt^2} = -\omega^2 y$$

or

$$\frac{d^2 y}{dt^2} + \omega^2 y = 0 \quad \ldots (3)$$

Here $d^2 y/dt^2$ represents the acceleration of the particle. Equation (3) represents the differential equation of simple harmonic motion.

It also shows that in any phenomenon where an equation similar to equation (3) is obtained, the body executes simple harmonic motion. The general solution of equation (3) is

$$y = a \sin(\omega t + \alpha).$$

Also the time period of a vibrating particle can be calculated from equation (3).

Numcrically

$$\omega = \sqrt{\frac{d^2 y/dt^2}{y}}$$

or

$$\omega = \sqrt{\frac{\text{Acceleration}}{\text{Displacement}}}$$

or

$$T = \frac{2\pi}{\omega} = 2\pi \sqrt{\frac{\text{Displacement}}{\text{Acceleration}}}.$$

1.4. Graphical Representation of SHM

Let P be a particle moving on the circumference of a circle of radius a. The foot of the perpendicular vibrates on the diameter YY'.

$$y = a \sin \omega t = a \sin 2\pi \frac{t}{T}$$

The displacement graph is a sine curve represented by *ABCDE* (Fig. 1.3).

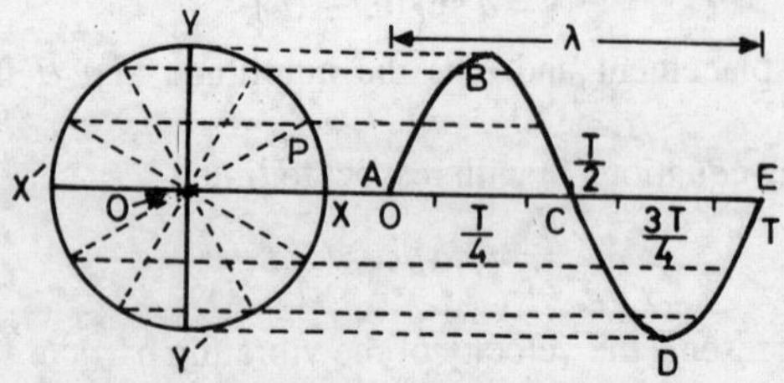

Fig. 1.3. Displacement Time Curve.

The motion of the particle *M* is simple harmonic.

The velocity of a particle moving with simple harmonic motion is

$$v = \frac{dy}{dt} = + a\,\omega \cos \omega t$$

The velocity–time graph is shown in Fig. 1.4. It is a cosine curve.

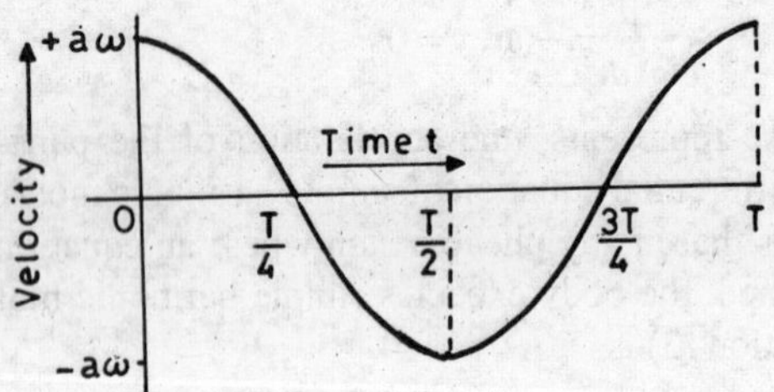

Fig. 1.4. Velocity — Time Curve.

The acceleration of a particle moving with simple harmonic motion is

$$\frac{d^2 y}{dt^2} = - a\,\omega^2 \sin \omega t$$

The acceleration—time graph is shown in Fig. 1.5. It is a negative sine curve.

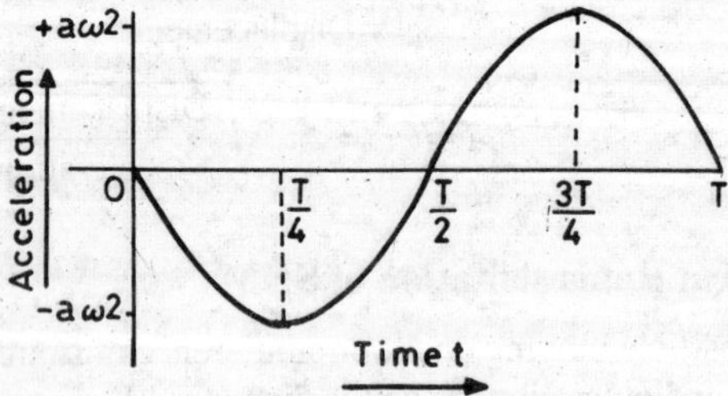

Fig. 1.5. Acceleration — Time Curve.

1.5. Average Kinetic Energy of a Vibrating Particle

The displacement of a vibrating particle is given by

$$y = a \sin(\omega t + \alpha)$$

$$v = \frac{dy}{dt} = a\omega \cos(\omega t + \alpha).$$

If m is the mass of the vibrating particle, the kinetic energy at any instant

$$= \frac{1}{2} m v^2 = \frac{1}{2} m \,.\, a^2 \omega^2 \cos^2 (\omega t + \alpha)\,.$$

The average kinetic energy of the particle in one complete vibration

$$= \frac{1}{T} \int_0^T \frac{1}{2} m a^2 \omega^2 \cos^2 (\omega t + \alpha)\, dt$$

$$= \frac{1}{T} \cdot \frac{m a^2 \omega^2}{4} \int_0^T 2 \cos^2 (\omega t + \alpha)\, dt$$

$$= \frac{m a^2 \omega^2}{4T} \int_0^T [1 + \cos 2(\omega t + \alpha)]\, dt$$

$$= \frac{m a^2 \omega^2}{4T} \left[\int_0^T dt + \int_0^T \cos 2(\omega t + \alpha)\, dt \right]$$

But $$\int_0^T \cos 2(\omega t + \alpha)\, dt = 0$$

$\therefore$ $$\text{Average K.E.} = \frac{m a^2 \omega^2}{4T} \cdot T + 0$$

$$= \frac{m a^2 \omega^2}{4} = \frac{m a^2 (4\pi^2 n^2)}{4}$$

$$= \pi^2 m a^2 n^2$$

where m is the mass of the vibrating particle, a is the amplitude of vibration and n is the frequency of vibration. Also, the average kinetic energy of a vibrating particle is directly proportional to the square of the amplitude.

1.6. Total Energy of a Vibrating Particle

$$y = a \sin(\omega t + \alpha)$$

$$\sin(\omega t + \alpha) = \frac{y}{a}$$

$$\cos(\omega t + \alpha) = \sqrt{1 - \frac{y^2}{a^2}} = \sqrt{\frac{a^2 - y^2}{a^2}}$$

$$= \frac{\sqrt{a^2 - y^2}}{a}$$

Velocity $\qquad v = a\,\omega \cos \omega\, t = \dfrac{a\omega \sqrt{a^2 - y^2}}{a}$

$$= \omega \sqrt{(a^2 - y^2)}$$

$\therefore$ The kinetic energy of the particle at the instant the displacement is y,

$$= \frac{1}{2}\, m\, v^2$$

$$= \frac{1}{2}\, m \,.\, \omega^2\, (a^2 - y^2)$$

Potential energy of the vibrating particle is the amount of work done in overcoming the force through a distance y.

Acceleration $\qquad = -\,\omega^2\, y$

Force $\qquad = -\,m\omega^2\, y$

(The –ve sign shows that the direction of the acceleration and force are opposite to the direction of motion of the virbrating particle.)

$$\therefore \qquad \text{P.E.} = \int_0^y m \,.\, \omega^2 y \,.\, dy$$

$$= m\omega^2 \cdot \frac{y^2}{2} = \frac{1}{2}\, m\omega^2\, y^2 .$$

Total energy of the particle at the instant the displacement is y

$$= \text{K.E} + \text{P.E.}$$

$$= \frac{1}{2}\, m\omega^2\, (a^2 - y^2) + \frac{1}{2}\, m\omega^2\, y^2$$

$$= \frac{1}{2}\, m\omega^2 \,.\, a^2$$

$$= \frac{1}{2}\, m\, (2\pi\, n)^2\, a^2$$

$$= \mathbf{2\pi^2\, m\, a^2\, n^2.}$$

As the average kinetic energy of the vibrating particle = $\pi^2\, ma^2n^2$, the average potential energy = $\pi^2\, ma^2n^2$, The total energy at any instant is a constant.

Example 1.1. *For a particle vibrating simple harmonically, the displacement is 12* cm *at the instant the velocity is 5* cm/s *and the displacement is 5* cm *at the instant the velocity is 12* cm/s. *Calculate (i) amplitude, (ii) frequency and (iii) time period.*

The velocity of a particle executing SHM,

$$v = \frac{dy}{dt} = \omega \sqrt{a^2 - y^2}$$

In the first case,

$$v_1 = \omega\sqrt{a^2 - y_1^{\,2}}$$

Here $\quad v_1 = 5$ cm/s, $y_1 = 12$ cm.

$$5 = \omega\sqrt{a^2 - 144} \qquad \ldots (1)$$

In the second case

$$v_2 = \omega\sqrt{a^2 - y_2^2}$$

Here $\quad v_2 = 12$ cm/s; $y_2 = 5$ cm

$$12 = \omega\sqrt{a^2 - 25} \qquad \ldots (2)$$

Dividing (2) by (1) and squaring

$$\frac{144}{25} = \frac{a^2 - 25}{a^2 - 144}$$

$$a = \mathbf{13\ cm}$$

The amplitude is **13 cm.**

Substituting the value of $a = 13$ cm in equation (1)

$$5 = \omega\sqrt{(13)^2 - 144}$$

or

$$\omega = 1 \text{ radian/s}$$

The frequency $n = \dfrac{\omega}{2\pi} = \dfrac{\mathbf{1}}{\mathbf{2\pi}}$ **hertz**

Time period $\quad T = \dfrac{1}{n} = \mathbf{2\pi}$ **seconds** .

Example 1.2. *Show that for a particle executing simple harmonic motion, its velocity at any instant is*

$$\frac{dy}{dt} = \omega\sqrt{a^2 - y^2}$$

The displacement,

$$y = a \sin \omega t \qquad \ldots (1)$$

The velocity at any instant is,

$$\frac{dy}{dt} = a\,\omega \cos \omega\, t \qquad \ldots (2)$$

From equation (1)

$$\sin \omega\, t = \frac{y}{a}$$

$$\cos \omega\, t = \sqrt{1 - \sin^2 \omega t}$$

$$\cos \omega t = \sqrt{1 - \frac{y^2}{a^2}}$$

$$\therefore \quad \frac{dy}{dt} = a\,\omega \sqrt{1 - \frac{y^2}{a^2}}$$

or

$$\frac{dy}{dt} = \omega \sqrt{a^2 - y^2}.$$

Example 1.3. *For a particle vibrating simple harmonically the displacement is 8* cm *at the instant the velocity is 6* cm/s *and the displacement is 6* cm *at the instant the velocity is 8* cm/s. *Calculate (i) amplitude, (ii) frequency and (iii) time period.*

The velocity of a particle executing SHM,

$$v = \frac{dy}{dt} = \omega \sqrt{a^2 - y^2}$$

In the first case,

$$v_1 = \omega \sqrt{a^2 - y_1^2}$$

Here $v_1 = 6$ cm/s, $y_1 = 8$ cm

$$\therefore \quad 6 = \omega \sqrt{a^2 - 64} \qquad \ldots (1)$$

In the second case,

$$v_2 = \omega \sqrt{a^2 - y_2^2}$$

Here $v_2 = 8$ cm/s $y_2 = 6$ cm ... (2)

$$\therefore \quad 8 = \omega \sqrt{a^2 - 36}$$

Dividing (2) by (1) and squaring

$$\frac{64}{36} = \frac{a^2 - 36}{a^2 - 64}$$

$$a = \mathbf{10\ cm.}$$

The amplitude of vibration = **10 cm**

Substituting the value of

$a = 10$ cm in equation (1)

$$6 = \omega \sqrt{100 - 64}$$

$$\omega = 1 \text{ radian/s}$$

Frequency $n = \frac{\omega}{2\pi} = \frac{\mathbf{1}}{\mathbf{2\pi}}$ **hertz**

Time period $T = \frac{1}{n} = \mathbf{2\pi}$ **seconds.**

Example 1.4. *The motion of a particle in simple harmonic motion is given by $x = a \sin \omega t$. If it has a speed u when the displacement is x_1 and speed v when the displacement is x_2, show that the amplitude of the motion is*

$$a = \left[\frac{v^2 x_1^2 - u^2 x_2^2}{v^2 - u^2}\right]^{\frac{1}{2}}$$

(Utkal, 1989)

Here $x = a \sin \omega t$

$$u = \frac{d x_1}{dt} = \omega \sqrt{a^2 - x_1^2} \qquad \ldots (i)$$

and

$$v = \frac{d x_2}{dt} = \omega \sqrt{a^2 - x_2^2} \qquad \ldots (ii)$$

Squaring and dividing

$$\frac{u^2}{v^2} = \frac{a^2 - x_1^2}{a^2 - x_2^2}$$

$$u^2 a^2 - u^2 x_2^2 = v^2 a^2 - v^2 x_1^2$$

$$a^2 [v^2 - u^2] = v^2 x_1^2 - u^2 x_2^2$$

$$a = \left[\frac{v^2 x_1^2 - u^2 x_2^2}{v^2 - u^2}\right]^{\frac{1}{2}} \qquad \ldots (iii)$$

Example 1.5. *Show that for a particle executing SHM, the instantaneous velocity is $\omega \sqrt{a^2 - y^2}$ and instantaneous acceleration is $-\omega^2 y$.*

For a particle executing SHM,

$$y = a \sin (\omega t + \alpha) \qquad \ldots (1)$$

The instantaneous velocity,

$$v = \frac{dy}{dt} = + a \omega \cos (\omega t + \alpha) \qquad \ldots (2)$$

From equation (1)

$$\sin (\omega t + \alpha) = \frac{y}{a}$$

$$\cos(\omega t+\alpha)=\sqrt{1-\sin^2(\omega t+\alpha)}$$

$$=\sqrt{1-\frac{y^2}{a^2}}$$

$$\therefore \quad v=a\omega\sqrt{1-\frac{y^2}{a^2}}$$

$$v=\omega\sqrt{a^2-y^2} \quad \ldots(3)$$

The instantaneous acceleration,

$$\frac{d^2y}{dt^2}=\frac{dv}{dt}=-a\omega^2\sin(\omega t+\alpha)$$

$$=-\omega^2[a\sin(\omega t+\alpha)]$$

$$=-\omega^2 y \quad \ldots(4)$$

Example 1.6. *A particle performs simple harmonic motion given by the equation*

$$y=20\sin[\omega t+\alpha]$$

If the time period is 30 seconds *and the particle has a displacement of 10* cm *at $t=0$, find (i) epoch ; (ii) the phase angle at $t=5$* seconds *and (iii) the phase difference between two positions of the particle 15* seconds *apart.*

Here $y=20\sin(\omega t+\alpha)$

$T=30$ s

$$\omega=\frac{2\pi}{T}=\frac{2\pi}{30}=\frac{\pi}{15}\text{ radians/s}$$

(i) At $t=0$, $y=10$ cm

$$\therefore \quad 10=20\sin\left(\frac{\pi}{15}\times 0+\alpha\right)$$

or $\sin\alpha=0{\cdot}5$

or $\sin\alpha=\frac{\pi}{6}$ radian

(ii) At $t=5$ s,

The phase angle $=(\omega t+\alpha)$

$$=\left(\frac{\pi}{15}\times 5+\frac{\pi}{6}\right)$$

$$=\frac{\pi}{2}$$

(iii) At $t = 0$

The phase angle $\theta_1 = \frac{\pi}{6}$

At $t = 15$

The phase angle $\theta_2 = (\omega t + \alpha)$

$$\theta_2 = \left(\frac{\pi}{15} \times 15 + \frac{\pi}{6}\right)$$

$$\theta_2 = \frac{7\pi}{6}$$

The phase difference $\theta_2 - \theta_1 = \frac{7\pi}{6} - \frac{\pi}{6} = \pi$ ***radians.***

Example 1.7. *A particle executes simple harmonic motion given by the equation*

$$y = 12 \sin\left(\frac{2\pi t}{10} + \frac{\pi}{4}\right)$$

Calculate (i) amplitude, (ii) frequency, (iii) epoch, (iv) displacement at t=1·25 s, *(v) velocity at t = 2·5*s *and (vi) acceleration at t = 5* s.

Here $$y = 12 \sin\left(\frac{2\pi t}{10} + \frac{\pi}{4}\right) \quad \ldots (1)$$

The displacement equation is

$$y = a \sin(\omega t + \alpha) \quad \ldots (2)$$

Comparing equations (1) and (2)

(i) Amplitude $a = 12$ units

(ii) $\omega = \frac{2\pi}{10}$

Frequency $n = \frac{\omega}{2\pi} = \frac{1}{10} =$ **0·1 hertz**

(iii) Epoch $\alpha = \frac{\pi}{4}$

(iv) When $t = 1{\cdot}25$ s

$$y = 12 \sin\left(\frac{2\pi \times 1{\cdot}25}{10} + \frac{\pi}{4}\right)$$

$$y = 12 \sin \pi/2$$

or $y =$ **12 units**

(v) At $t = 2{\cdot}5$ s

$$\text{Velocity} = \frac{dy}{dt} = a\omega \cos(\omega t + \alpha)$$

$$\frac{dy}{dt} = 12 \times \frac{2\pi}{10} \cos\left[\frac{2\pi}{10} \times 2{\cdot}5 + \frac{\pi}{4}\right]$$

$$\frac{dy}{dt} = \mathbf{-\ 5{\cdot}552\ units.}$$

The –ve sign shows that the velocity is directed towards the mean position.

(vi) At $t = 5$ s

$$\text{Acceleration} = \frac{d^2 y}{dt^2} = -a\omega^2 \sin(\omega t + \alpha)$$

$$\frac{d^2 y}{dt^2} = -12 \times \left(\frac{2\pi}{10}\right)^2 \sin\left(\frac{2\pi}{10} \times 5 + \frac{\pi}{4}\right)$$

$$= -0{\cdot}48\,\pi^2 \sin\left(\pi + \frac{\pi}{4}\right)$$

$$= \mathbf{3{\cdot}35\ units.}$$

Example 1.8. *A simple harmonic motion is represented by the equation*

$$y = 10 \sin\left(10\,t - \frac{\pi}{6}\right)$$

where y is measured in metres, *t in* seconds *and the phase angle in* radians. *Calculate :*

(i) *the frequency,*

(ii) *the time period,*

(iii *the maximum displacement,*

(iv) *the maximum velocity,*

(v) *the maximum acceleration, and*

(vi) *displacement, velocity and acceleration at time, t=0 and t=1* second.

Here $$y = 10 \sin\left(10\,t - \frac{\pi}{6}\right) \quad \ldots (1)$$

The displacement equation is

$$y = a \sin(\omega t + \alpha) \quad \ldots (2)$$

(i) From (1) and (2)

$$\omega = 10$$

But $$\omega = 2\,\pi n$$

$\therefore \qquad 2\pi n = 10$

or $\qquad n = \frac{10}{2\pi}$ hertz

$n = \mathbf{1{\cdot}6\ hertz.}$

(ii) Time period,

$$T = \frac{1}{n} = \frac{2\pi}{10}$$

or $\qquad T = 0{\cdot}63$ s

(iii) Maximum displacement,

$$a = \mathbf{10\ m}$$

(iv) Velocity, $\qquad \frac{dy}{dt} = a\omega \cos(\omega t + \alpha)$

Therefore, maximum velocity

$$\frac{dy}{dt} = a\omega$$

But $\qquad a = 10$ m

and $\qquad \omega = 10$

$\therefore \qquad \frac{dy}{dt} = 10 \times 10 = \mathbf{100\ m/s}$

(v) Acceleration, $\qquad \frac{d^2 y}{dt^2} = -a\omega^2 \sin(\omega t + \alpha)$

Maximum acceleration

$$\frac{d^2 y}{dt^2} = -a\omega^2$$

or $\qquad \frac{d^2 y}{dt^2} = -10 \times (10)^2 = \mathbf{-1{,}000\ m/s^2}$

– ve sign shows that the acceleration is directed towards the mean position.

(vi) From equation (1)

(a) At $\qquad t = 0$

$$y = 10 \sin\left(-\frac{\pi}{6}\right)$$

$$y = \mathbf{-5\ m}$$

$$\frac{dy}{dt} = a\omega \cos(\omega t + \alpha)$$

At $t = 0$

$$\frac{dy}{dt} = a\omega \cos \alpha$$

$$= 10 \times 10 \cos\left(-\frac{\pi}{6}\right)$$

$$= 100 \times 0{\cdot}866$$

$$= \mathbf{86{\cdot}6\ m/s}$$

$$\frac{d^2 y}{dt^2} = -a\omega^2 \sin \alpha$$

or

$$\frac{d^2 y}{dt^2} = -a\omega^2 \sin\left(-\frac{\pi}{6}\right)$$

or

$$\frac{d^2 y}{dt^2} = a\omega^2 \sin\left(\frac{\pi}{6}\right)$$

$$= 10 \times 100 \times 0{\cdot}5$$

$$= \mathbf{500\ m/s^2.}$$

(b) From equation (1)

At $t = 1$,

Displacement

$$y = 10 \sin\left(10 — \frac{\pi}{6}\right)$$

$$y = 10 \sin\left(\frac{60 - 3{\cdot}142}{6}\right)$$

$$y = 10 \sin\left(\frac{56{\cdot}858}{6}\right)$$

$$y = 10 \sin (3\pi) \quad \text{(approximately)}$$

or

$$y = 10 \sin \pi = 0.$$

Velocity

$$\frac{dy}{dt} = a\omega \cos\left(10 - \frac{\pi}{6}\right)$$

or

$$\frac{dy}{dt} = a\omega \cos \pi \quad \text{(approximately)}$$

$$= 10 \times 10 \times (-1)$$

$$= \mathbf{-100\ m/s.}$$

Acceleration

$$\frac{d^2 y}{dt} = -a\omega^2 \sin\left(10 - \frac{\pi}{6}\right)$$

$$= -a\omega^2 \sin \pi \quad \text{(approximately)}$$

$$= \mathbf{0.}$$

Example 1.9. *The equation of a progressive wave is given by*

$$y = 10 \sin\ (0{\cdot}5x - 200\ t)$$

where x and y are in cm *and t is in* second. *Calculate the amplitude, wavelength, velocity and frequency of the wave. Also calculate the maximum velocity of a particle of the medium.* (Bhagalpur, 1990)

$$y = 10 \sin (0{\cdot}5\ x - 200\ t)$$

$$y = -10 \sin (200\ t - 0{\cdot}5\ x) \quad \ldots (i)$$

The displacement equation is

$$y = a \sin (\omega t + \alpha) \quad \ldots (ii)$$

Comparing *(i)* and *(ii)*

$$a = -10 \text{ cm}$$

or

$$a = 10 \text{ cm} \quad \ldots (iii)$$

$$\omega = 200$$

$$\omega = 2\pi n$$

$$n = \frac{\omega}{2\pi}$$

$$= \frac{200}{2\pi} = \frac{100}{\pi} \text{ Hz} \quad \ldots (iv)$$

From equation (*i*)

$$y = -10 \sin (0{\cdot}5)(400\ t - x) \quad \ldots (v)$$

Also

$$y = a \sin \frac{2\pi}{\lambda} (vt - x) \quad \ldots (vi)$$

Comparing (*v*) and (*vi*)

$$v = \mathbf{400\ cm/s}$$

and

$$\frac{2\pi}{\lambda} = 0{\cdot}5$$

$$\lambda = \mathbf{4\pi\ cm}$$

$$\lambda = \mathbf{12{\cdot}56\ cm}$$

Maximum velocity

$$\frac{dy}{dt} = a\omega$$

$$= 10 \times 200$$

$$= \mathbf{2000\ cm/s}.$$

1.7. Energy of Vibration

For a vibrating system, the work done is equal to the product of the force and the displacement while the power required is the work done per second. Let the displacement of the vibrating system at any instant be given by

$$y = a \sin(\omega t - \phi)$$

and let the periodic force be

$$F = F_0 \sin \omega t$$

The work done for a displacement dy

$$dW = F \times dy$$

The total work done

$$W = \int F dy$$

$$W = \int F_0 \sin \omega t \ [a \cos(\omega t - \phi)] \ d(\omega t)$$

Work done per cycle of motion,

$$W = F_0 \, a \int_0^{2\pi} \sin \omega t \ \cos(\omega t - \phi) \ d(\omega t)$$

But $\quad \cos(\omega t - \phi) = \cos \omega t \cos \phi + \sin \omega t \ \sin \phi$

$$W = F_0 \, a \int_0^{2\pi} \sin \omega t \ (\cos \omega t \cos \phi + \sin \omega t \sin \phi) \ d(\omega t)$$

$$W = F_0 \, a \cos \phi \int_0^{2\pi} \sin \omega t \ \cos \omega t \ d(\omega t) + F_0 \, a \sin \phi \int_0^{2\pi} \sin^2 \omega t \ d(\omega t)$$

$$W = F_0 \, a \cos \phi \left[\frac{\sin^2 \omega t}{2} \right]_0^{2\pi} + F_0 \, a \sin \phi \left[\frac{\omega t}{2} - \frac{\sin 2\omega t}{4} \right]_0^{2\pi}$$

or $\quad W = [F_0 \, a \sin \phi] \, [\pi]$

$$W = \pi F_0 \, a \sin \phi.$$

Example 1.10. *The force and displacement of a simple dynamic system undergoing sinusoidal excitation are given by the equations*

$$F = 10 \sin \left(\frac{\pi t}{10} \right) \text{ newtons}$$

$$y = 0{\cdot}1 \sin\left(\frac{\pi t}{10} - \frac{\pi}{6}\right) \text{ metres}$$

Calculate the work done by the excitation force in (i) 20 seconds, *and (ii) 1* minute.

The work done per cycle by the excitation force is given by

$$W = \pi F_0 a \sin\phi$$

Here F_0 is the maximum force and a is the maximum displacement.

Here $$F = 10 \sin\left(\frac{\pi t}{10}\right) \quad \ldots (1)$$

$$y = 0{\cdot}1 \sin\left(\frac{\pi t}{10} - \frac{\pi}{6}\right) \quad \ldots (2)$$

Also $$F = F_0 \sin\omega t \quad \ldots (3)$$

and $$y = a(\sin\omega t - \phi) \quad \ldots (4)$$

From these equations

$$F_0 = 10 \text{ newtons}$$

$$a = 0{\cdot}1 \text{ metre}$$

$$\phi = \frac{\pi}{6}$$

and $$\omega = \frac{\pi}{10}.$$

Work done per cycle

$$W = \pi F_0 a \sin\phi$$

$$W = \pi \times 10 \times 0{\cdot}1 \times 0{\cdot}5$$

$$W = 0{\cdot}5\,\pi \text{ joules per cycle.}$$

For the angular frequency

$$\omega = \frac{\pi}{10} \text{ radians/s}$$

$\therefore$ Time period $$T = \frac{2\pi}{\omega}$$

$$= \frac{2\pi \times 10}{\pi}$$

$$= \mathbf{20 \text{ seconds}}$$

(*i*) As the time taken to complete one cycle is 20 seconds.

The work done in 20 seconds

= Work done per cycle

= 0·5 π joules.

(*ii*) In one minute, three cycles are completed.

Therefore, work done in one minute

$$= 3 \times 0{\cdot}5\,\pi \text{ joules}$$

$$= \mathbf{1{\cdot}5\,\pi \text{ joules.}}$$

Example 1.11. *Show that the mean kinetic and potential energies of non-dissipative simple harmonic vibrating systems are equal.*

For free vibration in the absence of damping, the displacement at any instant is given by

$$y = a \sin \omega t$$

$$\frac{dy}{dt} = a\,\omega \cos \omega t$$

$$\text{Kinetic energy} = \frac{1}{2} m \left(\frac{dy}{dt}\right)^2$$

$$\text{K.E.} = \frac{1}{2} m\,(a^2 \omega^2 \cos^2 \omega t)$$

or $$\text{K.E.} = \frac{1}{2} k\,a^2 \cos^2 \omega t \qquad \dots (1)$$

Here $$k = m\,\omega^2$$

or $$\omega^2 = \frac{k}{m}$$

Here k is the force per unit displacement

$$\text{Potential energy} = \frac{1}{2}\,k\,y^2$$

$$\text{P.E.} = \frac{1}{2}\,k\,a^2 \sin^2 \omega t \qquad \dots (2)$$

Total kinetic energy for one complete cycle

$$= \int_0^T \frac{1}{2} k\,a^2 \cos^2 \omega t\; dt$$

$$= \frac{1}{4} k a^2\,T \qquad \dots (3)$$

Total potential energy for one complete cycle

$$= \int_0^T \frac{1}{2} k\,a^2 \sin^2 \omega t\; dt$$

$$= \frac{1}{4}\,k\,a^2\,T \qquad \dots (4)$$

Hence the mean potential and kinetic energies are equal.

Example 1.12. *Write down the equation for a wave travelling along the negative Z direction and having an amplitude 0·01* m, *frequency 550* Hz *and speed 330* m/s. *How would the equation change if a wave with the same parameters was travelling along the positive Z direction. Justify your answer.*

[IAS]

Here $$y = a \sin \frac{2\pi}{\lambda} (vt - z)$$

The wave is travelling in the positive z-direction

Here $$a = 0{\cdot}01 \text{ m}$$

$$\nu = 550 \text{ Hz}$$

$$v = 330 \text{ m/s}$$

$$\lambda = \frac{v}{\nu} = \frac{330}{550}$$

$$\lambda = \mathbf{0{\cdot}6\ m}$$

$$\therefore \quad y = 0{\cdot}01 \sin \left(\frac{2\pi}{0{\cdot}6}\right) [330\ t - z] \quad \ldots (i)$$

The wave is travelling along + z direction

For wave travelling along – z direction

$$y = 0{\cdot}01 \sin \left(\frac{2\pi}{0{\cdot}6}\right) [330\, t + z] \quad \ldots \text{(ii)}$$

For $t = 0$, from equation (*i*)

$$y = -\,0{\cdot}01 \sin \left(\frac{2\pi}{0{\cdot}6}\right) z \quad \ldots (iii)$$

The wave is travelling in + z direction

Similarly for t=0, from equation (*ii*)

$$y = +\,0{\cdot}01 \sin \left(\frac{2\pi}{0{\cdot}6}\right) z \quad \ldots (iv)$$

The wave is travelling along – z direction.

1.8. Oscillations with One Degree of Freedom

A pendulum of a clock, a loaded spring and LC circuit have one degree of freedom. In the case of simple pendulum, the swing depends upon the angular dispacement made by the string with the vertical direction. In the case of a loaded spring, the displacement of the mass, and in the case of LC circuit the charge on the condenser plate describes the nature of the oscillations (Fig.1.6). These oscillations take place about the mean position. All these systems have one degree of freedom.

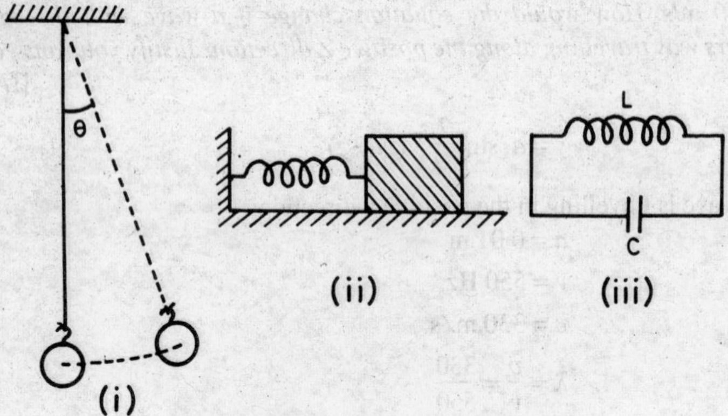

Fig. 1.6. Systems with one degree of freedom.

For oscillatory systems with one degree of freedom, the displacement of the "moving particle" depends upon the simple harmonic relation

$$y = a \sin(\omega t + \alpha)$$

In the case of a simple pendulum, at any instant of time t, y refers to the displacement of the bob from the mean position. In the case of a loaded spring at any instant of time t, y refers to the displacement of the oscillating mass from the mean position. In the case of the oscillatory LC circuit, at any instant of time t, y refers to the charge on the condenser.

An oscillatory system will continue to oscillate for an infinite time according to the equation

$$y = a \sin(\omega t + \alpha)\,.$$

Damped oscillations. In actual practice the oscillatory system experiences frictional or resistive forces. Due to these reasons the oscillations get damped. In the case of a pendulum, the amplitude decreases due to the resistance offered by air. In the case of LC circuit, the resistance of the circuit produces damping.

1.9. Linearity and Superposition Principle

The differential equation of SHM is given by

$$\frac{d^2 y}{dt^2} + \omega^2 y = 0$$

In this equation, $d^2 y/dt^2$ is proportional to $-y$. The time periods for a simple pendulum, mass and springs, bar pendulum, Kater's pendulum, LC circuit etc. are derived on the basis of this equation. In all these cases, the return force per unit displacement depends upon the term y. The equation does not contain the tems y^2, y^3 etc. The differential equations which do not contain the higher powers of y, such as y^2, y^3 . . . etc. terms are called linear equations.

Further this equation does not contain any term independent of y. Therefore, it is called a linear homogeneous equation. If, in a particular equation, higher powers of y *i.e.* y^2, y^3 . . . etc. terms are present, the equation is said to be non-linear. Moreover, if it contains terms independent of y also, it is said to be non-homogeneous and nonlinear equation. In general, it is not easy to solve non-linear equations.

The equation for a simple pendulum is

$$M\frac{d^2 y}{dt^2} = -Mg\sin\theta \qquad \ldots(1)$$

It is assumed that amplitude is small and $\sin\theta = \theta$.

Equation $M\dfrac{d^2 y}{dt^2} = -Mg\,\theta$ is linear.

However, $\sin\theta = \theta - \dfrac{\theta^3}{3!} + \dfrac{\theta^5}{5!} \ldots\ldots$

Taking this value of $\sin\theta$

$$M\frac{d^2 y}{dt^2} = -Mg\left(\theta - \frac{\theta^3}{3!} + \frac{\theta^5}{5!}\right) \qquad \ldots(2)$$

Equation (2) is non-linear for it contains terms of θ^3, θ^5 etc.

Linear homogeneous equations. One of the important properties of linear homogeneous equations is that the sum of any two solutions is a solution by itself. This property is not true in the case of non-linear equations.

Consider the differential equation,

$$\frac{d^2 y}{dt^2} = -\omega^2 y + Ay^2 + By^3 + Cy^4 \quad \text{etc.} \qquad \ldots(3)$$

If the values of the constants, A, B, C etc. in equation (3) are zero or they can be taken sufficiently near to zero, then equation (3) becomes

$$\frac{d^2 y}{dt^2} = -\omega^2 y \qquad \ldots(4)$$

This equation is linear and homogeneous.

If in equation (3), constants A, B, C etc. are not zero, then the equation is non-linear.

Suppose, y_1 is the first solution of equation (4) at some instant of time t_1 and y_2 is the second solution of equation (4) at some other instant of time t_2 (different displacement and velocity).

$$\therefore \qquad \frac{d^2 y_1}{dt^2} = -\omega^2 y_1 + A y_1^2 + B y_1^3 + C y_1^4 + \ldots \qquad \ldots (5)$$

$$\frac{d^2 y_2}{dt^2} = -\omega^2 y_2 + A y_2^2 + B y_2^3 + C y_2^4 + \ldots \qquad \ldots (6)$$

When the two instants are superimposed on each other, the resultant displacement is y. Here if superposition is true it is to be proved that $y = y_1 + y_2$.

$\therefore$ The equation for resultant displacement,

$$\frac{d^2 y}{dt^2} = \frac{d^2}{dt^2}(y_1 + y_2)$$

$$= -\omega^2 (y_1 + y_2) + A(y_1 + y_2)^2 + B(y_1 + y_2)^3 + C(y_1 + y_2)^4 + \ldots \qquad \ldots (7)$$

Adding equations (5) and (6)

$$\frac{d^2 y_1}{dt^2} + \frac{d^2 y_2}{dt^2} = -\omega^2 (y_1 + y_2) + A(y_1^2 + y_2^2) + B(y_1^3 + y_2^3) + C(y_1^4 + y_4^4) + \ldots \qquad \ldots (8)$$

The equations (7) and (8) are identical, only if

$$\frac{d^2}{dt^2}(y_1 + y_2) = \frac{d^2 y_1}{dt^2} + \frac{d^2 y_2}{dt^2} \qquad \ldots (9)$$

$$-\omega^2 (y_1 + y_2) = -\omega^2 y_1 - \omega^2 y_2 \qquad \ldots (10)$$

$$A(y_1^2 + y_2^2) = A(y_1 + y_2)^2 \qquad \ldots (11)$$

$$B(y_1^3 + y_2^3) = B(y_1 + y_2)^3 \qquad \ldots (12)$$

$$C(y_1^4 + y_2^4) = C(y_1 + y_2)^4 \qquad \ldots (13)$$

Equations (9) and (10) are true. But equations (11), (12) and (13) are true only, if

$$A = 0, \qquad B = 0, \qquad C = 0$$

When A, B, C etc. are zero, the equations become linear. Hence superposition principle is true only in the case of homogeneous linear equations. Also the sum of any two solutions is also a solution of the homogeneous linear equation.

All harmonic oscillators given in equations (9) and (10) obey superposition principle.

1.10. Simple Pendulum

A simple pendulum consists of a light string supporting a small sphere and fixed firmly at its upper end. An ideal simple pendulum should consist of a heavy particle suspended by means of a weightless, inextenisble, flexible string from a rigid support.

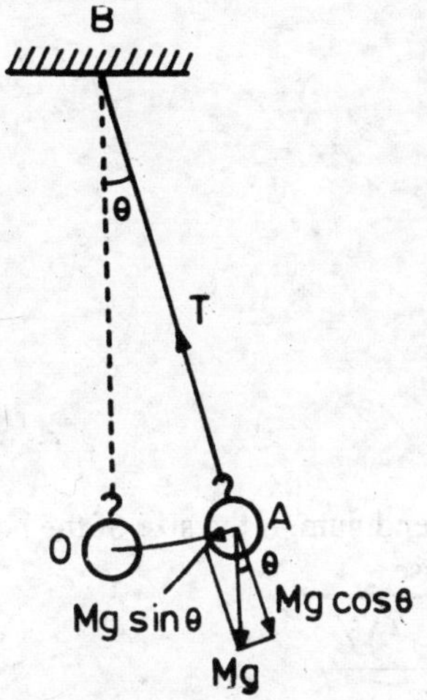

Fig. 1.7.

Let a pendulum be displaced from its mean positioin O and allowed to oscillate (Fig. 1.7). Suppose at any instant of time t, it is at A. The force acting upon the bob vertically downward = Mg. Resolve Mg into two rectangular components.

(1) Force along the string = $Mg \cos \theta$

(2) Force perpendicular to the string

$$= Mg \sin \theta$$

Let the tension in the string be T. The component $Mg \cos \theta$ balances the tension T

$$Mg \cos \theta = T$$

Thus the only force acting on the oscillating particle is $-Mg \sin \theta$.

$$\therefore \qquad F = -Mg \sin \theta$$

(– ve sign shows that the acceleration is directed towards the mean position)

According to Taylor's series of expansion

$$\sin \theta = \theta - \frac{\theta^3}{3!} + \frac{\theta^5}{5!} \cdots\cdots$$

For small angular displacements θ, $\quad \sin \theta = \theta$

Tangential force $\quad F = -Mg\theta$

The linear displacement $y = l\theta$

Acceleration $\quad \dfrac{d^2 y}{dt^2} = l \dfrac{d^2 \theta}{dt^2}$

$$\therefore \qquad \text{Force} = Ml \frac{d^2 \theta}{dt^2}$$

From Newton's second law

$$Ml \frac{d^2 \theta}{dt^2} = -Mg\,\theta$$

$$\frac{d^2 \theta}{dt^2} + \frac{g}{l}\theta = 0 \qquad \ldots (1)$$

This equation is similar to the equation of simple harmonic motion

$$\frac{d^2 y}{dt^2} + \omega^2 y = 0 \quad \ldots (2)$$

From (1) and (2)

$$\omega^2 = \frac{g}{l}$$

$$\omega = \sqrt{\frac{g}{l}}$$

Time period $T = \frac{2\pi}{\omega}$

$$T = 2\pi\sqrt{\frac{l}{g}} \quad \ldots (3)$$

[**Bob of large size.** In the case of a simple pendulum, if the size of the bob is large, a correction has to be applied. In this case

$$t = 2\pi\sqrt{\frac{l + (\frac{2}{5}r^2)/l}{g}}$$

Here $l + \left(\frac{\frac{2}{5}r^2}{l}\right)$ represents the equivalent length of a simple pendulum.

1.11. Compound Pendulum

A compound pendulum is a rigid mass capable of oscillating about a horizontal axis passing through any point of the mass. This point is called the point of suspension. In Fig. 1.8, G is the centre of gravity of the body and S is the point of suspension. At any instant of time, when the mass has been displaced, the force acting vertically downwards = Mg. At this position, the line SG makes an angle θ with the vertical and the restoring moment of this force about the point $S = Mg\, l \sin\theta$. This is the only moment which produces angular acceleration in the pendulum.

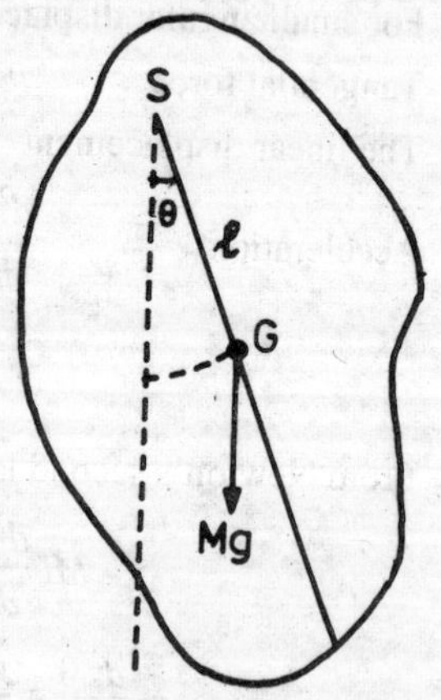

Fig. 1.8.

Let the moment of inertia of the pendulum about an axis passing through S and perpendicular to its length be I. If the angular acceleration at this instant is $\frac{d^2\theta}{dt^2}$, then

$$I\frac{d^2\theta}{dt^2} = -Mg\ l\ \sin\theta$$

[– ve sign shows that the force is directed towards the mean position].

$$\therefore \qquad I\frac{d^2\theta}{dt^2} = +Mg\ l\ \theta = 0$$

For small angular displacements, $\sin\theta = \theta$

$$\therefore \qquad I\frac{d^2\theta}{dt^2} + Mg\ l\,\theta = 0$$

The moment of inertia of the pendulum about an axis passing through S and perpendicular to its plane $= MK^2 + Ml^2$.

Here K is the radius of gyration *about an axis passing through the centre of gravity* of the pendulum.

$$\therefore \qquad M(K^2 + l^2)\frac{d^2\theta}{dt^2} + Mg\ l\,\theta = 0$$

$$\frac{d^2\theta}{dt^2} + \left(\frac{lg}{K^2 + l^2}\right)\theta = 0 \qquad \ldots (1)$$

This equation is similar to the equation of simple harmonic motion

$$\frac{d^2y}{dt^2} + \omega^2 y = 0 \qquad \ldots (2)$$

Here y refers to the angular displacement θ.

Comparing (1) and (2)

$$\omega^2 = \left(\frac{lg}{K^2 + l^2}\right)$$

Here ω is the angular frequency

$$\therefore \qquad \text{Time period} \qquad T = \frac{2\pi}{\omega}$$

$$T = 2\pi\sqrt{\frac{K^2 + l^2}{lg}} \qquad \ldots (3)$$

or

$$T = 2\pi\sqrt{\frac{(K^2/l) + l}{g}}$$

Here $(K^2/l + l)$ is called the equivalent length of the simple pendulum.

1.12. Bar Pendulum

A bar pendulum consists of a metal bar about 1 metre in length, 5 cm in breadth and 0·5 cm thick. It has a large number of holes as shown in Fig. 1.9. There are two knife edges which fit into the holes. A bar pendulum is a particular case of a compound pendulum. The time period is determined by fixing the knife edge in each hole. The distance of each hole from the centre of gravity is measured. A graph is drawn between the distance from the C.G. along the *x*-axis and the corresponding time period along the *y*-axis. The graph is as shown in Fig. 1.10.

From the graph the following results are obtained.

(1) On each side of the centre of gravity there are two positions for which the time periods are equal. It means there are four positions in the pendulum (two on each side of the C.G.) having the same time period. The points *A, B, C* and *D* in the graph have the same time period.

The time period of the bar pendulum,

$$t = 2\pi\sqrt{\frac{K^2 + l^2}{lg}}$$

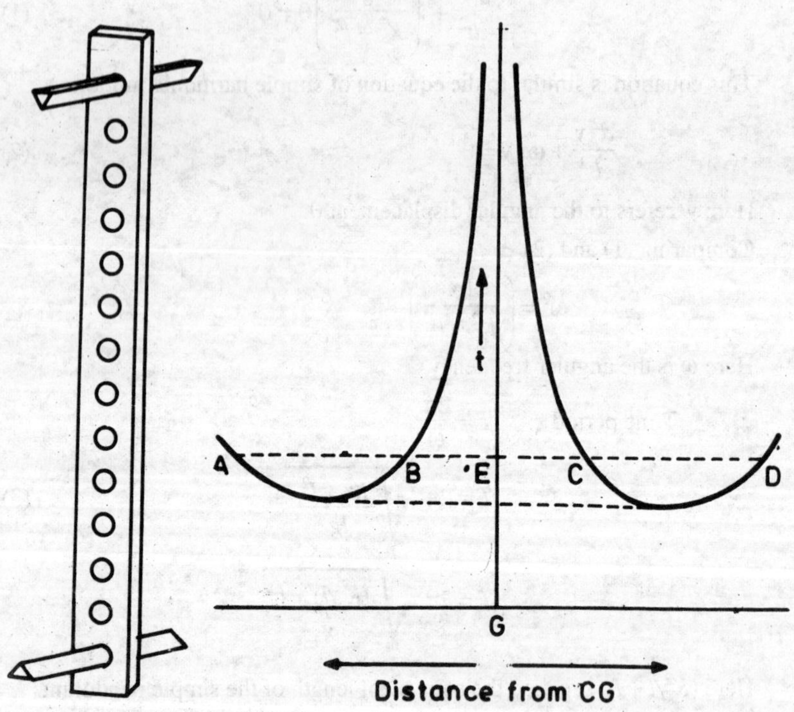

Fig. 1.9. Fig. 1.10.

Squaring

$$t^2 = 4\pi^2 \left(\frac{K^2 + l^2}{lg}\right)$$

or $$4\pi^2 l^2 - gt^2 l + 4\pi^2 K^2 = 0 \qquad \ldots (1)$$

It is a quadratic equation in l

$\therefore$ $$l = \frac{gt^2 \pm \sqrt{g^2 t^4 - 64\pi^4 K^2}}{8\pi^2} \qquad \ldots (2)$$

From equation (2), the two values of l are

$$l_1 = \frac{gt^2 + \sqrt{g^2 t^4 - 64\pi^4 K^2}}{8\pi^2} \qquad \ldots (3)$$

and $$l_2 = \frac{gt^2 - \sqrt{g^2 t^4 - 64\pi^4 K^2}}{8\pi^2} \qquad \ldots (4)$$

Thus two different points having the same value of t are obtained on each side of the bar pendulum.

(2) For a particular value of l, the time period is minimum, when

$$\sqrt{g^2 t^4 - 64\pi^4 K^2} = 0$$

$$l_1 = l_2 = \frac{gt^2}{8\pi^2} \qquad \ldots (5)$$

(3) It is seen from the graph that the time period increases enormously when the point of suspension approaches the centre of gravity.

From equation

$$t = 2\pi\sqrt{\frac{K^2 + l^2}{lg}}$$

When l is zero, t is infinite.

(4) From equations (3) and (4),

$$l_1 + l_2 = \frac{gt^2}{4\pi^2}$$

or $$t = 2\pi\sqrt{\frac{l_1 + l_2}{g}} \qquad \ldots (6)$$

Here $(l_1 + l_2)$ is the equivalent length of the simple pendulum. In the graph

$$EA = ED = l_1$$

$$EB = EC = l_2$$

$$l_1 + l_2 = EA + EC = AC$$

Also $l_1 + l_2 = EB + ED = BD$

The value of g can be determined from equation (6).

Also
$$t = 2\pi\sqrt{\frac{K^2/l_1 + l_1}{g}} \quad \ldots (7)$$

Comparing equations (6) and (7)

$$\frac{K^2}{l_1} = l_2$$

or
$$K^2 = l_1 l_2$$

$$K = \sqrt{l_1 \times l_2}$$

Also
$$K = \sqrt{EA \times EB} = \sqrt{ED \times EC}$$

Thus the value of radius of gyration K about the axis passing through the C.G. of a bar pendulum can be determined.

1.13. Points of Suspension and Oscillation are Interchangeable

In a bar pendulum, let the distance of A from the C.G. be l_1 and the distance of C be l_2. The time period when A is the point of suspension is t_1 and the time period when C is the point suspension is t_2.

The time periods are

$$t_1 = 2\pi\sqrt{\frac{K^2 + l_1^2}{l_1 g}} = 2\pi\sqrt{\frac{K^2/l_1 + l_1}{g}}$$

$$t_2 = 2\pi\sqrt{\frac{K^2 + l_2^2}{l_2 g}} = 2\pi\sqrt{\frac{K^2/l_2 + l_2}{g}}$$

But
$$\frac{K^2}{l_1} = l_2$$

$$\frac{K^2}{l_2} = l_1$$

i.e.
$$\frac{K^2}{l_1} + l_1 = \frac{K^2}{l_2} + l_2 = l_1 + l_2$$

$\therefore$
$$t_1 = t_2$$

Thus the points of suspension and oscillation are interchangeable. If A is the point of suspension, C is the point of oscillation. If C is the point of suspension, A is the point of oscillation.

1.14. Minimum Time Period

The time period of a compound pendulum,

$$t = 2\pi\sqrt{\frac{K^2 + l_l^2}{l_1 g}} \quad \ldots (1)$$

Here l_1 is the distance of the point of suspension from the centre of gravity and K is the radius of gyration about the C.G. Suppose the distance of the point of oscillation from the C.G. is l_2

$$\therefore \quad \frac{K^2}{l_1} = l_2 \quad \text{or} \quad K^2 = l_1 l_2 \quad \ldots (2)$$

For the time period to be minimum, the value of $\left(\frac{K^2}{l_1} + l_1\right)$ should be minimum.

Differeniating $\frac{K^2}{l_1} + l_1$ with respect to l_1

$$\frac{d}{dl_1}\left[\frac{K^2}{l_1} + l_1\right] = -\frac{K^2}{l_1^{\,2}} + 1 \quad \ldots (2a)$$

This value should be zero for the time period to be minimum.

$$\therefore \quad -\frac{K^2}{l_1^{\,2}} + 1 = 0$$

or $$K^2 = l_1^2$$

But $$K^2 = l_1 l_2$$

$$\therefore \quad l_1 l_2 = l_1^{\,2}$$

or $$l_2 = l_1 \quad \ldots (3)$$

It means that the time period will be minimum when the points of suspension and oscillation are equidistant from the centre of gravity.

[**Note.** Also t is maximum when l approaches zero, *i.e.* the point of suspension is at the centre of gavity, when

$$l = 0, \quad t = \infty.$$

Example 1.13. *A heavy circular ring of radius R oscillates in a vertical plane about a horizontal axis at a distance x from the centre. Show that the time period is minimum when x=R.*

The radius of gyration for a circular ring about an axis passing through the C.G. and perpendicular to its planes,

$$K^2 = R^2$$

or $$K = R$$

Here $$l_1 = x$$

For the time period to be minimum

$$l_1 = l_2$$

or $$K^2 = l_1 \times l_2 = l_1^2 = x^2$$

But $$K^2 = R^2$$

$\therefore$ $$x^2 = \mathbf{R^2} \qquad \text{or} \qquad x = \mathbf{R.}$$

Example 1.14. *A heavy uniform rod of length 90* cm *swings in a vertical plane about a horizontal axis passing through its one end. Calculate the position at which a concentrated mass may be placed so that the time of swing remains unaltered.*

Here, $$L = 90 \text{ cm}$$

Radius of gyration, $$K = \sqrt{\frac{L^2}{12}} = \frac{90}{\sqrt{12}} \text{ cm}$$

Distance of the point of suspension from the C.G.

$$l_1 = 45 \text{ cm}$$

Suppose the distance of the point of oscillation from the centre of gravity = l_2

$\therefore$ $$K^2 = l_1 \times l_2$$

$$l_2 = \frac{K^2}{l_1} = \frac{90 \times 90}{12 \times 45} = 15 \text{ cm}$$

The distance of the point of oscillation from the point of suspension

$$= l_1 + l_2 = 45 + 15 = \mathbf{60 \text{ cm.}}$$

Therefore, a concentrated mass has to be placed at a distance of 60 cm from the point of suspension so that the time period remains unaltered.

Example 1.15. *A uniform circular disc of diameter 20* cm *vibrates about a horizontal axis perpendicular to its plane and at a distance of 5* cm *from the centre. Calculate, the time period of oscillation and the equivalent length of the simple pendulum.*

Here $$R = 10 \text{ cm}$$

Radius of gyration $$K = \sqrt{\frac{R^2}{2}} = \sqrt{50} \text{ cm}$$

Distance of the point of suspension from the C.G.

$$= l = 5 \text{ cm}$$

$$g = 980 \text{ cm/s}^2$$

$$t = 2\pi\sqrt{\frac{K^2 + l^2}{lg}}$$

$$t = 2\pi\sqrt{\frac{50 + 25}{5 \times 980}}$$

$$t = \mathbf{0{\cdot}782 \text{ s.}}$$

Equivalent length of the simple pendulum=L

Here $$L = \frac{K^2 + l^2}{l}$$

$$L = \frac{50 + 25}{5} = \mathbf{15\ cm.}$$

Example 1.16. *A uniform square lamina of side 24* cm, *oscillates in a vertical plane about a horizontal axis perpendicular to the plane of the lamina and within its boundary. Calculate (i) the minimum time period of oscillation and (ii) the locus of the points of suspension about which the time period is minimum.*

Here $$l = 24 \text{ cm}$$

For a square lamina, the radius of gyration,

$$K = \frac{l}{\sqrt{6}} = \frac{24}{\sqrt{6}} \text{ cm}$$

$$K^2 = \frac{24 \times 24}{6} = 96$$

When the time period is minimum, the points of suspension and oscillation are equidistant from the centre of gravity.

$\therefore$ $$l_1 = l_2$$

$$K^2 = l_1 \times l_2 = l_1^2$$

or $$l_1^2 = l_2^2 = K^2 = 96$$

$$T = 2\pi \sqrt{\frac{K^2 + l_1^2}{l_1 g}}$$

$$T = 2\pi \sqrt{\frac{96 + 96}{\sqrt{96} \times 980}}$$

$$T = 2\pi \sqrt{\frac{192}{9{\cdot}8 \times 980}}$$

$$T = \frac{2\pi \times 8\sqrt{3}}{98}$$

$$T = 0{\cdot}8884 \text{ s}$$

$$l_1 = \sqrt{96} = \mathbf{9{\cdot}8\ cm.}$$

The locus of the points about which the time period is minimum is a circle of radius 9·80 cm with the C.G. as its centre.

Example 1.17. *A disc of radius 15* cm *is executing small oscillations about a point on its rim. Find the time period of oscillation.* [IAS]

Moment of inertia of the disc about an axis passing through its centre of gravity and perpendicular to its plane,

$$I = \frac{MR^2}{2}$$

$$MK^2 = \frac{MR^2}{2}$$

$$K^2 = \frac{R^2}{2}$$

Time period of oscillation

$$T = 2\pi\sqrt{\frac{K^2 + l^2}{lg}}$$

Here $l = R$

$$T = 2\pi\left[\frac{\left(\frac{R^2}{2}\right) + R^2}{Rg}\right]^{\frac{1}{2}}$$

$$T = 2\pi\left[\frac{3R}{2g}\right]^{\frac{1}{2}}$$

Here $R = 15 \text{ cm} = 0{\cdot}15 \text{ m}$

$$g = 9{\cdot}8 \text{ m/s}^2$$

$$T = 2\pi\left[\frac{3 \times 0{\cdot}15}{2 \times 9{\cdot}8}\right]^{\frac{1}{2}}$$

$$\mathbf{T = 0{\cdot}95 \text{ s.}}$$

1.15. Kater's Pendulum

A Kater's pendulum consists of a rigid metal bar having two knife edges whose positions can be altered along the length of the bar (Fig 1.11). M_1 and M_2 are two cylinders of the same size and shape. M_1 is made of wood and M_2 is made of metal say brass or iron. There is a third adjustable cylinder M_3 whose position can also be altered along the bar. On the outer surface of M_3 there is an adjustable screw whose position on the surface M_3 can be changed. This helps in the final adjustment of the time period.

The positions of knife edges K_1 and K_2 is fixed so that they face each other and are on either side of the centre of gravity of the pendulum. The time period of the pendulum about each knife edge is determined. In the beginning the difference in the time periods may be large. The position of the mass M_2 is adjusted suitably so that the difference in time periods decreases. Finally, by adjusting the positions of the masses M_3 and M_1, the difference in time periods is made extremely small (say 0·01 s).

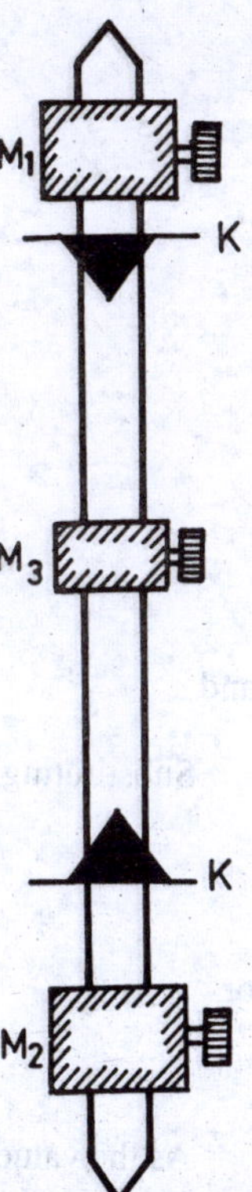

Fig. 1.11.

Let l_1 and l_2 be the distances of the knife edges from centre of gravity of the pendulum in the final adjusted positions. Let t_1 and t_2 be corresponding time periods about the two knife edges K_1 and K_2.

$$t_1 = 2\pi\sqrt{\frac{K^2 + l_1^2}{l_1 g}} \quad \ldots (1)$$

$$t_2 = 2\pi\sqrt{\frac{K^2 + l_2^2}{l_2 g}} \quad \ldots (2)$$

Squaring (1) and (2)

$$t_1^2 = 4\pi^2\left[\frac{K^2 + l_1^2}{l_1 g}\right] \quad \ldots (3)$$

$$t_2^2 = 4\pi^2\left[\frac{K^2 + l_2^2}{l_2 g}\right] \quad \ldots (4)$$

From equations (3) and (4)

$$t_1^2\, l_1 = \frac{4\pi^2}{g}\,[K^2 + l_1^2] \quad \ldots (5)$$

and

$$t_2^2\, l_2 = \frac{4\pi^2}{g}\,[K^2 + l_2^2] \quad \ldots (6)$$

Subtracting (6) from (5)

$$t_1^2\, l_1 - t_2^2\, l_2 = \frac{4\pi^2}{g}\,[l_1^2 - l_2^2]$$

or

$$\frac{4\pi^2}{g} = \frac{t_1^2\, l_1 - t_2^2\, l_2}{l_1^2 - l_2^2}$$

Using the method of partial fractions

$$\frac{4\pi^2}{g} = \frac{t_1^2 l_1 - t_2^2\, l_2}{(l_1^2 - l_2^2)} = \frac{A}{l_1 + l_2} + \frac{B}{l_1 - l_2} \quad \ldots (7)$$

$$= \frac{A(l_1 - l_2) + B(l_1 + l_2)}{(l_1^2 - l_2^2)}$$

$$= \frac{l_1(A + B) - l_2(A - B)}{(l_1^2 - l_2^2)}$$

$$\therefore \qquad A + B = t_1^2$$

$$A - B = t_2^2$$

$$\therefore \qquad A = \frac{t_1^2 + t_2^2}{2}$$

and

$$B = \frac{t_1^2 - t_2^2}{2}$$

Substituting these values in equation (7)

$$\frac{4\pi^2}{g} = \frac{t_1^2 + t_2^2}{2(l_1 + l_2)} + \frac{t_l^2 - t_2^2}{2(l_1 - l_2)}$$

or

$$g = \frac{8\pi^2}{\left(\frac{t_1^2 + t_2^2}{l_1 + l_2}\right) + \left(\frac{t_1^2 - t_2^2}{l_1 - l_2}\right)} \qquad \ldots (8)$$

As the values of l_1, l_2, t_1 and t_2 are known by experiment, the value of g can be determined, provided the position of the centre of gravity is accurately known.

However, as it is difficult to locate the position of the centre of gravity in a Kater's pendulum, the time periods t_1 and t_2 are adjusted to be very nearly equal so that in equation (8)

$$\frac{t_1^2 - t_2^2}{l_1 - l_2}$$

is negligibly small.

From equation (8)

$$g = \frac{8\pi^2}{\frac{t_1^2 + t_2^2}{l_1 + l_2}}$$

Taking $\qquad t_1 = t_2 = t \qquad l_1 + l_2 = L$

$$g = \frac{8\pi^2 L}{2t^2}$$

$$g = \frac{4\pi^2 L}{t^2} \quad ...(9)$$

Here L is the distance between the two knife edges. Equation (9) is similar to the equation of a simple pendulum.

[**Note.** When the positions of K_1, K_2, M_1, M_2 and M_3 are finally adjusted, then the time periods about each knife edge must be equal. The positions of M_1, M_2 and M_3 must be the same while determining the time period about each knife edge.]

1.16. Simple Harmonic Oscillations of a Mass between Two Springs

Consider two springs S_1 and S_2 each having a length l in the free position. Mass M is placed midway between the two springs on a frictionless surface [Fig. 1.12 *(i)*]. One end of the spring S_1 is attached to a rigid wall at A and the other end is attached to the mass M. Similarly one end of the spring S_2 is attached to a rigid wall at B and the other end is connected to the mass M. Here $AC = BC = L$. [Fig.1.12 *(ii)*]. At C the mass is equally pulled by both the springs and it is the equilibrium position.

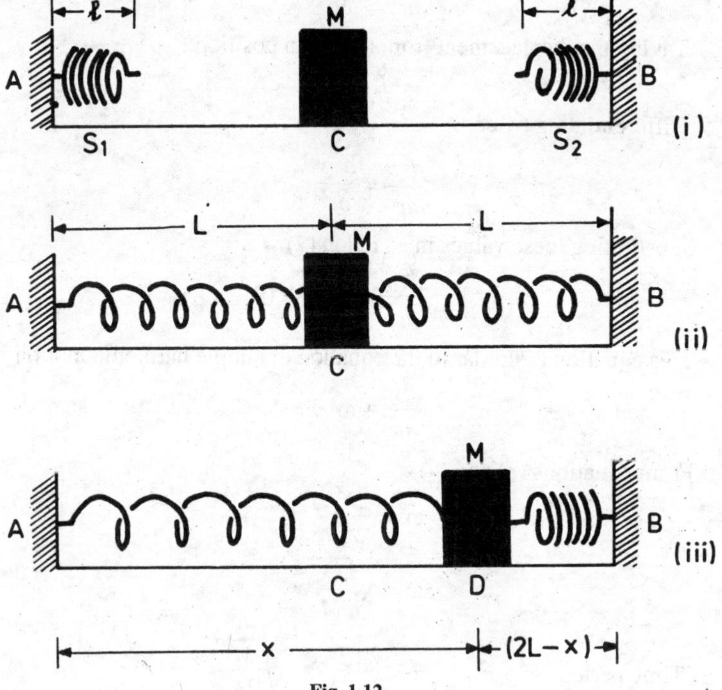

Fig. 1.12.

When the mass M is displaced from its equilibrium position and left, it executes simple harmonic oscillations. Let, at any instant, D be the displaced position of the mass M.

Here $AD = x$, and $BD = (2L - x)$

Let the tension per unit displacement in the spring be K. The displacement of the spring of S_1 is $(x - l)$ and it exerts a force $= K[x - l]$ in the direction DA. The displacement of the spring S_2 is $(2L - x - l)$ and it exerts a force $= K[2L - x - l]$ in the direction DB.

The resultant force on the mass M

$$= K[2L - x - l] - K[x - l] \text{ in the direction } DB$$

$$= -2K[x - L] \text{ in the direction } DB$$

According to Newton's second law of motion

$$F = M\frac{d^2x}{dt^2} = -2K[x - L] \quad \ldots (1)$$

$$\therefore \quad \frac{d^2x}{dt^2} = -\frac{2K}{M}[x - L]$$

or

$$\frac{d^2x}{dt^2} + \frac{2K}{M}(x - L) = 0 \quad \ldots (2)$$

Taking the displacement from the mean position

$$x - L = y$$

Differentiating twice,

$$\frac{d^2x}{dt^2} = \frac{d^2y}{dt^2}$$

Substituting these values in equation (2)

$$\frac{d^2y}{dt^2} + \frac{2K}{M}y = 0 \quad \ldots (3)$$

This equation is similar to the equation of simple harmonic motion

$$\frac{d^2y}{dt^2} + \omega^2 y = 0 \quad \ldots (4)$$

From equations (3) and (4)

$$\omega^2 = \frac{2K}{M}$$

$$\omega = \sqrt{\frac{2K}{M}}$$

Time period $$T = \frac{2\pi}{\omega} = 2\pi\sqrt{\frac{M}{2K}} \quad \ldots (5)$$

Thus, the mass M executes simple harmonic oscillations and the time period is given by equation (5). Knowing the values of M and K, the time period can be calculated.

1.17. Mass between Two Springs – Transverse Oscillations

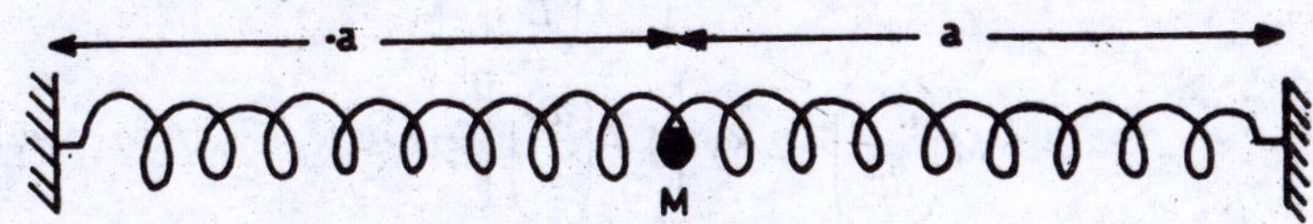

Fig. 1.13 (*a*)

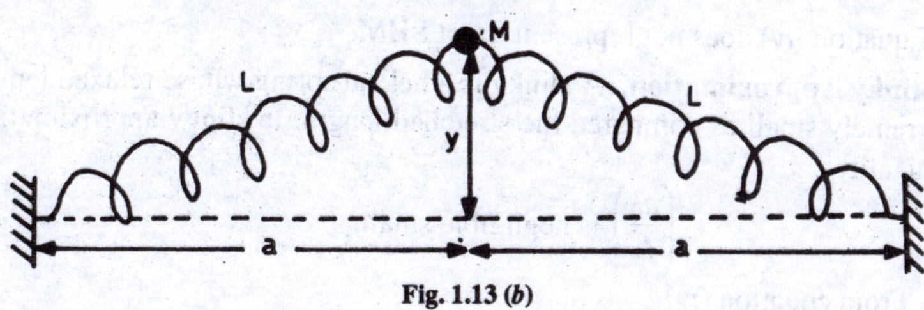

Fig. 1.13 (*b*)

Consider two springs each having a length a_0 in the free (relaxed) position. Mass M is placed midway between the two springs on a frictionless surface [(Fig. 1.13 (*a*)]. The length of each spring in the horizontal position is a. It is assumed that a_0 is extremely small as compared to a.

The mass M is displaced along the Y-axis through a displacement y. It is assumed that there is no displacement of the mass along X- or Z-axis.

In the equilibrium position, the tension in each spring is given by

$$T_0 = K(a - a_o) \qquad \ldots (i)$$

In the displaced position, each spring has a length L and tension in each spring is given by

$$T = K[L - a_o] \qquad \ldots (ii)$$

This tension acts along the axis of the spring.

The component along the Y-axis contributes the return force and the transverse oscillations are set up in the system :

The return force for each spring is $T \sin \theta$.

The net force acting on the mass due to both the springs along –ve y-axis is given by

$$F = -2T \sin \theta$$

Here $\sin\theta = \frac{y}{L}$

$$\therefore \quad F = -2K[L - a_0]\left(\frac{y}{L}\right)$$

$$F = -2Ky\left[1 - \left(\frac{a_0}{L}\right)\right] \qquad \ldots (iii)$$

$$\therefore \quad M\left(\frac{d^2y}{dt^2}\right) + 2Ky\left[1 - \left(\frac{a_0}{L}\right)\right] = 0 \qquad \ldots (iv)$$

$$\omega^2 = \left(\frac{2K}{M}\right)\left[1 - \left(\frac{a_0}{L}\right)\right] \qquad \ldots (v)$$

Equation (*iv*) does not represent exact SHM.

Slinky Approximation. A slinky is a helical spring whose relaxed length is extremely small as compared the stretched length. In slinky approximation the quantity

$\left(\frac{a_0}{L}\right)$ is negligible small.

From equation (*iv*)

neglecting $\left(\frac{a_0}{L}\right)$ we get

$$M\left(\frac{d^2y}{dt^2}\right) + 2Ky = 0$$

$$\frac{d^2y}{dt^2} + \left(\frac{2K}{M}\right)y = 0 \qquad \ldots (vi)$$

$$\omega = \left[\frac{2K}{M}\right]^{\frac{1}{2}} \qquad \ldots (vii)$$

$$\nu = \frac{1}{2\pi}\left[\frac{2K}{M}\right]^{\frac{1}{2}} \qquad \ldots (viii)$$

Time period $t = \frac{1}{\nu} = 2\pi\left[\frac{M}{2K}\right]^{\frac{1}{2}}$

The solution of equation (*vi*) is

$$y = A\sin(\omega t + \phi) \qquad \ldots (ix)$$

Also from equation (*i*)

$$K = \frac{T_0}{a\left[1-\left(\frac{a_0}{a}\right)\right]}$$

$\frac{a_0}{a}$ is negligibly small

$$\therefore \qquad K = \frac{T_0}{a}$$

Substituting this value of K in equation (*vii*)

$$\omega = \left[\frac{2\,T_o}{Ma}\right]^{\frac{1}{2}} \qquad \ldots (x)$$

and

$$\nu = \frac{1}{2\pi}\left[\frac{2\,T_0}{Ma}\right]^{\frac{1}{2}} \qquad \ldots (xi)$$

Time period

$$T = \frac{1}{\nu} = 2\pi\left[\frac{Ma}{2\,T_0}\right]^{\frac{1}{2}} \qquad \ldots (xii)$$

It may be noted that equation (*ix*) has no restriction on the amplitude A. Even for large amplitude there will be perfect linearity of the return force. This holds good only for slinky approximation.

The frequency is the same for both longitudinal and transverse oscillations.

Example. 1.18 *Show that for a mass connected between two identical springs,*

$$\frac{\omega_{long}}{\omega_{trans}} = \frac{1}{\left[1-\frac{a_0}{a}\right]^{\frac{1}{2}}}$$

For longitudinal oscillations

$$\omega_{long} = \left[\frac{2\,K}{M}\right]^{\frac{1}{2}} \qquad \ldots (i)$$

For transverse oscillations

$$\omega_{trans} = \left[\frac{2\,K}{M}\left[1-\left(\frac{a_0}{a}\right)\right]\right]^{\frac{1}{2}} \qquad \ldots (ii)$$

$$\therefore \qquad \frac{\omega_{long}}{\omega_{trans}} = \frac{1}{\left[1-\frac{a_0}{a}\right]^{\frac{1}{2}}} \qquad \ldots (iii)$$

1.18. Simple Harmonic Oscillations of a Loaded Spring

Consider a spring S whose upper end is fixed to a rigid support and the lower end is attached to a mass M (Fig.1.14). In the equilibrium position, the mass is at A. When the mass is displaced downwards and left, it oscillates simple harmonically in the vertical direction.

Suppose at any instant the mass is at B. The distance $AB = y$. Let the tension per unit displacement of the spring be K.

Force exerted by the spring=Ky

According to Newton's second law

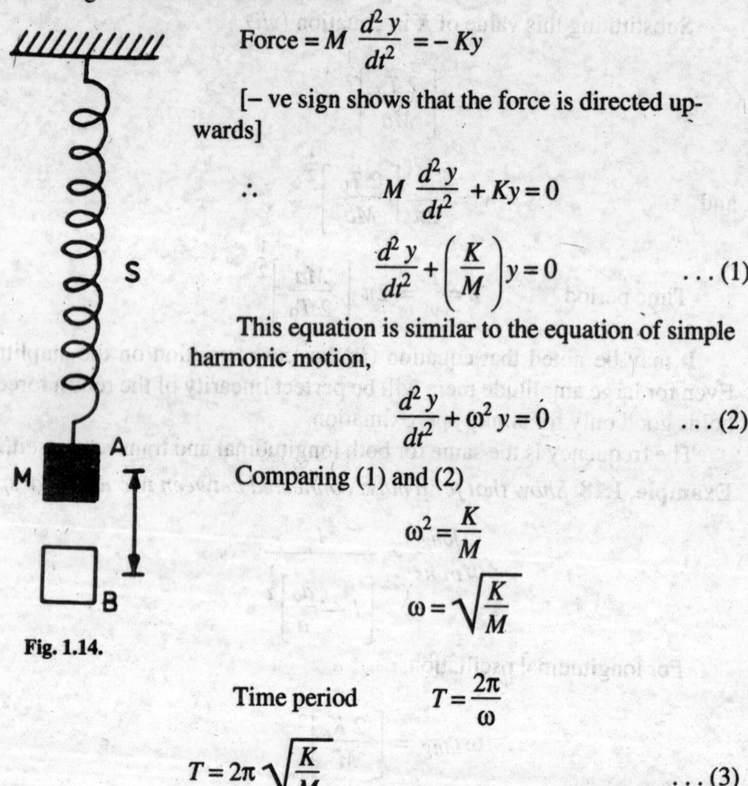

Fig. 1.14.

$$\text{Force} = M\frac{d^2y}{dt^2} = -Ky$$

[– ve sign shows that the force is directed upwards]

$$\therefore \qquad M\frac{d^2y}{dt^2} + Ky = 0$$

$$\frac{d^2y}{dt^2} + \left(\frac{K}{M}\right)y = 0 \qquad \ldots (1)$$

This equation is similar to the equation of simple harmonic motion,

$$\frac{d^2y}{dt^2} + \omega^2 y = 0 \qquad \ldots (2)$$

Comparing (1) and (2)

$$\omega^2 = \frac{K}{M}$$

$$\omega = \sqrt{\frac{K}{M}}$$

Time period $\qquad T = \dfrac{2\pi}{\omega}$

or

$$T = 2\pi\sqrt{\frac{K}{M}} \qquad \ldots (3)$$

Knowing the values of M and K, the value of T can be calculated.

Determination of K. To determine the value of tension per unit displacement of the spring, a small mass m is attached to the free end of the spring. The increase in length of the spring is noted. Let it be x.

Then, $\qquad K = \left(\dfrac{mg}{x}\right)$

Substituting the value of K in equation (3),

$$T = 2\pi\sqrt{\frac{Mx}{mg}} \qquad \ldots (4)$$

It is to be noted that $\frac{mg}{x}$ is constant for a given spring.

Example 1.19. *A spring is hung vertically and loaded with a mass of 100* grams *and allowed to oscillate. Calculate (i) the time period and (ii) the frequency of oscillation. When the spring is loaded with 200* grams *it extends by 10* cm.

Here

$$M = 100 \text{ grams}$$
$$m = 200 \text{ grams}$$
$$x = 10 \text{ cm}$$
$$g = 980 \text{ cm/s}^2$$

(*i*)

$$T = 2\pi\sqrt{\frac{Mx}{mg}}$$

$$T = 2\pi\sqrt{\frac{100 \times 10}{200 \times 980}}$$

$$T = \frac{2\pi}{14} = \mathbf{0{\cdot}449\ s}$$

(*ii*) Frequency

$$n = \frac{1}{T} = \frac{1}{0{\cdot}449}$$

$$n = \mathbf{2{\cdot}22\ hertz.}$$

Example 1.20. *The scale of a spring balance reading from 0 — 10* kg *is 0·25* m. *A body suspended from the balance oscillates with a frequency of $\frac{10}{\pi}$* hertz. *Calculate the mass of the body attached to the spring.*

Here

$$m = 10 \text{ kg}$$
$$x = 0{\cdot}25 \text{ m}$$
$$M = ?$$
$$g = 9{\cdot}8 \text{ m/s}^2$$
$$n = \frac{10}{\pi} \text{ hertz}$$

or

$$T = \frac{1}{n} = \frac{\pi}{10} \text{ s}$$

$$T = 2\pi\sqrt{\frac{Mx}{mg}}$$

or
$$M = \frac{T^2 mg}{4\pi^2 x}$$

$$M = \frac{\pi^2 \times 10 \times 9{\cdot}8}{100 \times 4\pi^2 \times 0{\cdot}25}$$

$$\mathbf{M = 0{\cdot}98\ kg.}$$

Example 1.21. *A body of mass 0·5* kg *is suspended from a spring of negligible mass and it stretches the spring by 0·07*m. *For a displacement of 0·03* m *it is given a downward velocity 0·4* m/s. *Calculate (i) the time period, (ii) the frequency and (iii) amplitude of vibration of the spring.*

Here
$$m = 0{\cdot}5\ \text{kg}$$
$$x = 0{\cdot}07\ \text{m}$$
$$M = 0{\cdot}5\ \text{kg}$$
$$g = 9{\cdot}8\ \text{m/s}^2$$

(*i*)
$$T = 2\pi\sqrt{\frac{Mx}{mg}}$$

$$T = 2\pi\sqrt{\frac{0{\cdot}5 \times 0{\cdot}07}{0{\cdot}5 \times 9{\cdot}8}}$$

$$\mathbf{T = 0{\cdot}5311\ s.}$$

(*ii*) Frequency
$$n = \frac{1}{T} = \frac{1}{0{\cdot}5311}$$

$$n = 1{\cdot}882\ \text{hertz}$$

Angular frequency $\omega = \frac{2\pi}{T} = \frac{2\pi}{0{\cdot}5311} = \mathbf{11{\cdot}8\ radians.}$

(*iii*) Amplitude

$$y = a \sin \omega t$$

$$\frac{dy}{dt} = a\,\omega \cos \omega t$$

$$\frac{dy}{dt} = a\omega\sqrt{1 - \frac{y^2}{a^2}}$$

$$\frac{dy}{dt} = \omega\sqrt{a^2 - y^2}$$

Here
$$\frac{dy}{dt} = 0{\cdot}4\ \text{m/s}$$

$$y = 0{\cdot}03\ \text{m}$$

and $\omega = 11{\cdot}8$ radians/s

$$\therefore \quad 0{\cdot}4 = 11{\cdot}8\sqrt{a^2 - (0{\cdot}03)^2}$$

By simplification

$$\therefore \quad a = \mathbf{0{\cdot}04526\ m.}$$

Example 1.22. *A uniform spring of force constant k is cut into two pieces of equal length. What is the force constant of each piece ?*

Suppose, Force $= F$

Increase in length $= l$

$$k = \frac{F}{l} \qquad \ldots (i)$$

In the second, the length is half and increase in length for the force $F = \frac{l}{2}$

$$\therefore \quad k_1 = \frac{F}{l/2}$$

$$k_1 = 2\left[\frac{F}{l}\right]$$

$$k_1 = 2\,\mathrm{k} \qquad \ldots (ii)$$

Force constant for each piece = **2 k.**

Example 1.23. *A uniform spring of force constant k is cut into two pieces whose lengths are in the ratio of 1 : 3. Calculate the force constant of each piece.*

Here Force $= F$

Increase in length $= l$

$$k = \frac{F}{l} \qquad \ldots (i)$$

When the springs are cut in the ratio 1 : 3,

Increase in length for the first piece for force $F = \frac{l}{4}$

Increase in length for the second piece for force $F = \frac{3l}{4}$

$$\therefore \quad k_1 = \frac{F}{l/4} = 4\left[\frac{F}{l}\right]$$

$$k_1 = \mathbf{4}\,k \qquad \ldots (ii)$$

and

$$k_2 = \frac{F}{3l/4} = \frac{4}{3}\left[\frac{F}{l}\right]$$

$$k_2 = \frac{4k}{3} \qquad \ldots (iii)$$

Example 1.24. *Two identical springs each of force constant k are connected as shown in Fig 1.15. Find the force constant of the system.*

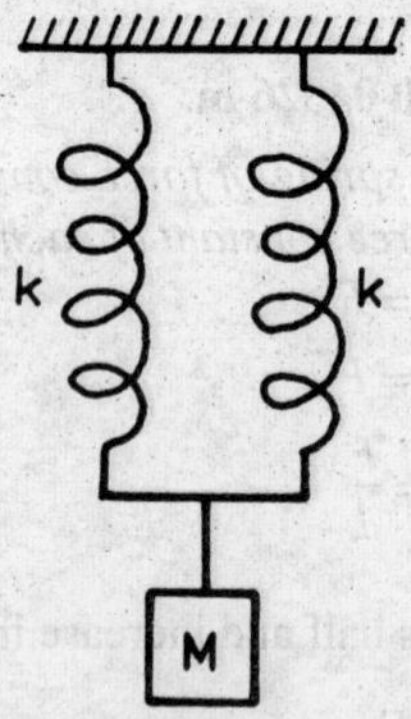

Fig. 1.15.

When force F is applied to each spring separately, increase in length $= l$

$$\therefore \qquad k = \frac{F}{l} \qquad \ldots (i)$$

When force F is applied to the system, force $F/2$ acts on each spring and increase in length of the system $= \dfrac{l}{2}$

Force constant of the system.

$$k_1 = \frac{F}{l/2} = 2\left[\frac{F}{l}\right]$$

$$k_1 = 2\,\mathbf{k}.$$

Example 1.25. *Two identical springs each of force constant k are connected as shown in Fig. 1.16. Calculate the force constant of the system.*

When force F is applied to each spring separately, increase in length $= l$

$$\therefore \qquad k = \frac{F}{l} \qquad \ldots (i)$$

When force F is applied to the system, each spring extends by l and increase in length of the system $= 2\,l$

$$\therefore \qquad k_1 = \frac{F}{2l} = \frac{l}{2}\left[\frac{F}{l}\right]$$

$$k_1 = \frac{\mathbf{k}}{\mathbf{2}} \qquad \ldots (ii)$$

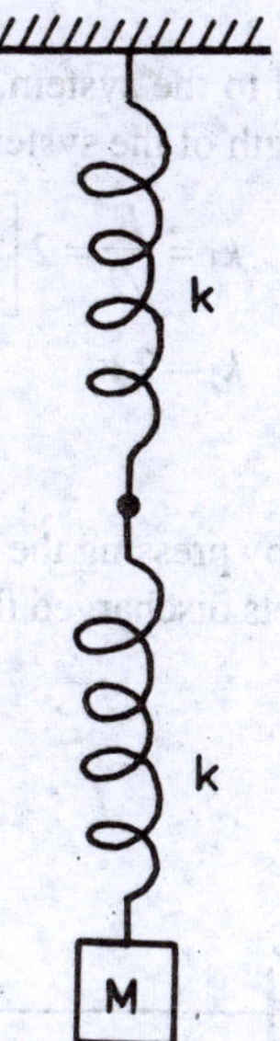

Fig. 1.16.

Example 1.26. *Two identical springs each of force constant k are connected as shown in Fig. 1.17. Find the force constant of the system.*

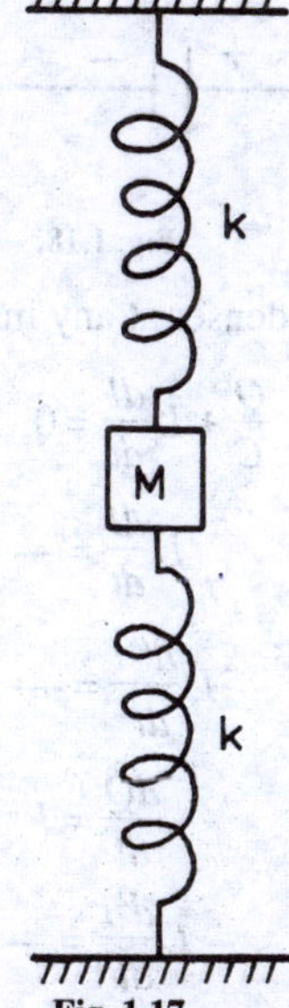

Fig. 1.17.

When force F is applied to each spring separately, increase in length = l

$$\therefore \qquad k = \frac{F}{l} \qquad \ldots (i)$$

When force F is applied to the system, change in length in each spring is $l/2$. Therefore change in length of the system = $l/2$.

$$\therefore \qquad k_1 = \frac{F}{l/2} = 2\left[\frac{F}{l}\right]$$

$$k_1 = \mathbf{2\,k}.$$

1.19. LC Circuit

A condenser is charged by pressing the morse key and when the morse key is released, the condenser gets discharged through the inductance L (Fig.1.18).

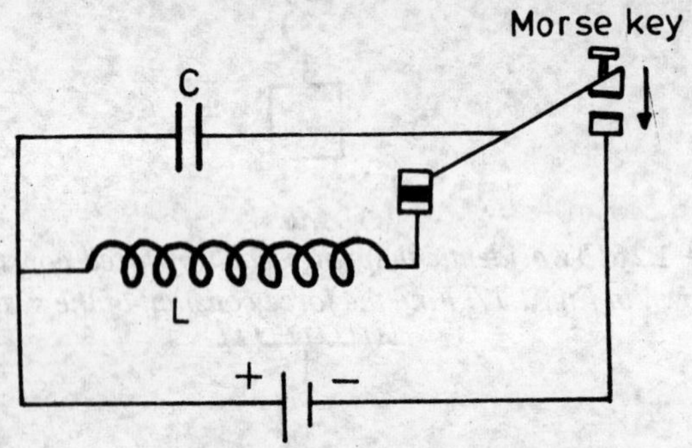

Fig. 1.18.

Let the charge on the condenser at any instant of time be Q.

$$\therefore \qquad \frac{Q}{C} + L\frac{dI}{dt} = 0$$

$$L\frac{dI}{dt} = -\frac{Q}{C}$$

$$L\frac{d^2 I}{dt^2} = -\frac{1}{C}\frac{dQ}{dt}$$

But
$$\frac{dQ}{dt} = I$$

$$\therefore \qquad L\frac{d^2 I}{dt^2} = -\frac{I}{C}$$

$$\frac{d^2 I}{dt^2} + \left(\frac{1}{LC}\right) I = 0 \qquad \ldots(1)$$

This equation is similar to the equation of simple harmonic motion,

$$\frac{d^2y}{dt^2} + \omega^2 y = 0 \qquad \ldots (2)$$

From equations (1) and (2)

$$\omega^2 = \frac{1}{LC}$$

or $$\omega = \sqrt{\frac{1}{LC}}$$

Time period $$T = \frac{2\pi}{\omega} = 2\pi\sqrt{LC}$$

Frequency $$n = \frac{1}{T} = \frac{1}{2\pi\sqrt{LC}}$$

This shows that in a circuit containing an inductance and a capacitance, the discharge is oscillatory and undamped.

The general solution of equation (1) is given by

$$I = A \sin \omega t + \alpha$$

When $$t = 0$$

$$I = I_0$$

and when $$\alpha = \pi/2, A = I_0$$

$\therefore$ $$I = I_0 \sin(\omega t + \pi/2)$$

or $$I = I_0 \sin\left(\frac{t}{\sqrt{LC}} + \frac{\pi}{2}\right) \qquad \ldots (3)$$

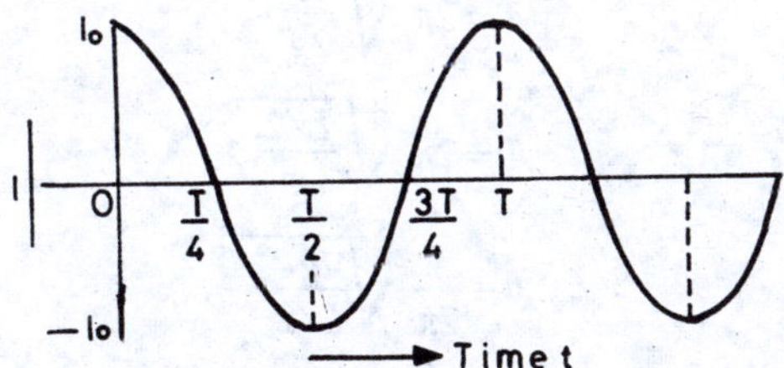

Fig. 1.19. Current—Time Graph.

Equation (3) shows that the discharge is oscillatory as represented in Fig.1.19. This circuit is of importance in wireless.

1.20. U-Tube

In a U-tube containing a liquid, the time period of oscillation depends upon the force constant K and the length of the liquid column. Suppose, the liquid is disturbed and the displacement is Y. (Fig.1.20).

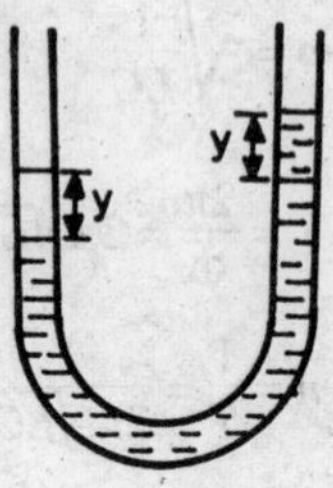

Fig. 1.20.

Force per unit displacement,

$$K = \frac{\text{Force}}{y} = \frac{2 \times y A\rho g}{y} = 2A\rho g$$

Here A is the area of cross-section and ρ is the density of the liquid.

Mass of the vibrating liquid

$$m = AL\rho$$

Here L is the total length of the liquid column

But $$T = 2\pi \sqrt{\frac{m}{K}}$$

$\therefore$ $$T = 2\pi \sqrt{\frac{AL\rho}{2A\rho g}}$$

$$T = 2\pi \sqrt{\frac{L}{2g}}$$

$$T = 2\pi \sqrt{\frac{L/2}{g}} \quad \ldots (i)$$

or $$\nu = \frac{1}{2\pi} \sqrt{\frac{g}{L/2}} \quad \ldots (ii)$$

Example 1.27 *Nine kilograms of mercury is poured into a glass U-tube of uniform internal diameter of 1.2cm. It oscillates freely about its equilibrium position. Calculate the period of oscillation.* [IAS,1990]

Diameter of the tube,

$$d = 1{\cdot}2 \text{ cm} = 1{\cdot}2 \times 10^{-2} \text{ m}$$

Area of cross section

$$A = \frac{\pi d^2}{4} = \frac{3{\cdot}14 \times (1{\cdot}2\times10^{-2})^2}{4}$$

$$A = 1{\cdot}1304 \times 10^{-4} \text{ m}^2$$

Density of mercury

$$\rho = 13{\cdot}6\times10^3 \text{ kg/m}^3$$

$$M = 9 \text{ kg}$$

Total length of liquid column $= L$

$$M = \rho A L$$

$$L = \frac{M}{\rho A}$$

$$= \frac{9}{13{\cdot}6 \times 10^3 \times 1{\cdot}1304 \times 10^{-4}}$$

$$= \mathbf{5{\cdot}854 \text{ m.}}$$

Suppose, the liquid is displaced through y,

Force per unit displacement,

$$K = \frac{\text{Force}}{y} = \frac{2 \times yA\rho g}{y}$$

$$K = 2A\rho g$$

$$M = AL\rho$$

$$\frac{M}{K} = \frac{AL\rho}{2A\rho g} = \frac{L}{2g}$$

Time period.

$$T = 2\pi\sqrt{\frac{M}{K}}$$

$$= 2\pi\sqrt{\frac{L}{2g}}$$

$$= 2\pi\sqrt{\frac{5{\cdot}854}{2 \times 9{\cdot}8}}$$

$$= \mathbf{3{\cdot}434 \text{ s.}}$$

1.21. Air Chamber with a Neck

Suppose a ball of mass m oscillates in the neck of a chamber containing air (Fig.1.21). The volume of air = V. Area of cross section of the neck = A. Displacement of the ball = y. The bulk modulus of elasticity of air = E. Volume of air displaced by the ball =yA.

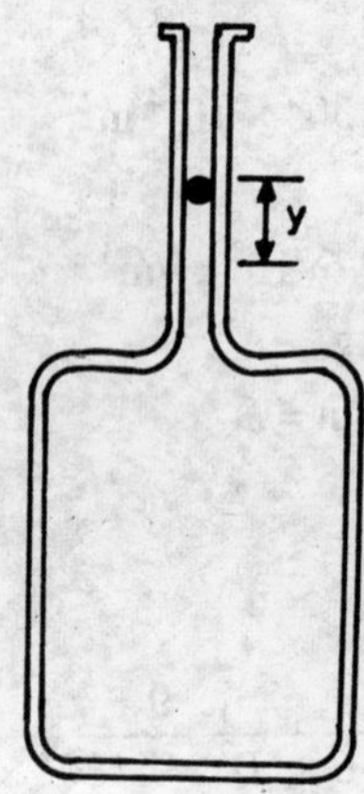

Fig. 1.21.

Elasticity, $$E = \frac{P}{\frac{yA}{V}}$$

$$p = \frac{yAE}{V}$$

or $$F = -pA = -\frac{yA^2E}{V}$$

Force constant, $$K = \frac{F}{-Y} = \frac{A^2E}{V}$$

∴ $$T = 2\pi\sqrt{\frac{m}{K}}$$

$$T = 2\pi\sqrt{\frac{mV}{A^2E}} \quad \ldots (i)$$

$$\nu = 2\pi\sqrt{\frac{A^2E}{mV}} \quad \ldots (ii)$$

Here, mass of air in the neck is negligibly small as compared to the mass of the ball.

However, if there is no ball in the neck and air column of length L in the neck oscillates due to sudden change in pressure, then

$$m = AL\rho$$

Substituting this value of m in equation (i)

$$T = 2\pi\sqrt{\frac{AL\rho V}{A^2E}}$$

$$T = 2\pi\sqrt{\frac{L\rho V}{AE}} \qquad \ldots \text{(iii)}$$

$$v = \frac{1}{2\pi}\sqrt{\frac{AE}{L\rho V}} \qquad \ldots (iv)$$

1.22. Free Oscillations with Two Degrees of Freedom

There are many interesting examples of systems possessing free oscillations with two degrees of freedom. One of the simplest examples is a double pendulum viz., one pendulum attached to another pendulum, [Fig. 1.22 (i)]

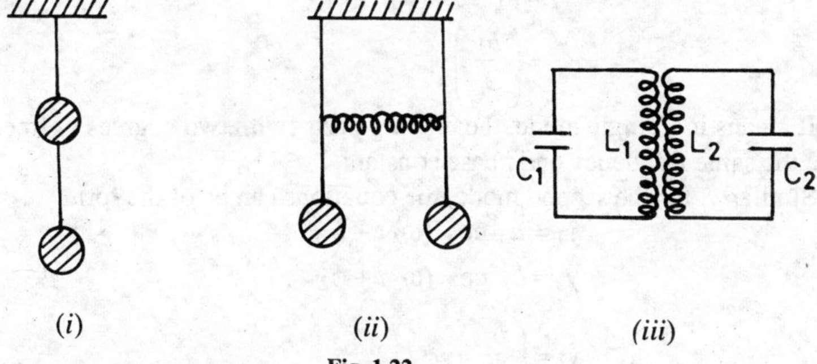

Fig. 1.22.

Two pendulums coupled by a spring also possess oscillations with two degrees of freedom [Fig. 1.22 (ii)]. Similarly two coupled LC circuits also correspond to a system with two degrees of freedom [Fig. 1.22 (iii)].

For a system with two degrees of freedom, the configuration can be described by taking two variables, y_1 and y_2. In the case of a double pendulum, the two variables y_1 and y_2 correspond to the positions of the pendulum in two perpendicular horizontal directions. In the case of pendulums coupled with a spring, y_1 and y_2 represent the positions of the pendulum. In the case of coupled LC circuits, y_1 and y_2 represent the charges on the condensers at any instant of time.

The motion of such a system with two degrees of freedom has a complicated motion as no individual part possesses simple harmonic motion. However, for linear equations of motion, the combined motion is the result of the superposition of two simple harmonic motions. The component simple harmonic motions are called normal modes. With the help of suitable starting conditions viz., y_1, y_2, dy_1/dt and dy_2/dt, it is possible to make the system oscillate with only one mode or the other. Here, the modes are not coupled whereas the moving parts are coupled.

1.23. Properties of a Mode

When an oscillating system possesses only one mode, each moving part executes simple harmonic motion. All the parts oscillate with the same frequency and pass through the mean position at the same time. In such cases, the equations for a single mode cannot be in the form

$$y_1 = a\cos\omega t$$

and $$y_2 = b\sin\omega t \qquad \text{(different phase constants)}$$

or $$y_1 = a\cos\omega_1 t$$

and $$y_2 = b\cos\omega_2 t \qquad \text{(different frequencies)}$$

For the mode 1, the displacements at any instant can be of the form

$$y_1 = a_1\cos(\omega_1 t + \phi_1) \qquad \ldots (1)$$

$$y_2 = b_1\cos(\omega_1 t + \phi_1) \qquad \ldots (2)$$

i.e. $$y_2 = \left(\frac{b_1}{a_1}\right) y_1 \qquad \ldots (3)$$

It means in a single mode, the moving parts (with two degrees of freedom) have the same frequency and phase constant.

Similarly, for the second mode, the equations can be of the form

$$y_1 = a_2\cos(\omega_2 t + \phi_2) \qquad \ldots (4)$$

$$y_2 = b_2\cos(\omega_2 t + \phi_2) \qquad \ldots (5)$$

or $$y_2 = \left(\frac{b_2}{a_2}\right) y_1 \qquad \ldots (6)$$

Each mode has its own characteristic frequency. The general motion of the system having two modes can be written as

$$y_1 = a_1\cos(\omega_1 t + \phi_1) + a_2\cos(\omega_2 t + \phi_2) \qquad \ldots (7)$$

and $$y_2 = b_1\cos(\omega_1 t + \phi_1) + b_2\cos(\omega_2 t + \phi_2) \qquad \ldots (8)$$

It is to be noted that in each mode, the system has a characteristic configuration or shape governed by the ratio of the amplitudes. For mode 1, this ratio is (a_1/b_1) and for mode 2, it is (a_2/b_2). It means in both the modes the ratio y_1/y_2 is constant. It can be either positive or negative depending on the instant of time.

1.24. Two Coupled LC Circuits

Consider a system consisting of coupled circuits (Fig 1.23). If the rate of change of current in circuit 1, is $\frac{dI_1}{dt}$, then the induced emf across the

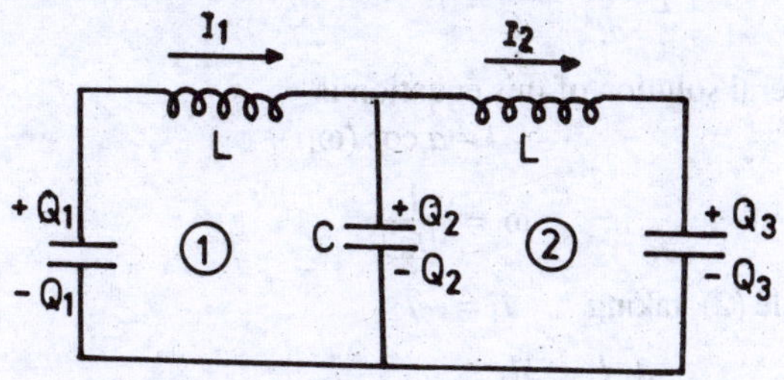

Fig. 1.23.

inductance L is $L\frac{dI_1}{dt}\cdot$ The positive charge Q_1 on the left hand side condenser increases the current I_1 and the charge Q_2 on the middle condenser tends to decrease I_1.

$$\therefore \qquad L\frac{dI_1}{dt} = \frac{Q_1}{C} - \frac{Q_2}{C} \qquad \ldots (1)$$

Similarly for the circuit 2,

$$L\frac{dI_2}{dt} = \frac{Q_2}{C} - \frac{Q_3}{C} \qquad \ldots (2)$$

Differentiating equations (1) and (2)

$$L\frac{d^2 I_1}{dt^2} = \frac{1}{C}\frac{dQ_1}{dt} - \frac{1}{C}\frac{dQ_2}{dt} \qquad \ldots (3)$$

and

$$L\frac{d^2 I_2}{dt^2} = \frac{1}{C}\frac{dQ_2}{dt} - \frac{1}{C}\frac{dQ_3}{dt} \qquad \ldots (4)$$

If the charge is conserved,

$$\frac{dQ_1}{dt} = -I_1$$

$$\frac{dQ_2}{dt} = I_1 - I_2$$

$$\frac{dQ_3}{dt} = I_2$$

Substituting these values in equations (3) and (4)

$$L\frac{d^2 I_1}{dt^2} = -\frac{I_1}{C} + \frac{(I_2 - I_1)}{C} \qquad \ldots (5)$$

$$L\frac{d^2 I_2}{dt^2} = \frac{-(I_2 - I_1)}{C} - \frac{I_2}{C} \qquad \ldots (6)$$

For mode 1, taking $I_1 = I_2$

$$L\frac{d^2 I_1}{dt^2} + \frac{I_1}{C} = 0$$

The general solution of this equation is

$$I = a \cos(\omega_1 t + \phi_1)$$

Here $\omega_1 = \frac{1}{\sqrt{LC}}$ ... (7)

For mode (2), taking $I_1 = -I_2$

$$L\frac{d^2 I_1}{dt^2} + \frac{3I_1}{C} = 0$$

The general solution of this equation is

$$I_1 = b \cos(\omega_2 t + \phi_2)$$

Here $\omega_2 = \sqrt{\frac{3}{LC}}$ ... (8)

In mode 1, the middle condenser does not acquire any charge. This condenser can even be removed without affecting the motion of the charges. Also in this mode, the charges Q_1 and Q_3 are always of equal magnitude but of opposite sign.

In the case of mode 2, the charges Q_1 and Q_3 are equal in magnitude and sign. The charge Q_2 has twice this magnitude but of opposite sign.

Example 1.28. *A particle of mass 10 g has in a potential given by*

$$V_{(x)} = 32\,x^2 + 0{\cdot}2$$

where x is in metres *and V_x in* joules. *Write down the equation of motion and solve it. What is the frequency of oscillation?* (IAS, 1991)

$$V_x = 32\,x^2 + 0{\cdot}2 \qquad \ldots (i)$$

Here V_x is potential energy in joules and x in metres.

Also potential energy of an oscillator

$$V_x = \tfrac{1}{2} K x^2 + \text{constant} \qquad \ldots (ii)$$

Comparing (*i*) and (*ii*)

$$\frac{1}{2} K = 32$$

$$\mathbf{K = 64\ N/m}$$

Mass, $M = 10\text{ g} = 10^{-2}\text{ kg}$

K is force constant

$$\omega = \sqrt{\frac{K}{M}}$$

$$\omega = \sqrt{\frac{64}{10^{-2}}} = \mathbf{80\ rad/s}$$

$$\nu = \frac{\omega}{2\pi} = \frac{80}{2\pi}\ \text{Hz}$$

$$\nu = \mathbf{12{\cdot}72\ Hz}$$

Equation of motion is

$$\frac{d^2x}{dt^2} + \omega^2 x = 0 \qquad \ldots (iii)$$

$$\frac{d^2x}{dt^2} + 6400\,x = 0$$

$$x = A \sin \omega t$$

$$x = A \sin 80\,t \qquad \ldots (iv)$$

1.25. Two Dimensional Harmonic Oscillator

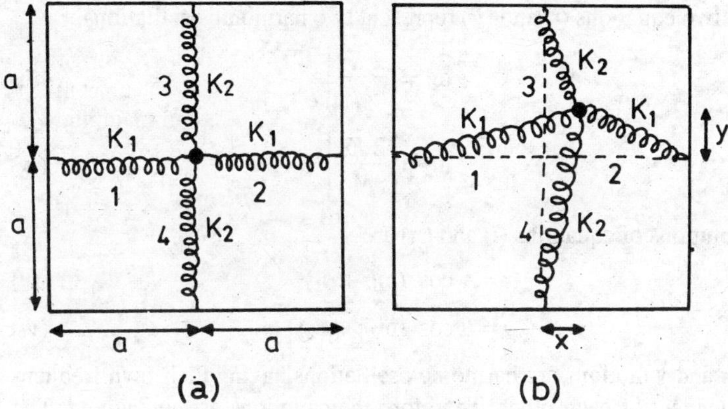

Fig. 1.24.

A mass M is connected by four springs as shown in Fig. 1.24. The springs with spring constant K_1 displace the mass along x-axis and the springs with spring constant K_2 displace the mass along y-axis.

Consider that the mass M moves in the x-y plane with oscillations of very small amplitude. Here, length of the each spring a is very large as compared to displacement x or y. Suppose, frist the mass M is displaced in the X -direction by displacement x. The spring 1 is elongated and spring 2 in compressed. In this case the return force,

$$F_x = -2K_1 x$$

and $$F_y = 0$$

Further, displace the mass M by a small distance y along + Y direction [Fig. 1.24 (b)]. The springs 1 and 2 are extended by a small amount which is neglected. The spring 3 gets compressed and spring 4 gets elongated. It is assumed that F_x and F_y are not independent due to the small values of x and y.

$$\therefore \quad F_x = -2\,K_1 x$$

$$F_y = -2K_2\, y$$

$$\therefore \quad M\left(\frac{d^2x}{dt^2}\right) = -2K_1 x$$

$$M\left(\frac{d^2y}{dt^2}\right) = -2K_2\, y$$

$$\frac{d^2x}{dt^2} + \left(\frac{2K_1}{M}\right)x = 0 \qquad \ldots (i)$$

$$\frac{d^2y}{dt^2} + \left(\frac{2K_2}{M}\right)y = 0 \qquad \ldots (ii)$$

These two equations (i) and (ii) represent two harmonic oscillations,

$$\therefore \quad \omega_1 = \left[\frac{2\,K_1}{M}\right]^{1/2} \qquad \ldots (iii)$$

$$\omega_2 = \left[\frac{2\,K_2}{M}\right]^{1/2} \qquad \ldots (iv)$$

The solutions of equations (i) and (ii) are

$$x = A\cos\ (\omega_1 t + \phi_1) \qquad \ldots (v)$$

$$y = B\cos\ (\omega_2\, t + \phi_2) \qquad \ldots (vi)$$

Here x and y motions are harmonic oscillations having their own frequencies *independent* of each other. Therefore their motions are *uncoupled*. The motion along x-axis corresponds to one normal mode of frequency

$$\nu_1 = \frac{1}{2\pi}\sqrt{\frac{2K_1}{M}} \qquad \ldots (vii)$$

and motion along y-axis corresponds to second normal mode of frequency

$$\nu_2 = \frac{1}{2\pi}\sqrt{\frac{2K_2}{M}} \qquad \ldots (viii)$$

1.26. Two Coupled Pendulums

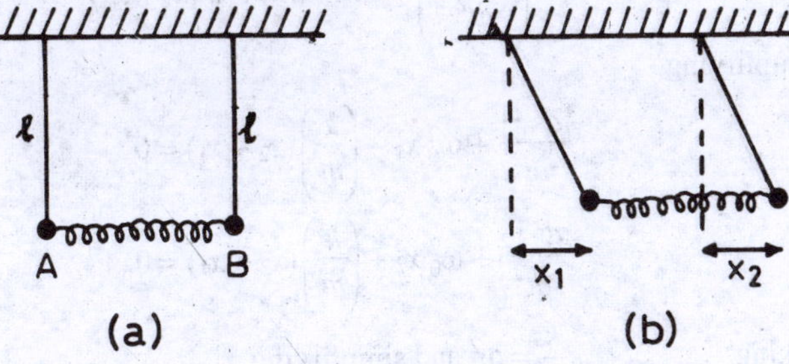

Fig. 1.25.

A and *B* are two simple pendulums suspended from a rigid support. Each pendulum has a length *l*. Mass of the bob of each pendulum is *m*. The two masses are attached to a spring of force constant *k* and relaxed length *l*. [Fig. 1.25 (*a*)]

The bobs are displaced slightly from their equilibrium position [Fig. 1.25 (*b*)]. The two pendulums begin to vibrate. Let x_1 and x_2 be the displacement of the bobs at time *t*.

The spring is either stretched or compressed depending on whether x_2 is greater than x_1 or x_2 is less than x_1. The tension in the spring is $k(x_2 - x_1)$. If x_2 is greater than x_1, the tension will act against the acceleration of the bob *B* but in favour of the acceleration of bob *A*.

For small oscillations in a plane, the equations of motion of the two pendulums are given by

$$m\left(\frac{d^2 x_1}{dt^2}\right) = -\left(\frac{mg}{l}\right) x_1 + k(x_2 - x_1) \qquad \ldots (i)$$

and

$$m\left(\frac{d^2 x_2}{dt^2}\right) = -\left(\frac{mg}{l}\right) x_2 - k(x_2 - x_1) \qquad \ldots (ii)$$

These equations do not represent S.H.M since the accelerations are not proportional to the displacement of the respective bobs.

In case the spring is absent, $k = 0$ and here the two pendulums will vibrate simple harmonically with angular frequency

$$\omega_0 = \left[\frac{g}{l}\right]^{\frac{1}{2}}$$

Substituting this value in equations (*i*) and (*ii*)

$$m\left(\frac{d^2 x_1}{dt^2}\right) = -m\omega_0^2 x_1 + k(x_2 - x_1) \qquad \ldots (iii)$$

and $$m\left(\frac{d^2 x_2}{dt^2}\right) = -m\,\omega_0^2 x_2 - k(x_2 - x_1) \qquad \ldots (iv)$$

Simplifying

$$\frac{d^2 x_1}{dt^2} + \omega_0^2 x_1 - \left(\frac{k}{m}\right)(x_2 - x_1) = 0$$

and $$\frac{d^2 x_2}{dt^2} + \omega_0^2 x_2 + \left(\frac{k}{m}\right)(x_2 - x_1) = 0$$

Taking $\frac{k}{m} = \omega^2$ and simplifying

$$\frac{d^2 x_1}{dt^2} + (\omega_0^2 + \omega^2)\,x_1 - \omega^2 x_2 = 0 \qquad \ldots (v)$$

$$\frac{d^2 x_2}{dt^2} + (\omega_0^2 + \omega^2)\,x_2 - \omega^2 x_1 = 0 \qquad \ldots (vi)$$

These two equations cannot be solved separately because the equation (*v*) contains the term x_2 and the equation (*vi*) contains the term x_1. This means that the motion of *A* is only confined to *A* alone but *B* is also affected and *vice versa.* The above equations can be solved simultaneously.

Adding equations (*v*) and (*vi*)

$$\frac{d^2}{dt^2}\left[x_1 + x_2\right] + \omega_0^2\,(x_1 + x_2) = 0 \qquad \ldots (vii)$$

Subtracting equation (*vi*) from equation (*v*)

$$\frac{d^2}{dt^2}\left[x_1 - x_2\right] + \left(\omega_0^2 + 2\,\omega^2\right)\left(x_1 - x_2\right) = 0 \qquad \ldots (viii)$$

Taking $x_1 + x_2 = y_1$

and $x_1 - x_2 = y_2$

$$\omega_0^2 = \omega_1^2$$

$$\left[\omega_0^2 + 2\omega^2\right] = \omega_2^2$$

Rewriting these equations,

$$\frac{d^2 y_1}{dt^2} + \omega_1^2 y_1 = 0 \qquad \ldots (ix)$$

$$\frac{d^2 y}{dt^2} + \omega_2^2 y_2 = 0 \qquad \ldots (x)$$

The possible solutions are

$$y_1 = C \cos \omega_1 t \qquad \ldots (xi)$$

$$y_2 = D \cos \omega_2 t \qquad \text{... (xii)}$$

The values of C and D depend upon the initial conditions. Equations (*xi*) and (*xii*) represent two independent oscillations. They give another description of the normal modes in terms of the variables y_1 and y_2 respectively. These variables y_1 and y_2 are called **normal coordinates.** Changes in the values of y_1 and y_2 occur independently.

In terms of the original coordinates x_1 and x_2 the solutions are

$$x_1 = \frac{1}{2}\left[y_1 + y_2 \right]$$

$$x_1 = \frac{1}{2}\left[C \cos \omega_1 t + D \cos \omega_2 t \right] \qquad \text{... (xiii)}$$

and

$$x_2 = \frac{1}{2}\left[y_1 - y_2 \right]$$

$$x_2 = \frac{1}{2}\left[C \cos \omega_1 t - D \cos \omega_2 t \right] \qquad \text{... (xiv)}$$

Case 1. If $C = 0$

$$x_1 = \frac{1}{2}\left[D \cos \omega_2 t \right]$$

and

$$x_2 = -\frac{1}{2}\left[D \cos \omega_2 t \right]$$

It means both the pendulums oscillate with the angular frequency ω_2.

Case 2. If $D = 0$

$$x_1 = \frac{1}{2}\left[C \cos \omega_1 t \right]$$

and

$$x_2 = -\frac{1}{2}\left[C \cos \omega_1 t \right]$$

It means both the pendulums oscillate with angular frequency ω_1.

Here ω_1 and ω_2 are the individual **normal modes** and are called **normal angular frequencies**. It means the characteristic of a normal angular frequency is that both A and B can oscillate with the same angular frequency.

Conclusion

(*i*) When $x_1 = x_2$, the angular frequency is ω_1. In this case both the pendulums vibrate with the same angular frequency and in the same direction

$$\omega_1 = \left[\frac{g}{l} \right]^{\frac{1}{2}}$$

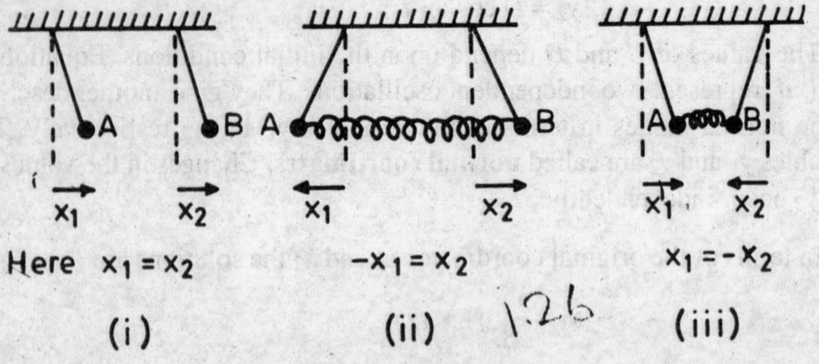

A B A B A B

x_1 x_2 x_1 x_2 x_1 x_2

Here $x_1 = x_2$ $-x_1 = x_2$ $x_1 = -x_2$

(i) (ii) (iii)

Fig. 1.26.

(*ii*) When $\quad x_1 = -x_2$

or $\quad x_1 = -x_2$ [Fig 1.26 (*ii*) and (*iii*)] the angular frequency is ω_1.

where
$$\omega_2 = \left[\omega_1^2 + 2\omega^2\right]^{\frac{1}{2}}$$

$$\omega_2 = \left[\omega_1^2 + 2\left(\frac{k}{m}\right)\right]^{\frac{1}{2}}$$

$$\omega_2 = \left[\left(\frac{g}{l}\right) + 2\left(\frac{k}{m}\right)\right]^{\frac{1}{2}}$$

It means, when the pendulums are vibrating with angular frequency ω_2 they are out of phase. Here the sping is either elongated or compressed [Fig. 1.26 (*ii*) and (*iii*)].

However when the pendulums are vibrating with angular frequency ω_0, they are in phase [Fig 1.26 (*i*)]

Hence, for mode 1,

$$\omega_1^2 = \omega_0^2 = \left[\frac{g}{l}\right] \qquad \ldots (i)$$

for mode 2

$$\omega_2^2 = \left[\omega_1^2 + (2\,k/m)\right]$$

$$\omega_2^2 = \left[\omega_0^2 + (2\,k/m)\right]$$

1.27. Energy Exchange (Beats) between Two Coupled Identical Oscillators

Consider two pendulums *A* and *B* coupled by spring of spring constant *k*. There are two possible modes of vibration given by

$$\omega_1^2 = \left(\frac{g}{l}\right)$$

and $$\omega_2^2 = \left[\frac{g}{l} + (2k/m)\right]$$

The general motion of pendulum is obtained by superposing the two modes.

$$x_A = x_1 + x_2$$

Similarly for the pendulum B,

$$x_B = x_1 - x_2$$

The displacement equations are

$$x_A = x_1 + x_2$$

$$= A_1 \cos(\omega_1 t + \phi_1) + A_2 \cos(\omega_2 t + \phi_2) \quad \ldots (i)$$

and $$x_B = x_1 - x_2$$

$$= A_1 \cos(\omega_1 t + \phi_1) - A_2 \cos(\omega_2 t + \phi_2) \quad \ldots (ii)$$

As the two pendulums are identical both the Oscillations will have approximately equal amplitude and produce strong beats.

Taking $A_1 = A_2 = A$ and $\phi_1 = \phi_2 = 0$

$$x_A = A\,[\cos\omega_1 t + \cos\omega_2 t] \quad \ldots (iii)$$

and $$x_B = A\,(\cos\omega_1 t - \cos\omega_2 t) \quad \ldots (iv)$$

Also, the velocities of the two pendulums will be

$$\left(\frac{dx_A}{dt}\right) = -A\,(\omega_1 \sin\omega_1 t + \omega_2 \sin\omega_2 t) \quad \ldots (v)$$

$$\left(\frac{dx_B}{dt}\right) = -A\,(\omega_1 \sin\omega_1 t - \omega_2 \sin\omega_2 t) \quad \ldots (vi)$$

At any instant of time, $t = 0$

$$x_A = 2A$$

$$x_B = 0$$

$$\frac{dx_A}{dt} = 0$$

$$\frac{dx_B}{dt} = 0$$

It means at $t = 0$, the bob A is given a displacement $2A$ and the bob B has zero displacement.

Both the bobs are released at this instant.

Gradually, it is noticed that the amplitude of pendulum A decreases and that of the pendulum B increases . Eventually the pendulum A comes to rest and pendulum B has amplitude $2A$. It means whole of the energy of pendulum A is transferred to pendulum B. The process repeats and energy of B is transferred back to A and *vice versa.*

One complete round trip of energy transfer from A to B and back from B to A is called a beat. The beat period is the time for the round trip of the energy transfer and

$$\text{Frequency of beats} = \frac{1}{\text{beat period}}$$

Also
$$\omega_m = \left(\frac{\omega_2 - \omega_1}{2}\right)$$

and
$$\omega_a = \left(\frac{\omega_2 + \omega_1}{2}\right)$$

Adding and subtracting

$$\omega_2 = \omega_m + \omega_a$$
$$\omega_1 = \omega_a - \omega_m$$

Substituting these values of ω_1 and ω_2 in equations *(iii)* and *(iv)*

$$x_A = A\left[\cos(\omega_a - \omega_m)\,t + \cos(\omega_a + \omega_m)\,t\right]$$
$$= 2A\cos\omega_a t\cos\omega_m t$$
$$= (2A\cos\omega_m t)\cos\omega_a t$$
$$= A_m\cos\omega_a t \qquad (\text{Taking } 2A\cos\omega_m t = A_m)$$

Similarly

$$x_B = A\left[\cos(\omega_a - \omega_m)\,t - \cos(\omega_a + \omega_m)\,t\right]$$
$$= 2A\sin\omega_m t\sin\omega_a t$$
$$= B_m\sin\omega_a t \qquad [\text{Here } B_m = 2A\sin\omega_m t\,]$$

The oscillation amplitudes A_m and B_m are taken as constant for fast oscillations. The energy transferred between the weak coupling spring and the pendulums is neglected. In fact for a weak spring there is never a significant amount of stored energy.

In one fact oscillation cycle, the pendulum A is considered as a harmonic oscillator of angular frequency W_A with constant amplitude A_m The average kinetic energy of pendulum A

$$E_A = \frac{1}{2}\,m\,\omega_a^2\,A_m^2$$
$$= 2m\,A^2\,\omega_a^2\cos^2\omega_m t$$

For the pendulum B,

$$E_B = 2m\, A^2\, \omega_0^2 \sin^2 \omega_m t$$

The total energy of both the pendulums

$$E = E_A + E_B$$

$$= 2m\, A^2\, \omega_0^2 (\sin^2 \omega_m t + \cos^2 \omega_m t)$$

$$= \mathbf{2m\, A^2\, \omega_0^2}$$

It means that the total energy is constant and is transferred between the two pendulums back and forth at the beat frequency.

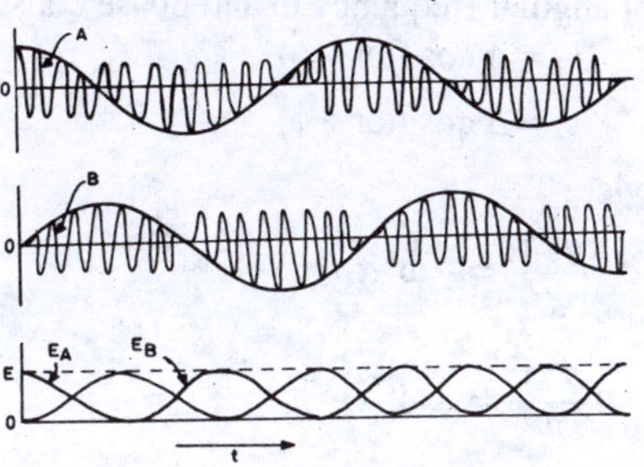

Fig. 1.27

1.28. Two Coupled Masses—Normal Modes of Longitudinal Oscillations

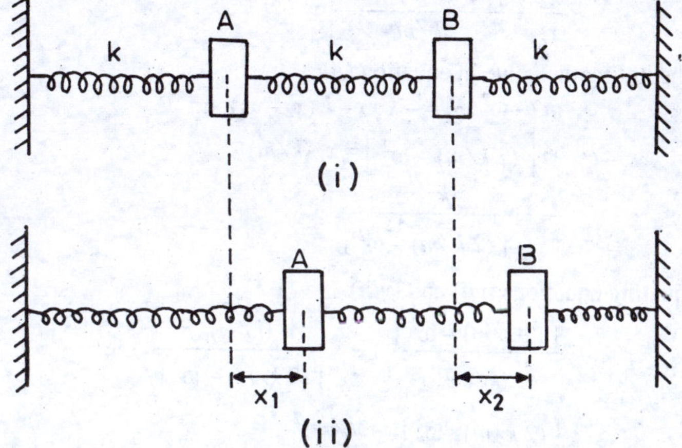

Fig. 1.28.

Consider two masses A and B each of mass m and three springs each of spring constant k. These are connected as shown in Fig. 1.28 (*i*). Each spring in its equilibrium position has a length a. Consider that at any instant of time t, the displacement of the mass A is x_1 and displacement of mass B is x_2 [Fig. 1.28 (*ii*)].

Consider that $x_2 > x_1$. The equations of motion for the general configuration are

$$m\left(\frac{d^2 x_1}{dt^2}\right) = -k x_1 + k\left(x_2 - x_1\right) \qquad \ldots (i)$$

$$m\left(\frac{d^2 x_2}{dt^2}\right) = -k x_2 - k\left(x_2 - x_1\right) \qquad \ldots (ii)$$

To determine the angular frequencies of the normal modes, assume that the normal modes exist at angular frequency ω and phase constant φ. It means

$$x_1 = A \cos(\omega t + \phi) \qquad \ldots (iii)$$

$$x_2 = B \cos(\omega t + \phi) \qquad \ldots (iv)$$

Differentiating twice,

$$\frac{d^2 x_1}{dt^2} = -\omega^2 x_1 \qquad \ldots (v)$$

and

$$\frac{d^2 x_2}{dt^2} = -\omega^2 x_2 \qquad \ldots (vi)$$

Substituting these values in equation (*i*)

$$m(-\omega^2 x_1) = -k x_1 + k(x_2 - x_1)$$

$$\therefore \quad x_2(k/m) = x_1\left[(2k/m) - \omega^2\right]$$

$$\frac{x_2}{x_1} = \frac{[(2k/m) - \omega^2]}{k/m} \qquad \ldots (vii)$$

Substituting the value in equation (*ii*)

$$m(-\omega^2 x_2) = -k x_2 - k(x_2 - x_1)$$

$$x_2\left[(2k/m) - \omega^2\right] = x_1 (k/m)$$

$$\frac{x_2}{x_1} = \frac{(k/m)}{[(2k/m) - \omega^2]} \qquad \ldots (viii)$$

Equating equations (*vii*) and (*viii*)

$$\frac{[(2k/m) - \omega^2]}{(k/m)} = \frac{(k/m)}{[(2k/m) - \omega^2]}$$

$$\therefore \quad [(2k/m) - \omega^2]^2 = (k/m)^2$$

$$(2k/m) - \omega^2 = \pm\, k/m \qquad \ldots (ix)$$

From equation (*ix*) two values of ω are obtained

$$\omega_1^2 = \left(\frac{2k}{m}\right) - \frac{k}{m} = \frac{k}{m}$$

and $$\omega_2^2 = \left(\frac{2k}{m}\right) + \frac{k}{m} = \frac{3k}{m}$$

$$\therefore \quad \omega_1 = [k/m]^{\frac{1}{2}} \quad \ldots (x)$$

and $$\omega_2 = [3k/m]^{\frac{1}{2}} \quad \ldots (xi)$$

Here $\omega_2 > \omega_1$

Equations (x) and (xi) give the angular frequencies of the two normal modes.

Mode 1. To obtain the shape of the configuration, for mode 1, the value of $\omega_1 = [k/m]^{\frac{1}{2}}$ can be substituted in equation (*vii*) or (*viii*). The value

$$\frac{x_2}{x_1} = \frac{(2k/m) - (k/m)}{k/m} = 1$$

$$\therefore \quad x_2 = x_1 \quad \text{[Fig .1 29 (i)]}$$

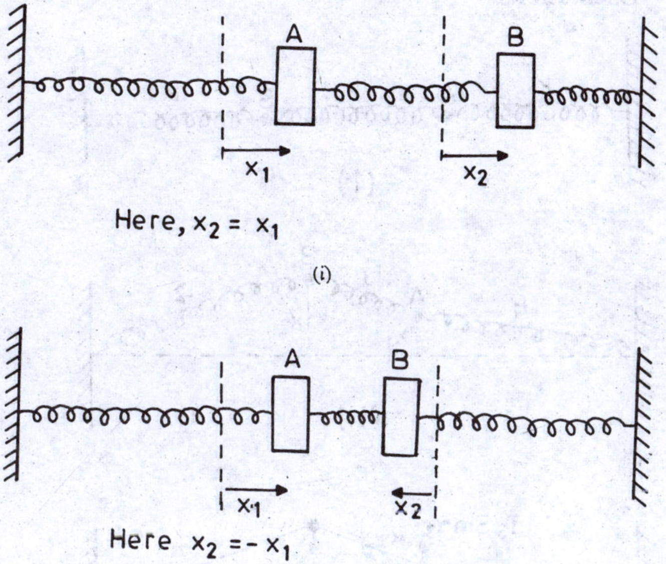

(*i*)

(*ii*)

Fig. 1.29

Mode 2. To obtain the shape of the configuration for mode 2, the value of $\omega_2 = [3k/m]^{\frac{1}{2}}$ can be substituted in equation (*vii*) or (*viii*)

$$\frac{x_2}{x_1} = \frac{(2k/m) - (3k/m)}{k/m}$$

$$\frac{x_2}{x_1} = -1$$

$$x_2 = -x_1 \qquad \text{[Fig. 1.29 (}ii\text{)]}$$

Fig. 1.29 (*i*) represents the configuration of mode 1 and Fig. 1.29 (*ii*) represents the configuration of mode 2 . The angular frequency of mode 2 is more than the angular frequency of mode 1.

Also $\omega_2^2 = (3\,k/m)$

$$\omega_2^2 = 3\,\omega_1^2$$

$$\omega_2 = (3)^{1/2}\,\omega_1$$

$$\omega_2 = 1{\cdot}732\,\omega_1$$

Consequently $\omega_2 > \omega_1$

1.29. Two Coupled Masses – Normal Modes of Transverse Oscillations

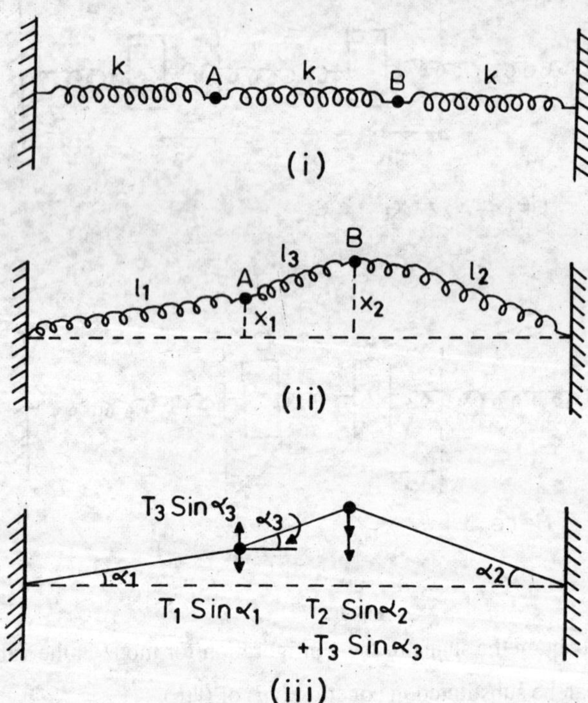

Fig. 1.30.

Consider two masses A and B each of mass m and three springs each of spring constant k. The relaxed length of each spring is a_0. The springs are connected as shown in Fig. 1.30 *(i)*. Each spring is extended from a_0 to a. The equilibrium tension in each spring $T_0 = k(a - a_0)$.

Fig. 1.30 *(ii)* shows the configuration of the system in transverse oscillations.

At any instant of time t, the force acting on each mass is shown in Fig. 1.30 *(iii)*. The new lengths of the springs are l_1, l_2 and l_3.

The tensions in the springs are,

$$T_1 = k[l_1 - a_0]$$

$$T_2 = k[l_2 - a_0]$$

$$T_3 = k[l_3 - a_0]$$

The restoring force acting on mass $A = -T_1 \sin \alpha_1 + T_3 \sin \alpha_3$ ... *(i)*

The restoring force acting on mass $B = -T_2 \sin \alpha_2 - T_3 \sin \alpha_3$ ... *(ii)*

Here α_1, α_2 and α_3 are the angles subtended by the spring on the x-axis.

The equation of motions are

$$m\left(\frac{d^2 x_1}{dt^2}\right) = -T_1 \sin \alpha_1 + T_3 \sin \alpha_3$$

$$= -T_1\left(\frac{x_1}{l_1}\right) + T_3\left(\frac{x_2 - x_1}{l_3}\right) \qquad \ldots (iii)$$

Also

$$m\left(\frac{d^2 x_2}{dt^2}\right) = -T_2 \sin \alpha_2 - T_3 \sin \alpha_3$$

$$= -T_2\left(\frac{x_2}{l_2}\right) - T_3\left(\frac{x_2 - x_1}{l_3}\right) \qquad \ldots (iv)$$

Substituting the values of T_1, T_2 and T_3 in equations *(iii)* and *(iv)* we get,

$$m\left(\frac{d^2 x_1}{dt^2}\right) = -\frac{k(l_1 - a_0)x_1}{l_1} + \frac{k(l_3 - a_0)(x_2 - x_1)}{l_3}$$

$$= -k\left[1 - \left(\frac{a_0}{l_1}\right)\right]x_1 + k\left[1 - \left(\frac{a_0}{l_3}\right)\right]\left(x_2 - x_1\right) \qquad \ldots (v)$$

Similarly

$$m\left(\frac{d^2 x_2}{dt^2}\right) = -\frac{k(l_2 - a_0)x_2}{l_2} + \frac{k(l_3 - a_0)(x_2 - x_1)}{l_3}$$

$$= -k\left[1 - \left(\frac{a_0}{l_2}\right)\right]x_2 - k\left[1 - \left(\frac{a_0}{l_3}\right)\right]\left(x_2 - x_1\right) \qquad \ldots (vi)$$

Normal Modes – Slinky Approximation

In the slinky approximation

$$a_0 \ll a \text{ as } a \text{ is} < l, \text{ therefore}$$

$$a_0 \leq l$$

or
$$\frac{a_0}{l} \leq 1$$

It means $\frac{a_0}{l}$ can be neglected as compared to 1.

Therefore neglecting $\frac{a_0}{l_1}, \frac{a_0}{l_2}$ and $\frac{a_0}{l_3}$ in equations *(v)* and *(vi)*

$$m\left(\frac{d^2 x_1}{dt^2}\right) = -k\, x_1 + k\left(x_2 - x_1\right) \quad \ldots (vii)$$

and
$$m\left(\frac{d^2 x_2}{dt^2}\right) = -k\, x_2 - k\left(x_2 - x_1\right) \quad \ldots (viii)$$

To determine the angular frequencies of the normal modes, assume that normal modes exist at angular frequency ω and phase constant ϕ. It means

$$x_1 = A \cos(\omega t + \phi_1) \quad \ldots (ix)$$

$$x_2 = A \cos(\omega t + \phi_2) \quad \ldots (x)$$

Differentiating twice

$$\frac{d^2 x_1}{dt^2} = -\omega^2 x_1 \quad \ldots (xi)$$

$$\frac{d^2 x_2}{dt^2} = -\omega^2 x_2 \quad \ldots (xii)$$

Substituting these values in equation *(vii)*

$$m(-\omega^2 x_1) = -k\, x_1 + k\,(x_2 - x_1)$$

$$x_2\,(k/m) = x_1[(2\,k/m) - \omega^2]$$

$$\frac{x_2}{x_1} = \frac{[(2k/m) - \omega^2]}{k/m} \quad \ldots (xiii)$$

Substituting the values in equation *(viii)*

$$m(-\omega^2 x_2) = -kx_2 - k\,(x_2 - x_1)$$

$$x_2\,[(2k/m) - \omega^2] = x_1\,(k/m)$$

$$\frac{x_2}{x_1} = \frac{k/m}{[(2\,k/m) - \omega^2]} \quad \ldots (xiv)$$

Equating equations *(xiii)* and *(xiv)*

$$\frac{(2k/m)-\omega^2}{k/m} = \frac{k/m}{(2\,k/m)-\omega^2}$$

$$[(2k/m)-\omega^2\,]^2 = (k/m)^2$$

$$\therefore \quad (2k/m)-\omega^2 = \pm\,(k/m) \qquad \ldots (xv)$$

From equation *(xv)* two values of ω are obtained

$$\omega_1^2 = (2\,k/m)-(k/m) = k/m$$

and $$\omega_2^2 = (2\,k/m)+(k/m) = (3\,k/m)$$

$$\therefore \quad \omega_1 = [\,k/m\,]^{1/2} \qquad \ldots (xvi)$$

$$\omega_2 = [\,3\,k/m\,]^{1/2} \qquad \ldots (xvii)$$

Here $\omega_2 > \omega_1$

Equations *(xvi)* and *(xvii)* give the angular frequencies of the two normal modes.

Mode 1. To obtain the shape of configuration for mode 1, the value $\omega_1 = [k/m]^{1/2}$ is substituted is equation *(xiii)* or *(xiv)*

$$\therefore \quad \frac{x_2}{x_1} = \frac{(2\,k/m)-(k/m)}{k/m} = 1$$

$$\therefore \quad x_2 = x_1 \quad \ldots \quad [\text{Fig. 1.31 } (i)\,]$$

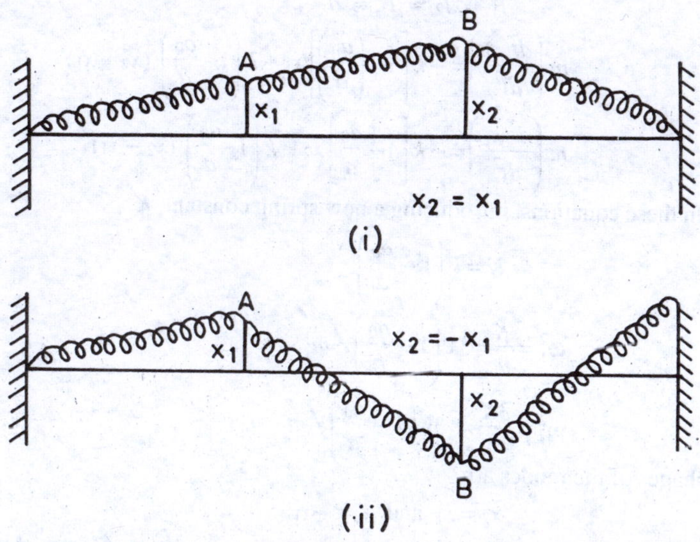

Fig. 1.31.

Mode 2. To obtain the shape of the configuration for mode 2, the value of $\omega_2^2 = 3\,k/m$ is substituted in equations *(xiii)* or *(xiv)*

$$\frac{x_2}{x_1} = \frac{(2k/m) - (3\,k/m)}{k/m} = -1$$

$\therefore \qquad x_2 = -x_1 \qquad \ldots$ [Fig. 1.31 *(ii)*]

Fig. 1.31 *(i)* represents the configuration of mode 1 and Fig. 1.31 *(ii)* represents the configuration of mode 2.

The angular frequency of mode 2 is more than the angular frequency of mode 1.

Also $\qquad \omega_2^2 = (3\,k/m)$

$$\omega_2^2 = 3\,\omega_1^2$$

$$\omega_2 = [\,3\,]^{1/2}\,\omega_1$$

$$\boldsymbol{\omega_2 = 1{\cdot}732\,\omega_1}$$

Consequently $\qquad \omega_2 > \omega_1$.

Conclusion. Therefore *in the slinky approximation, the angular frequencies of normal modes of transverse Oscillations are the same as those of longitudinal Oscillations*. It means **there is a form of degeneracy.**

1.30. Small Oscillations Approximation

In the small oscillations a_0 is not negligibly small as compard to a and the system remains close to the equilibrium position. In that case

$$l_1 \approx l_2 \approx l_3 \approx a$$

$$\therefore \qquad m\left(\frac{d^2 x_1}{dt^2}\right) = -k\left[1-\left(\frac{a_0}{a}\right)\right]x_1 + k\left(1-\frac{a_0}{a}\right)(x_2 - x_1)$$

and
$$m\left(\frac{d^2 x_2}{dt^2}\right) = -k\left[1-\frac{a_0}{a}\right]x_2 - k\left(1-\frac{a_0}{a}\right)(x_2 - x_1)$$

In these equations, introducing a new spring constant, K

$$K = k\left(1-\frac{a_0}{a}\right)$$

and
$$\omega_1^2 = \frac{K}{m} = k\left(1-\frac{a_0}{a}\right)\Big/ m \qquad \ldots (i)$$

and
$$\omega_2^2 = \frac{3K}{m} = 3k\left(1-\frac{a_0}{a}\right)\Big/ m \qquad \ldots (ii)$$

and shapes of the modes are

$$x_2 = x_1 \text{ and } x_2 = -x_1.$$

It clearly shows that in small transverse oscillations the angular frequencies are smaller that the angular frequencies in longitudinal oscillations.

1.31. Normal Modes of Transverse Oscillations of many (N) Coupled Oscillations

A piece of a string or fluid volume contains a large number of particles bound (coupled) to each other by forces of cohesion. Here, a case of a flexible string is considered. The mass of the string is taken as negligible small and N particles are at a distance l from each other.

The total length of the string $= L$

Number of particles $= N$

$\therefore \qquad L = (N+1)\, l$

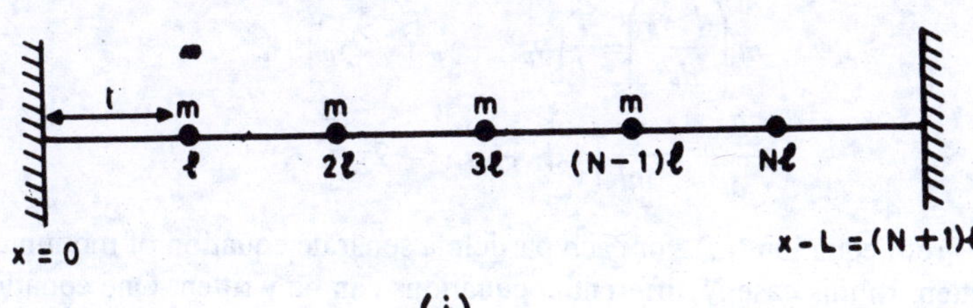

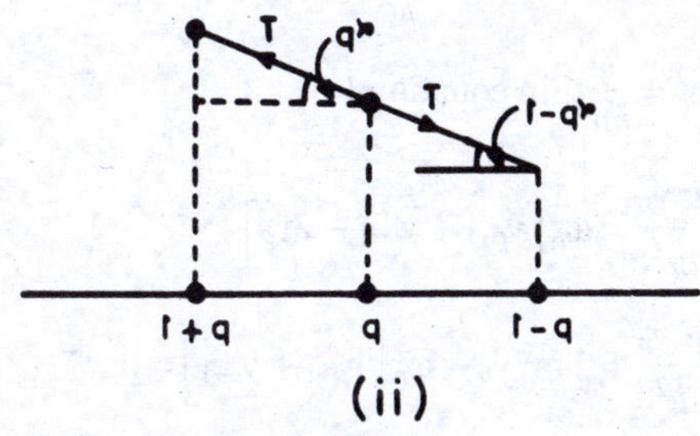

Fig. 1.32

The mass of each particle is m and equilibrium tension is T.

Fig. 1.32 (*i*) shows N particles. Fig. 1.32 (*ii*) shows the configuration of particles at any instant of time t during their transverse oscillations. It is assumed that the amplitude of these oscillations is small. Consider three particles $(p-1)$, p and $(p+1)$. Here

$$p = 1, 2, 3, \; .\,.\,.\,. \; (N-1), N.$$

The displacements are y_{p-1} y_p, y_{p+1} respectively.

Here $\qquad \sin\, \alpha_{p-1} = \dfrac{y_p - (y_{p-1})}{l}$

and $\qquad \sin \alpha_p = \dfrac{y_{p+1} - y_p}{l}$

Here displacement is small as compared to l.

The resultant y component of the force on the particle p,

$$F_p = -T\sin\alpha_{p-1} + T\sin\alpha_p$$

$$= -\left(\frac{T}{l}\right)\left(y_p - y_{p-1}\right) + \frac{T}{l}\left(y_{p+1} - y_p\right)$$

$$F_p = \frac{T}{l}\left[y_{p+1} + y_{p-1} - 2y_p\right] \qquad \ldots (i)$$

Therefore, for the *pth* particle, the equation of motion is

$$m\left(\frac{d^2 y_p}{dt^2}\right) = \frac{T}{l}\left[y_{p+1} + y_{p-1} - 2y_p\right] \qquad \ldots (ii)$$

$$\frac{d^2 y_p}{dt^2} = \frac{T}{ml}\left[y_{p+1} + y_{p-1} - 2y_p\right] \qquad \ldots (iii)$$

From equation *(iii)*, for each particle a separate equation of motion can be written. In this case N differential equations can be written. One equation for each value of p form 1 to N. Also $y_0 = 0$ and $y_{N+1} = 0$.

Normal Modes

Taking $\omega_0^2 = \dfrac{T}{ml}$ in equation *(iii)*

$$\frac{d^2 y_p}{dt^2} = \omega_0^2\left[y_{p+1} + y_{p-1} - 2y_p\right]$$

$$\frac{d^2 y_p}{dt^2} + 2\,\omega_0^2 y_p - \omega_0^2\left[y_{p+1} + y_{p-1}\right] = 0 \qquad \ldots (iv)$$

(i) **For p = N = 1,**

$$y_{p-1} = y_0 = 0 \text{ and } y_{p+1} = y_{N+1} = 0$$

$$\frac{d^2 y_1}{dt^2} + 2\,\omega_0^2 y_1 = 0$$

This represents a transverse harmonic motion of angular frequency

$$(\sqrt{2})\,\omega_0 = (\sqrt{2})\left(\frac{T}{ml}\right)^{\frac{1}{2}} = \left[\frac{2T}{ml}\right]^{\frac{1}{2}}$$

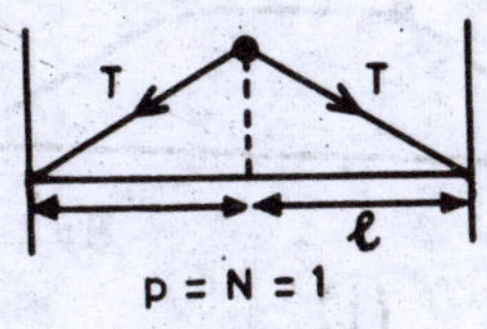

Fig. 1.33.

(ii) **For N = 2**

There are two differential equations for values of $p = 1$ and $p = 2$

$$\frac{d^2 y_1}{dt^2} + 2\,\omega_0^2 y_1 - \omega_0^2 y_2 = 0 \qquad \ldots (v)$$

$$\frac{d^2 y_2}{dt^2} + 2\,\omega_0^2 y_2 - \omega_0^2 y_1 = 0 \qquad \ldots (vi)$$

Adding equations *(v)* and *(vi)*

$$\left[\frac{d^2 y_1}{dt^2} + \left(\frac{d^2 y_2}{dt^2}\right)\right] + \omega_0^2 (y_1 + y_2) = 0 \qquad \ldots (vii)$$

Subtracting *(vi)* from *(v)*

$$\left[\frac{d^2 y_1}{dt^2} - \left(\frac{d^2 y_2}{dt^2}\right)\right] + (3\,\omega_0^2)(y_1 - y_2) = 0 \qquad \ldots (viii)$$

$\therefore$ $$\omega_1^2 = \omega_0^2 \qquad \ldots (ix)$$

and $$\omega_2^2 = 3\,\omega_0^2 \qquad \ldots (x)$$

$\therefore$ $$\omega_1 = \omega_0 \qquad \ldots (xi)$$

and $$\omega_2 = (\sqrt{3})\,\omega_0 \qquad \ldots (xii)$$

Therefore for a two particle system there are two normal modes and their configuration is shown in Fig. 1.34.

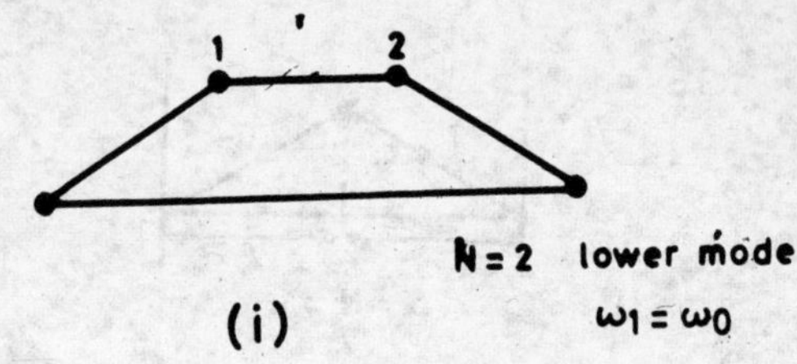

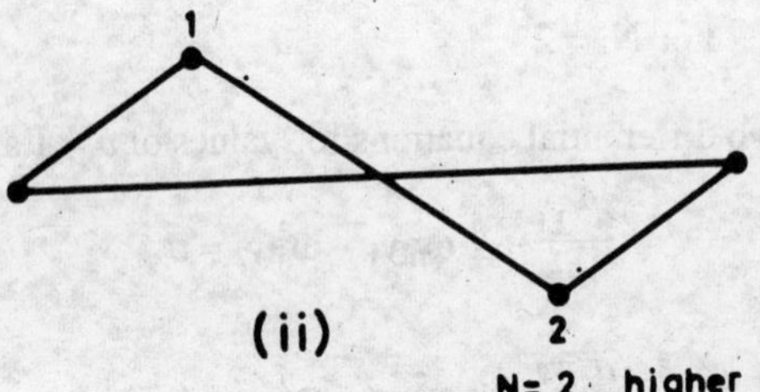

Fig. 1.34

∴ $$\omega_1 = \left[\frac{T}{ml}\right]^{\frac{1}{2}}$$

and $$\omega_2 = \left[\frac{3T}{ml}\right]^{\frac{1}{2}}$$

Consequently $\omega_2 > \omega_1$.

(iii)* General Solution for any Value of *N

Suppose that each particle viberate with the same angular frequency ω and amplitude A_p where $p = 1, 2 \ldots N$. The solution is

$$y_p = A_p \cos \omega t \qquad \ldots (xiii)$$

$$\frac{d^2 y_p}{dt^2} = -\omega^2 y_p \qquad \ldots (xiv)$$

The general equation is

$$\frac{d^2 y_p}{dt^2} + 2\omega_0^2 y_p - \omega_0^2 (y_{p+1} + y_{p-1}) = 0$$

Substituting the values

$$(-\omega^2 + 2\omega_0^2) y_p - \omega_0^2 (y_{p+1} + \dot{y}_{p-1}) = 0$$

Therefore for $p = 1$

$$(-\omega^2 + 2\omega_0^2) A_1 - \omega_0^2 (A_2 + A_0) = 0$$

For $p = 2$

$$(-\omega^2 + 2\omega_0^2) A_2 - \omega_0^2 (A_3 + A_1) = 0$$

and in general

$$(-\omega^2 + 2\omega_0^2) A_p - \omega_0^2 (A_{p+1} + A_{p-1}) = 0 \qquad \dots (xv)$$

$$\therefore \quad \frac{A_{p+1} + A_{p-1}}{A_p} = \frac{(-\omega^2 + 2\omega_0^2)}{\omega_0^2} \qquad \dots (xvi)$$

For a particular value of ω, the right hand side of equation (*xvi*) is constant. Therefore the left hand side must be a constant.

The value assigned to A_p should be such that

$$A_0 = 0 \text{ and } A_{N+1} = 0$$

Taking,

$$A_p = C \sin p\theta \qquad \dots (xvii)$$

$$A_{p-1} + A_{p+1} = C[\sin(p-1)\theta + \sin(p+1)\theta]$$

$$= 2C \sin p\theta \cos\theta$$

But $\quad C \sin p\theta = A_p \qquad \dots (xviii)$

$$\therefore \quad A_{p-1} + A_{p+1} = 2A_p \cos\theta$$

or

$$\frac{A_{p-1} + A_{p+1}}{A_p} = 2\cos\theta \qquad \dots (xix)$$

The right hand side of equation (*xviii*) is constant and independent of *p*.

cos θ is contant if

$$(N+1)\theta = n\pi$$

where $n = 1, 2, 3.....$

$$\theta = \left(\frac{n\pi}{N+1}\right) \qquad \dots (xx)$$

$$\therefore \quad A_p = C \sin\left(\frac{pn\pi}{N+1}\right) \qquad \dots (xxi)$$

From equation (*xvi*), (*xix*) and (*xx*), we get

$$\frac{A_{p+1} + A_{p-1}}{A_p} = \left(\frac{-\omega^2 + 2\omega_0^2}{\omega_0^2}\right) = 2\cos\left(\frac{n\pi}{N+1}\right)$$

$$\therefore \quad \omega_n^2 = 2\omega_0^2\left[1 - \cos\left(\frac{n\pi}{N+1}\right)\right]$$

$$\omega_n^2 = 4\omega_0^2 \sin^2\left[\frac{n\pi}{2(N+1)}\right]$$

$$\omega_n = 2\omega_0 \sin\left[\frac{n\pi}{2(N+1)}\right] \qquad \ldots (xxii)$$

Here $\omega_0 = \sqrt{\dfrac{T}{ml}}$

and $n = 1, 2, 3 \ldots N.$

$$\therefore \quad \omega_n = 2\left[\frac{T}{ml}\right]^{\frac{1}{2}} \sin\left[\frac{n\pi}{2(n+1)}\right] \qquad \ldots (xxiii)$$

Special Cases :

Here $\omega = 2\,\omega_0 \sin\left[\dfrac{n\pi}{2(N+1)}\right]$

and $A_p = C \sin\left(\dfrac{Pn\pi}{N+1}\right)$

(1) Single Oscillator

Here $N=1, \quad n=1, p=1$

$$\therefore \quad \omega = 2\,\omega_0 \sin\left(\frac{\pi}{4}\right)$$

$$= (\sqrt{2})\,\omega_0$$

$$= \left[\frac{2T}{ml}\right]^{\frac{1}{2}}$$

and $A = C \sin\left(\dfrac{\pi}{2}\right)$

$$A = C$$

(2) Two Coupled Oscillators

Here $N = 2, \quad p = 1$ and 2

$n = 1$ and 2

For first mode, $n = 1,\ p = 1,$ and $p = 2$

$$\omega_1 = 2\,\omega_0 \sin\left(\frac{\pi}{6}\right)$$

$$\omega_1 = \omega_0$$

$$\omega_1 = \left[\frac{T}{ml}\right]^{\frac{1}{2}}$$

$$A_1 = C \sin (\pi/3)$$

$$A_1 = \left(\frac{\sqrt{3}}{2}\right) C$$

$$A_2 = C \sin \left(\frac{2\pi}{3}\right) = \left(\frac{\sqrt{3}}{2}\right) C$$

For second mode,

$$n = 2, \ p = 1, \text{ and } p = 2$$

$$\omega_2 = 2\,\omega_0 \sin \left(\frac{\pi}{3}\right)$$

$$\omega_2 = (\sqrt{3})\,\omega_0$$

$$\omega_2 = \sqrt{\frac{3T}{ml}}$$

$$A_1 = C \sin \left(\frac{2\pi}{3}\right) = \left(\frac{\sqrt{3}}{2}\right) C$$

$$A_2 = C \sin \left(\frac{4\pi}{3}\right) = -\left(\frac{\sqrt{3}}{2}\right) C$$

(3) For $N = 5, \ n = 1, 2, 3, 4, 5$

When $n = 1$

$$\omega_1 = 2\,\omega_0 \sin \left(\frac{\pi}{12}\right)$$

$$\omega_1 = 2\left(\sqrt{\frac{T}{ml}}\right) \sin \left(\frac{\pi}{12}\right)$$

When $n = 2$,

$$\omega_2 = 2\,\omega_0 \sin \left(\frac{\pi}{6}\right)$$

$$\omega_2 = \sqrt{\frac{T}{ml}}$$

When $n = 3$,

$$\omega_3 = 2\,\omega_0 \sin\left(\frac{\pi}{4}\right)$$

$$\omega_3 = \sqrt{\frac{2\,T}{ml}}$$

When $n = 4$,

$$\omega_4 = 2\,\omega_0 \sin\left(\frac{\pi}{3}\right)$$

$$= \sqrt{3}\,\omega_0$$

$$= \sqrt{\frac{3T}{ml}}$$

When $n = 5$,

$$\omega_5 = 2\,\omega_0 \sin\left(\frac{5\pi}{12}\right).$$

1.32. Normal Modes of Longitudinal Oscillations of many (N) Coupled Oscillators

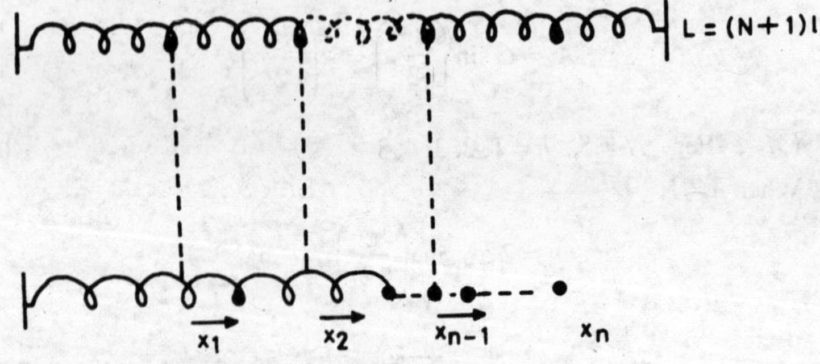

Fig. 1.35.

Consider N particle each of mass m and connected by $(N+1)$ spring. The relaxed length of each spring is l

$$\therefore \qquad L = (N + l)\,l$$

Let the displacement of the particles be $x_1, x_2 \ldots\ldots x_{n-1}$ and x_n from their equilibrium position.

The equation of motion for pth particle is given by

$$\frac{d^2 x_p}{dt^2} = \left(\frac{k}{m}\right)\left(x_{p+1} - x_p\right) - \left(\frac{k}{m}\right)\left(x_p - x_{p-1}\right)$$

$$= \left(\frac{k}{m}\right)\left[x_{p+1} + x_{p-1} - 2\,x_p\right] \qquad \ldots (i)$$

Taking $\omega_0^2 = \dfrac{k}{m}$

$$\frac{d^2 x_p}{dt^2} = -2\omega_0^2 x_p + \omega_0^2\left(x_{p+1} + x_{p-1}\right) \qquad \ldots (ii)$$

It can be solved similar to transverse vibrations of N coupled oscillators. The angular frequency of nth normal mode is given by

$$\omega_n = 2\,\omega_0 \sin\left[\frac{n\pi}{2\,(N+1)}\right] \qquad \ldots (iii)$$

where $\omega_0 = \left[\dfrac{k}{m}\right]^{\frac{1}{2}}$

The amplitudes of individual particles, in a given mode *i.e.* for given value of n is given by

$$A_p = C\ \sin\left(\frac{p\,n\pi}{N+1}\right)$$

Here $p = 1, 2, 3, \ldots\ldots N.$

Example 1.29. *Two blocks of masses m_1 and m_2 connected together by a spring of constant k, are resting on a smooth horizontal surface as shown in Fig. 1.36. Obtain an expression for the natural frequency of the system.*

[Delhi (Hons) 1991]

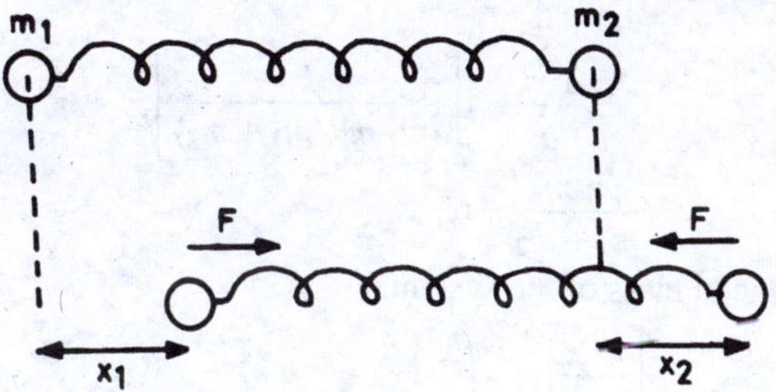

Fig. 1.36.

Suppose x_1 and x_2 are the displacement of the masses from their equilibrium positions. Here $x_2 > x_1$. It means the spring is stretched and increase in its length is $(x_2 - x_1)$.

The tension in the spring,

$$F = k\,(x_2 - x_1)$$

This force acting on the two masses will be equal and opposite in direction (Fig. 1.36).

The equations of motion are

$$m_1\left(\frac{d^2 x_1}{dt^2}\right)=k\,(x_2-x_1) \qquad \ldots (i)$$

and

$$m_2\left(\frac{d^2 x_2}{dt^2}\right)=-\,k\,(x_2-x_1) \qquad \ldots (ii)$$

Multiply equation *(i)* by m_2 and equation *(ii)* by m_1

$$m_1\,m_2\left(\frac{d^2 x_1}{dt^2}\right)=m_2\,k\,(x_2-x_1) \qquad \ldots (iii)$$

$$m_1\,m_2\left(\frac{d^2 x_2}{dt^2}\right)=-\,m_1\,k\,(x_2-x_1) \qquad \ldots (iv)$$

Subtract equation *(iii)* form equation *(iv)*

$$m_1\,m_2\left[\left(\frac{d^2 x_2}{dt^2}\right)-\left(\frac{d^2 x_1}{dt^2}\right)\right]=-\,k(m_1+m_2)\,(x_2-x_1) \qquad \ldots (v)$$

Take $\qquad (x_2-x_1)=x$

and $\qquad \dfrac{d^2 x_2}{dt^2}-\dfrac{d^2 x_1}{dt^2}=\dfrac{d^2 x}{dt^2}$

Subsituting these values in equation (v) we get

$$m_1\,m_2\left[\frac{d^2 x}{dt^2}\right]=-\,k(m_1+m_2)\,x$$

$$\frac{d^2x}{dt^2}=-\left[\frac{k}{(m_1\,m_2/m_1+m_2)}\right]x \qquad \ldots (vi)$$

Take $\qquad \dfrac{m_1\,m_2}{m_1+m_2}=\mu$

Here μ is reduced mass of the system.

$$\left[\frac{d^2 x}{dt^2}\right]=-\left(\frac{k}{\mu}\right)x \qquad \ldots (vii)$$

Equation (*vii*) represents S.H.M.

$$\therefore \qquad \omega^2=\left(\frac{k}{\mu}\right)$$

Angular frequency $\omega = \left[\frac{k}{\mu}\right]^{\frac{1}{2}}$ ... (*viii*)

Frequency $\nu = \frac{\omega}{2\pi} = \frac{1}{2\pi}\left[\frac{k}{\mu}\right]^{\frac{1}{2}}$ ... (*ix*)

Thus the system executes S.H.M. with a frequency

$$\nu = \frac{1}{2\pi}\left[\frac{k}{\mu}\right]^{\frac{1}{2}}$$

Here $\mu = \left[\frac{m_1 m_2}{m_1 + m_2}\right]$

μ is less than m_1 or m_2.

Example 1.30. *Calculate the frequency of oscillation of hydrogen molecule if its force constant is 4·8 × 10*^{2} N/m *and mass of hydrogen atom is 1·67 × 10*$^{-27}$ kg.

(Bombay, 1990)

Here
$$k = 4{\cdot}8 \times 10^2 \text{ N/m}$$
$$m_1 = 1{\cdot}67 \times 10^{-27} \text{ kg}$$
$$m_2 = 1{\cdot}67 \times 10^{-27} \text{ kg}$$
$$m_1 = m_2 = m$$

Reduced mass, $\mu = \left[\frac{m_1 \times m_2}{m_1 + m_2}\right] = \frac{m^2}{2m} = \left(\frac{m}{2}\right)$

$$\mu = \frac{m}{2} = \frac{1{\cdot}67 \times 10^{-27}}{2} = 0{\cdot}835 \times 10^{-27} \text{ kg}$$

$$\nu = \frac{1}{2\pi}\sqrt{\frac{k}{\mu}}$$

$$\nu = \frac{1}{2\pi}\left[\frac{4{\cdot}8 \times 10^2}{0{\cdot}835 \times 10^{-27}}\right]^{\frac{1}{2}}$$

$$\nu = \mathbf{1{\cdot}2 \times 10^{14} \text{ Hz}}.$$

Example 1.31. *Three masses m_1, m_2, m_3 are connected with equal length springs as shown in the diagram. Each coupling spring has constant k. If $m_1 = m_3$, set up equations of motion for the masses which oscillate along the line joining their centres. From these equations, prove that there will be only two modes of oscillations for this system.*

Calculate the ratio of frequencies of these two modes, taking $m_1 = m_3 = 16$ units and $m_2 = 12$ units.

(Delhi Hons., 1990)

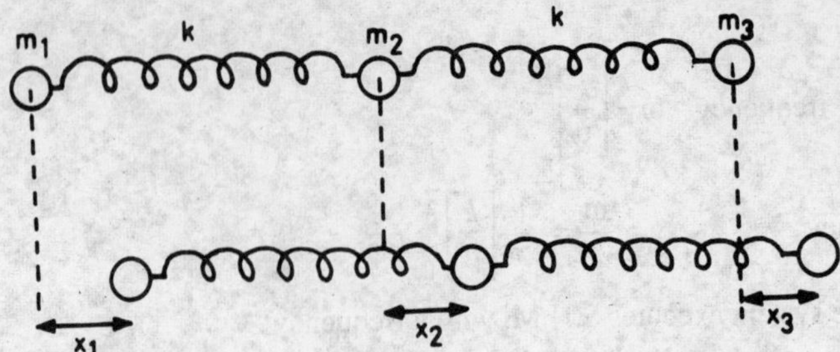

Fig. 1.37.

As shown in Fig. 1.37, suppose the displacements of the three masses are x_1, x_2 and x_3 respectively.

Here $$x_3 > x_2 > x_1$$

The equations of motion are

$$m_1\left(\frac{d^2 x_1}{dt^2}\right) = k\,(x_2 - x_1) \qquad \ldots (i)$$

$$m_2\left(\frac{d^2 x_2}{dt^2}\right) = k(x_3 - x_2) - k\,(x_2 - x_1) \qquad \ldots (ii)$$

$$m_3\left(\frac{d^2 x_3}{dt^2}\right) = -\,k\,(x_3 - x_2) \qquad \ldots (iii)$$

Here $$m_1 = m_3 \text{ (given)}$$

Substituting the value $m_1 = m_3$ and subtracting equation (*i*) from equation (*iii*), we get

$$m_1\left[\frac{d^2 x_3}{dt^2} - \frac{d^2 x_1}{dt^2}\right] = -\,k\,[x_3 - x_1] \qquad \ldots (iv)$$

Taking $$x_3 - x_1 = x \text{ and } \left(\frac{d^2 x_3}{dt^2} - \frac{d^2 x_1}{dt^2}\right) = \frac{d^2 x}{dt^2}$$

$$\frac{d^2 x}{dt^2} = -\left(\frac{k}{m_1}\right)x \qquad \ldots (v)$$

Equation (*v*) gives one of the normal mode of the system having angular frequency ω_1.

Here $$\omega_1^2 = \frac{k}{m_1}$$

Angular frequency,

$$\omega_1 = \left[\frac{k}{m_1}\right]^{\frac{1}{2}} \qquad \ldots (vi)$$

To determine the other modes, consider that

$$x_1 = A\cos(\omega t + \theta)$$
$$x_2 = B\cos(\omega t + \theta)$$
$$x_3 = C\cos(\omega t + \theta)$$

and

$$\frac{d^2 x_1}{dt^2} = -\omega^2 x_1$$
$$\frac{d^2 x_2}{dt^2} = -\omega^2 x_2$$
$$\frac{d^2 x_3}{dt^2} = -\omega^2 x_3$$

Substituting there values in equations (*i*), (*ii*), and (*iii*)

$$-m_1\omega^2 x_1 = k(x_2 - x_1) \qquad \ldots (vii)$$
$$-m_2\omega^2 x_2 = k(x_3 + x_1 - 2x_2) \qquad \ldots (viii)$$
$$-m_3\omega^2 x_3 = k(x_3 - x_2) \qquad \ldots (ix)$$

Eliminating x_1, x_2, and x_3 from these equations, we get

$$\omega_2^2 = \omega^2 = k\left[\frac{m_2 + 2m_1}{m_1 m_2}\right] \qquad \ldots (x)$$

From equations (*vi*) and (*x*)

$$\therefore \quad \frac{\omega_2^2}{\omega_1^2} = \left(\frac{m_2 + 2m_1}{m_2}\right)$$

$$\frac{\omega_2}{\omega_1} = \left(\frac{m_2 + 2m_1}{m_2}\right)^{\frac{1}{2}}$$

$$\frac{2\pi\nu_2}{2\pi\nu_1} = \left(\frac{m_2 + 2m_1}{m_2}\right)^{\frac{1}{2}}$$

$$\therefore \quad \frac{\nu_2}{\nu_1} = \left(\frac{m_2 + 2m_1}{m_2}\right)^{\frac{1}{2}} \qquad \ldots (xi)$$

As $m_1 = m_3 = 16$ units

$m_2 = 12$ units

$$\frac{\nu_2}{\nu_1} = \left(\frac{12 + 2 \times 16}{12}\right)^{\frac{1}{2}} = \left[\frac{44}{12}\right]^{\frac{1}{2}}$$

$$= [3{\cdot}67]^{1/2}$$

$$= \mathbf{1{\cdot}915}.$$

Note. This is the case of CO_2 molecule.

Example 1.32. *Two equal masses m are connected with two identical massless springs each of spring constant k. Show that the angular frequencies of two normal modes in the vertical direction are given by*

$$\omega_1^2 = (3+\sqrt{5})\left(\frac{k}{2m}\right)$$

and
$$\omega_2^2 = (3-\sqrt{5})\left(\frac{k}{2m}\right).$$

Also calculate the ratio of the amplitudes of the two masses in each separate mode.

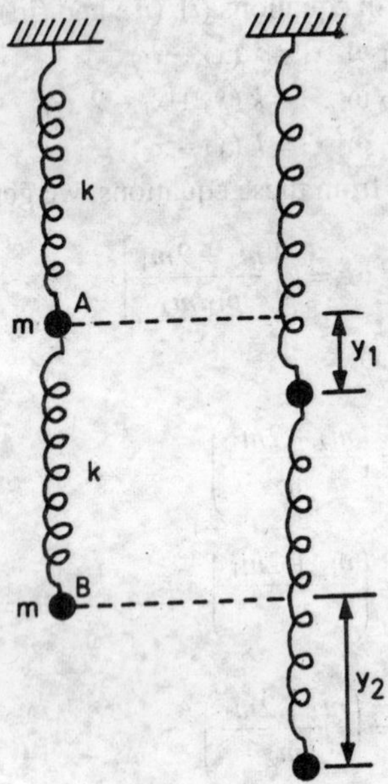

Fig. 1. 38.

The displacements of the masses be y_2 and y_1.

The gravitational forces acting on the two masses are not taken into consideration because they are independent of displacements and do not contribute to the restoring forces.

Restoring force acting on mass $A = -ky_1 + k(y_2 - y_1)$

Restoring force acting on mass $B = -k(y_2 - y_1)$

Equations of motions are

$$m\left(\frac{d^2y_1}{dt^2}\right) = -ky_1 + k(y_2 - y_1) \qquad \ldots (i)$$

and $$m\left(\frac{d^2 y_2}{dt^2}\right) = -k\,(y_2 - y_1) \qquad \ldots (ii)$$

$$\frac{d^2 y_1}{dt^2} = -\left(\frac{2k}{m}\right) y_1 + \left(\frac{k}{m}\right) y_2 \qquad \ldots (iii)$$

$$\frac{d^2 y_2}{dt^2} = -\left(\frac{k}{m}\right) y_2 + \left(\frac{k}{m}\right) y_1 \qquad \ldots (iv)$$

For a normal mode of angular frequency ω and phase constant φ,

$$y_1 = A\cos(\omega t + \phi)$$
$$y_2 = B\cos(\omega t + \phi)$$

and $$\frac{d^2 y_1}{dt^2} = -\omega^2 y_1$$
$$\frac{d^2 y_2}{dt^2} = -\omega^2 y_2$$

Substituting these values in equations (*iii*) and (*iv*)

$$-\omega^2 y_1 = -\left(\frac{2k}{m}\right) y_1 + \left(\frac{k}{m}\right) y_2$$

$$\frac{y_1}{y_2} = \frac{k/m}{(2\,k/m) - \omega^2} \qquad \ldots (v)$$

and $$-\omega^2 y_2 = -\left(\frac{k}{m}\right) y_2 + \frac{k}{m} y_1$$

$$\frac{y_1}{y_2} = \frac{(k/m) - \omega^2}{k/m} \qquad \ldots (vi)$$

Equating right hand sides of equations (*v*) and (*vi*)

$$\frac{(k/m)}{(2\,k/m - \omega^2)} = \frac{(k/m) - \omega^2}{(k/m)}$$

$$[(2k/m) - \omega^2]\ [(k/m) - \omega^2] = (k/m)^2$$

$$\omega^4 - (3k/m)\,\omega^2 + \frac{k^2}{m^2} = 0 \qquad \ldots (vii)$$

This is a quadratic equation

The solutions are

$$\omega_1^2 = (3 + \sqrt{5})\ (k/2m) \qquad \ldots (viii)$$

and $$\omega_2^2 = (3 - \sqrt{5})\ (k/2m) \qquad \ldots (ix)$$

Here ω_1 is greater than ω_2.

ω_1 and ω_2 are the angular frequencies of the two modes.

Amplitude Ratio

(1) For higher angular frequency

$$\omega_1^2 = (3+\sqrt{5})\ (k/2m)$$

$$\frac{A}{B} = \frac{y_1}{y_2} = \frac{k/m - \omega^2}{k/m}$$

Here $\omega^2 = \omega_1^2$

$$\therefore \quad \frac{A}{B} = \frac{[(k/m) - (3+\sqrt{5})\cdot(k/2m)]}{(k/m)}$$

$$= \frac{-(\sqrt{5}+1)}{2} \qquad \ldots (x)$$

The negative sign shows that this mode is out of phase.

(2) For lower angular frequency

$$\omega_2^2 = (3-\sqrt{5})\ (k/2m)$$

Here $\omega^2 = \omega_2^2$

$$\frac{A}{B} = \frac{y_1}{y_2}\ \frac{[(k/m) - (3-\sqrt{5})\ (k/2m)]}{(k/m)}$$

$$= \frac{(\sqrt{5})-1}{2}.$$

Example 1.33. *Two identical pendulums, each of mass m and length l are connected by a spring as shown in Fig. 1.39. The spring has force constant k.*

Calculate the angular frequencies and frequencies of the normal modes of the system.

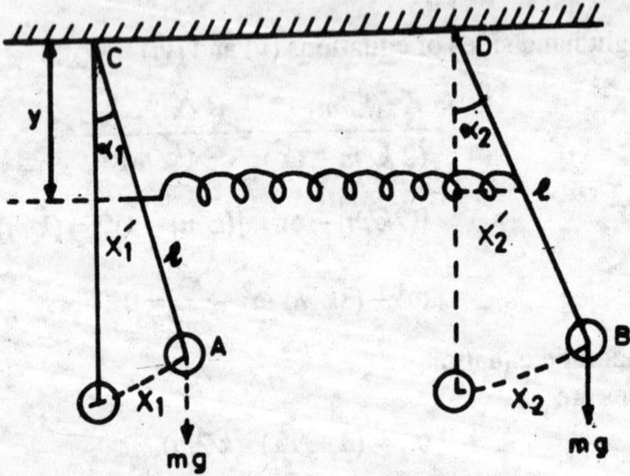

Fig. 1.39.

Let x_1 and x_2 be the displacements of the two bobs. As shown in Fig. 1.39, the change in length of the spring is $x_2' - x_1'$.

From geometry of the figure,

$$\frac{x_1'}{x_1} = \frac{y}{l} \text{ and } \frac{x_2'}{x_2} = \frac{y}{l}$$

$$x_1' = \left(\frac{y}{l}\right) x_1 \text{ and } x_2' = \left(\frac{y}{l}\right) x_2$$

$$x_2' - x_1' = \frac{y}{l}(x_2 - x_1)$$

The force in the spring is F, where

$$F = k(x_2' - x_1')$$

$$F = k\left(\frac{y}{l}\right)(x_2 - x_1)$$

Clockwise moment of mass A about the point C

$$= ml\left(\frac{d^2 x_1}{dt^2}\right) + (mg \sin \alpha_1)\, l \qquad \ldots (i)$$

Anticlockwise moment of mass A about the point C

$$= F \times y$$

$$= k\left(\frac{y^2}{l}\right)(x_2 - x_1) \qquad \ldots (ii)$$

For small oscillations $\quad \sin \alpha_1 = \alpha_1 = \dfrac{x_1}{l}$

Equating (*i*) and (*ii*)

$$ml\left(\frac{d^2 x_1}{dt^2}\right) + mg\left(\frac{x_1}{l}\right) l = k\left(\frac{y^2}{l}\right)(x_2 - x_1)$$

$$\frac{d^2 x_1}{dt^2} = -\left(\frac{g}{l}\right) x_1 + \left(\frac{ky^2}{ml^2}\right)(x_2 - x_1) \qquad \ldots (iii)$$

Proceeding in the same way, for the mass B,

$$\frac{d^2 x_2}{dt^2} = -\left(\frac{g}{l}\right) x_2 - \left(\frac{ky^2}{ml^2}\right)(x_2 - x_1) \qquad \ldots (iv)$$

From equations (*iii*) and (*iv*)

$$\frac{d^2 x_1}{dt^2} = -\left[\left(\frac{g}{l}\right) + \frac{ky^2}{ml^2}\right] x_1 + \left(\frac{ky^2}{ml^2}\right) x_2 \qquad \ldots (v)$$

$$\frac{d^2 x_2}{dt^2} = -\left[\left(\frac{g}{l}\right) + \frac{ky^2}{ml^2}\right] x_2 + \left(\frac{ky^2}{ml^2}\right) x_1 \qquad \ldots (vi)$$

Taking

$$\left(\frac{g}{l} + \frac{ky^2}{ml^2}\right) = P$$

and

$$\left(\frac{ky^2}{ml^2}\right) = Q$$

$$\frac{d^2 x_1}{dt^2} = -P\, x_1 + Q\, x_2 \qquad \ldots (vii)$$

$$\frac{d^2 x_2}{dt^2} = -P\, x_2 + Q\, x_1 \qquad \ldots (viii)$$

In a normal mode of angular frequency ω and phase ϕ

$$x_1 = A \cos(\omega t + \phi)$$
$$x_2 = B \cos(\omega t + \phi)$$

Differentiating twice,

$$\frac{d^2 x_1}{dt^2} = -\omega^2 x_1$$

and

$$\frac{d^2 x_2}{dt^2} = -\omega^2 x_2$$

Substituting the values in equations (*vii*) and (*viii*) we get,

$$-\omega^2 x_1 = -P\, x_1 + Q\, x_2$$

$$\frac{x_1}{x_2} = \left(\frac{Q}{P - \omega^2}\right) \qquad \ldots (ix)$$

and

$$-\omega^2 x_2 = -P\, x_2 + Q\, x_1$$

$$\frac{x_1}{x_2} = \left(\frac{P - \omega^2}{Q}\right) \qquad \ldots (x)$$

Equating right hand sides of equation (*xi*) and (*x*)

$$\left(\frac{Q}{P - \omega^2}\right) = \left(\frac{P - \omega^2}{Q}\right)$$

$$Q^2 = (P - \omega^2)^2$$
$$(P - \omega^2) = \pm Q$$
$$\omega^2 = P \pm Q$$
$$\omega = [P \pm Q]^{1/2}$$

The angular frequencies of the two modes are

(*i*) $$\omega_1 = [P + Q]^{1/2}$$

$$\omega_1 = \left[\frac{g}{l} + \left(\frac{2ky^2}{ml^2}\right)\right]^{\frac{1}{2}} \quad \ldots (xi)$$

(*ii*) $$\omega_2 = [P - Q]^{1/2}$$

$$\omega_2 = \left[\frac{g}{l}\right]^{\frac{1}{2}} \quad (xii)$$

The frequencies are

(*i*) $$\nu_1 = \frac{\omega_1}{2\pi} = \frac{1}{2\pi}\left[\frac{g}{l} + \left(\frac{2ky^2}{ml^2}\right)\right]^{\frac{1}{2}}$$

(*ii*) $$\nu_2 = \frac{\omega_2}{2\pi} = \frac{1}{2\pi}\left[\frac{g}{l}\right]^{\frac{1}{2}}$$

Example 1.34. *A double pendulum is formed by suspending two pendulums one below the other. (Fig 1.40). The masses of the two bobs are equal and length of the each pendulum is the same.*

Calculate the (*i*) angular frequencies (*ii*) frequencies of the two normal modes for small oscillations.

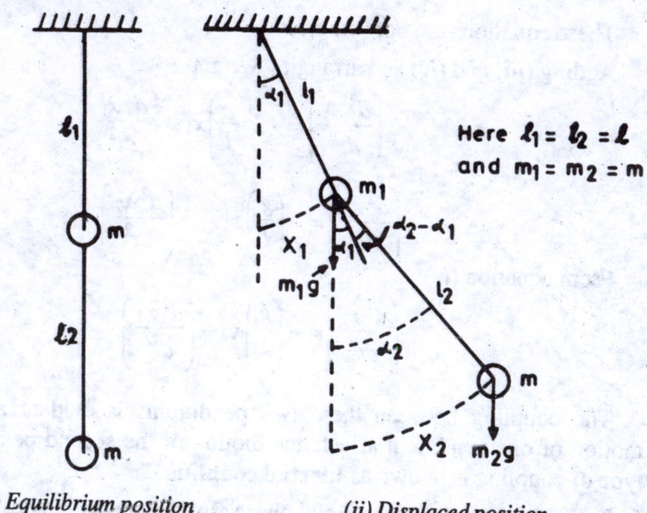

(*i*) *Equilibrium position* (*ii*) *Displaced position*

Fig . 1.40.

Fig. 1.39 (*ii*) represents the displaced position. The equations of motion are

$$m_1\left(\frac{d^2 x_1}{dt^2}\right) = -m_1 g \sin \alpha_1 + m_2 g \sin(\alpha_2 - \alpha_1) \qquad \ldots(i)$$

$$m_2\left(\frac{d^2 x_2}{dt^2}\right) = -m_2 g \sin \alpha_2 - m_1\left(\frac{d^2 x_1}{dt^2}\right)\cos(\alpha_2 - \alpha_1) \qquad \ldots(ii)$$

For small displacement

$$\sin \alpha_1 = \alpha_1 = \frac{x_1}{l_1}$$

$$\sin \alpha_2 = \alpha_2 = \frac{x_2}{l_2}$$

$$\sin(\alpha_2 - \alpha_1) = \alpha_2 - \alpha_1 = \frac{x_2}{l_2} - \frac{x_1}{l_1}$$

$\cos(\alpha_2 - \alpha_1) = 1$ (For small value of $\alpha_2 - \alpha_1$)

But $l_1 = l_2 = l$ and $m_1 = m_2 = m$

Substituting these values in equations *(i)* and *(ii)* and simplifying

$$\left(\frac{d^2 x_1}{dt^2}\right) = -\left(\frac{g}{l}\right)x_1 + \frac{g}{l}(x_2 - x_1) \qquad \ldots(iii)$$

$$\left(\frac{d^2 x_2}{dt^2}\right) = -\left(\frac{g}{l}\right)x_2 - \left(\frac{d^2 x_1}{dt^2}\right) \qquad \ldots(iv)$$

These equations are not symmetric

Adding *(iii)* and *(iv)* and arranging we get

$$2\left(\frac{d^2 x_1}{dt^2}\right) = -2\left(\frac{g}{l}\right)x_1 - \left(\frac{d^2 x_2}{dt^2}\right)$$

$$\left(\frac{d^2 x_1}{dt^2}\right) = -\left(\frac{g}{l}\right)x_1 - \frac{1}{2}\left(\frac{d^2 x_2}{dt^2}\right) \qquad \ldots(v)$$

From equation *(iv)*

$$\left(\frac{d^2 x_2}{dt^2}\right) = -\left(\frac{g}{l}\right)x_2 - \left(\frac{d^2 x_1}{dt^2}\right) \qquad \ldots(vi)$$

The coupling between these two pendulums is evident and clear. The motion of one pendulum affects the motion of the second pendulum. Such a type of coupling is known as **Inertial coupling.**

Take angular frequency ω and phase ϕ in the normal mode,

$$x_1 = A_1 \cos(\omega t + \phi)$$
$$x_2 = A_2 \cos(\omega t + \phi)$$

Differentiating twice,

$$\frac{d^2 x_1}{dt^2} = -\omega^2 x_1$$

$$\frac{d^2 x_2}{dt^2} = -\omega^2 x_2$$

Substituting these values in equations (*v*) and (*vi*)

$$-\omega^2 x_1 = -\left(\frac{g}{l}\right) x_1 + \frac{1}{2}\omega^2 x_2$$

$$\left(\frac{g}{l} - \omega^2\right) x_1 = \frac{1}{2}\omega^2 x_2$$

$$\frac{x_1}{x_2} = \frac{\omega^2}{2\left[\frac{g}{l} - \omega^2\right]} \qquad \ldots (vii)$$

and

$$-\omega^2 x_2 = -\left(\frac{g}{l}\right) x_2 + \omega^2 x_1$$

$$\frac{x_1}{x_2} = \frac{\left(\frac{g}{l} - \omega^2\right)}{\omega^2} \qquad \ldots (viii)$$

Equating right hand sides of equations (*vii*) and (*viii*)

$$\frac{\omega^2}{2\left(\frac{g}{l} - \omega^2\right)} = \frac{\frac{g}{l} - \omega^2}{\omega^2}$$

or

$$\omega^4 = 2\left(\frac{g}{l} - \omega^2\right)^2$$

$$\omega^4 = 2\left(\frac{g}{l}\right)^2 + 2\omega^4 - \left(\frac{4g}{l}\right)\omega^2$$

$\therefore$

$$\omega^4 - \left(\frac{4g}{l}\right)\omega^2 + 2\left(\frac{g}{l}\right)^2 = 0 \qquad \ldots (ix)$$

It is a quadratic equation. Its solution is

$$\omega^2 = 2\left(\frac{g}{l}\right) \pm (\sqrt{2})\,\frac{g}{l}$$

$$= (2 \pm \sqrt{2})\left(\frac{g}{l}\right)$$

$$\omega = \left[(2 \pm \sqrt{2})\left(\frac{g}{l}\right)\right]^{\frac{1}{2}}$$

Angular frequencies

(*i*) $$\omega_1 = \left[(2 + \sqrt{2})\left(\frac{g}{l}\right)\right]^{\frac{1}{2}} \qquad \ldots (x)$$

(*ii*) $$\omega_2 = \left[(2 - \sqrt{2})\left(\frac{g}{l}\right)\right]^{\frac{1}{2}} \qquad \ldots (xi)$$

Frequencies

(*i*) $$\nu_1 = \frac{\omega_1}{2\pi} = \frac{1}{2\pi}\left[(2 + \sqrt{2})\left(\frac{g}{l}\right)\right]^{\frac{1}{2}} \qquad \ldots (xii)$$

(*ii*) $$\nu_2 = \frac{\omega_2}{2\pi} = \frac{1}{2\pi}\left[(2 - \sqrt{2})\left(\frac{g}{l}\right)\right]^{\frac{1}{2}} \qquad \ldots (xiii)$$

Equations (*x*) and (*xi*) represent the angular frequencies of the two normal modes and equations (*xii*) and (*xiii*) represent the frequencies of the two normal modes.

EXERCISES

1. Explain simple harmonic motion and discuss its characteristics.
2. Show that for a body vibrating simple harmonically the time period is given by

$$t = 2\pi\sqrt{\frac{\text{displacement}}{\text{acceleration}}}$$

3. Calculate the average kinetic energy and the total energy of a body executing simple harmonic motion.
4. Give examples of the systems that oscillate with one degree of freedom. Explain the term damped oscillations.
5. Show that the superposition principle is valid only in the case of homogeneous, linear vibrations.
6. Obtain the expression for the time period of a simple pendulum. Also derive the expression for the time period for a bob of large size.

7. Give the theory of a compound pendulum and derive an expression for its time period.
8. Show that in the case of a compound pendulum, the points of suspension and oscillation are interchangeable.
9. Give the theory of Kater's reversible pendulum.
10. Discuss the case of simple harmonic oscillations of a mass held between two linear springs.
11. Show that the time period of oscillation of a loaded spring is

$$t = 2\pi\sqrt{\frac{Mx}{mg}}.$$

12. Discuss the *LC* circuit and calculate the expression for the frequency of oscillations.
13. Discuss with examples free oscillations of a system with two degrees of freedom.
14. Discuss the two normal modes of oscillations of coupled *LC* circuits.
15. How will you determine the value of acceleration due to gravity using a compound pendulum.
16. Derive an expression for the time period of a compound pendulum and show that there are four collinear points on a compound pendulum about which the period of oscillation is the same. Give Bessel's computed time of Kater's pendulum.
17. A particle vibrates simple harmonically with an amplitude of 13 cm. The time period of oscillation is 2π seconds. Calculate the velocity of the vibrating particle at the instant the displacement is 5 cm. Also calculate the frequency of oscillation.

[**Ans.** (*i*) 12 cm/s ; (*ii*) $\frac{1}{2\pi}$ hertz]

18. A simple periodic wave disturbance with an amplitude of 8 units, travels a line of particles in the positive *x* direction. At a given instant, the displacement of a particle 10 cm from the origin is 6 units, and that of a particle 25 cm from the origin is 4 units, both particles being in positive displacement. What is the wavelength of the disturbance.

[**Ans.** 290 cm.]

19. At time $t = 0$, a train of waves has the form

$$y = 4 \sin 2\pi\left(\frac{x}{100}\right)$$

The velocity of the wave is 30 cm/s. Find the equation giving the waveform at a time t=2 s.

Hint : $$y = a \sin 2\pi \left[\frac{t}{T} - \frac{x}{\lambda}\right]$$

When $t = 0$

$$y = a \sin 2\pi \left(\frac{x}{\lambda}\right)$$

But $$y = 4 \sin 2\pi \left(\frac{x}{100}\right)$$

$\therefore$ $\lambda = 100$ cm

At time $t = 2\text{s}, x = 2 \times 30 = 60$ cm

$\therefore$ $$y = a \sin 2\pi \left[\frac{x}{100} - \frac{60}{100}\right]$$

$$y = a \sin 2\pi \left[\frac{x}{100} - 0{\cdot}6\right].$$

20. A spring is hung vertically and loaded with of mass of 400 grams and made to oscillate. Calculate (*i*) the time period and (*ii*) the frequency of oscillation. When the spring is loaded with 100 grams it extends by 5 cm.

[Ans. (*i*) 0·898 s, (*ii*) 1·11 hertz]

21. A scale of a spring balance reading 0 → 5 kg is 120·5 cm. A body suspended from the balance oscillates with a frequency $5/\pi$ hertz. Calculate the mass of the body attached to the spring.

[Ans. M = 3·92 kg]

22. (*a*) In case of a two-dimensional harmonic oscillator as shown in Fig. 1.41 below obtain the equations of the two normal modes of oscillation.

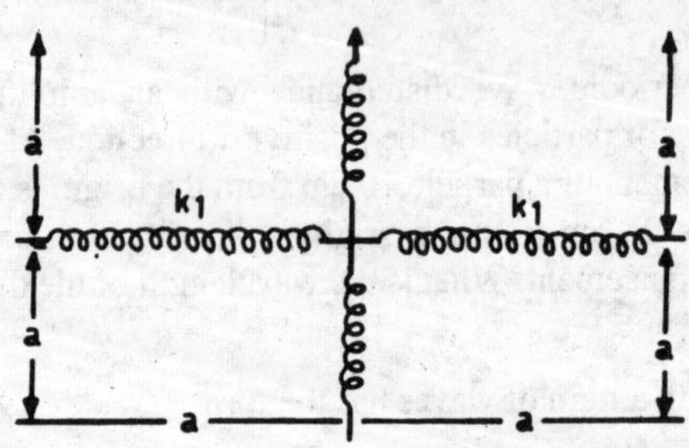

Fig. 1.41.

(*b*) Discuss the two normal modes of oscillations of two coupled *L-C* circuits. [*Delhi (Hons) 1991*]

23. (*a*) A mass M is attached with two springs each of spring constant K and relaxed length a_0. Show that the frequency of transverse oscil lations is the same for slinky approximated small scale oscillations.

(*b*) Explain normal modes and normal coordinates.

[*Delhi (Hons.) 1991*]

24. Two blocks of mass m_1 and m_2 connected together by a spring of constant K, are resting on *a* smooth horizontal surface as shown in Fig. 1.42 below.

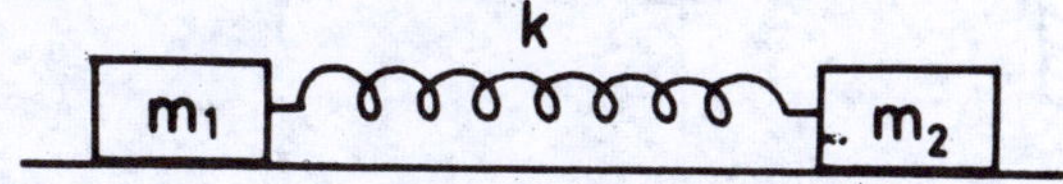

Fig. 1.42.

Obtain an expression for the natural frequencies of the oscillation.

[*Delhi (Hons.) 1991*]

25. A system consists of two identical pendulums, each having mass m, suspended on a light rigid rod of length l. The masses are connected by a massless spring of force constant K. The natural length of the spring is the distance between the masses when neither is displaced from equilibrium.

(*a*) Obtain the normal mode frequencies of vibration of the system.

(*b*) Assuming weak coupling, give the theory of beat formation between the two normal modes.

[*Delhi (Hons) 1992*]

26. A uniform stretched string under tension T and fixed at both ends has N identical particles, each of mass m, with a constant spacing l between two neighbouring particles. Show that the frequency of the nth normal mode is given by

$$\omega_n = 2\sqrt{\frac{T}{m\,.\,l}}\ \sin\left[\frac{n\pi}{2(N+1)}\right].$$

Also show that the maximum number of normal modes is N.

[*Delhi (Hons) 1992*]

27. What are normal coordinates and normal modes of a coupled system. What is their significance?

[*Delhi (Hons) 1993*]

28. In the system given in Fig. 1.43, *A* and *B* are two masses m_a, m_b respectively and springs have constants k_1, k_2 and k_3 as shown. The spring k_3 provides the couples. The masses are pulled along a frictionless horizontal surface as shown is the figure and then released. Find the angular frequencies of vibration of the system in the possible normal modes.

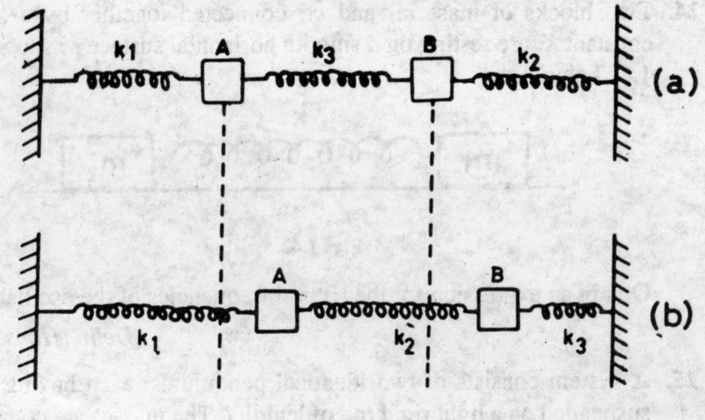

Fig. 1.43

[Delhi (Hons) 1993]

CHAPTER 2

Lissajous' Figures

2.1. Lissajous' Figures

When a particle is influenced simultaneously by two simple harmonic motions at right angles to each other, the resultant motion of the particle traces a curve. These curves are called Lissajous' figures. The shape of the curve depends on the time period, phase difference and the amplitude of the two constituent vibrations. Lissajous' figures are helpful in determining the ratio of the time periods of two vibrations and to compare the frequencies of two tuning forks.

2.2. Composition of Two Simple Harmonic Motions in a Straight Line

Analytical method. Let the two simple harmonic vibrations be represented by the equations

$$y_1 = a_1 \sin(\omega t + \alpha_1) \quad \ldots (1)$$

and

$$y_2 = a_2 \sin(\omega t + \alpha_2) \quad \ldots (2)$$

where y_1 and y_2 are the displacements of a particle due to the two vibrations, a_1 and a_2 are the amplitudes of the two vibrations and α_1 and α_2 are the epoch angles. Here, the two vibrations are assumed to be of the same frequency and hence ω is the same for both. The resultant displacement y of the particle is given by

$$\begin{aligned} y &= y_1 + y_2 \\ &= a_1 \sin(\omega t + \alpha_1) + a_2 \sin(\omega t + \alpha_2) \\ &= a_1 (\sin \omega t \cos \alpha_1 + \cos \omega t \sin \alpha_1) \\ &\qquad + a_2 (\sin \omega t \cos \alpha_2 + \cos \omega t \sin \alpha_2) \\ y &= (a_1 \cos \alpha_1 + a_2 \cos \alpha_2) \sin \omega t \\ &\qquad + (a_1 \sin \alpha_1 + a_2 \sin \alpha_2) \cos \omega t \quad \ldots (3) \end{aligned}$$

Since the amplitudes a_1 and a_2 and the angles α_1 and α_2 are constant, the coefficients of $\sin \omega t$ and $\cos \omega t$ in equation (3) can be substituted by $A \cos \phi$ and $A \sin \phi$

$$A \cos \phi = a_1 \cos \alpha_2 + a_2 \cos \alpha_2 \qquad \ldots (4)$$

and
$$A \sin \phi = a_1 \sin \alpha_1 + a_2 \sin \alpha_2 \qquad \ldots (5)$$

Squaring and adding equations (4) and (5)

$$A^2 (\sin^2 \phi + \cos^2 \phi) = a_1^2 (\sin^2 \alpha_1 + \cos^2 \alpha_1) + a_2^2 (\sin^2 \alpha_2 + \cos^2 \alpha_2)$$
$$+ 2a_1 a_2 [\cos \alpha_1 \cos \alpha_2 + \sin \alpha_1 \sin \alpha_2]$$

$$\therefore \quad A^2 = a_1^2 + a_2^2 + 2a_1 a_2 \cos (\alpha_1 - \alpha_2) \qquad \ldots (6)$$

Dividing equation (5) by (4)

$$\tan \phi = \frac{a_1 \sin \alpha_1 + a_2 \sin \alpha_2}{a_1 \cos \alpha_1 + a_2 \cos \alpha_2} \qquad \ldots (7)$$

Equations (6) and (7) give the values of A and ϕ in terms of a_1, a_2, α_1 and α_2.

$$\therefore \quad y = A \cos \phi \sin \omega t + A \sin \phi \cos \omega t$$
$$y = A \sin (\omega t + \phi) \qquad \ldots (8)$$

Equation (8) is similar to the original equations (1) and (2). The amplitude of the resultant vibrations is A and epoch angle is ϕ. The time period of the resultant vibration is the same as the original vibrations. The values of A and ϕ are given by equations (6) and (7). Thus, the resultant of two simple harmonic vibrations of the same period and acting in the same line is also a simple harmonic vibration with a resultant amplitude A and epoch angle ϕ.

Special case. $\alpha_1 = \alpha_2 = \alpha$, then

$$A = a_1 + a_2 \text{ and } \phi = \alpha$$

or
$$y = (a_1 + a_2) \sin (\omega t + \alpha) \qquad \ldots (9)$$

Graphical method. Let OP and OQ represent the radius vectors at any instant (Fig. 2.1).

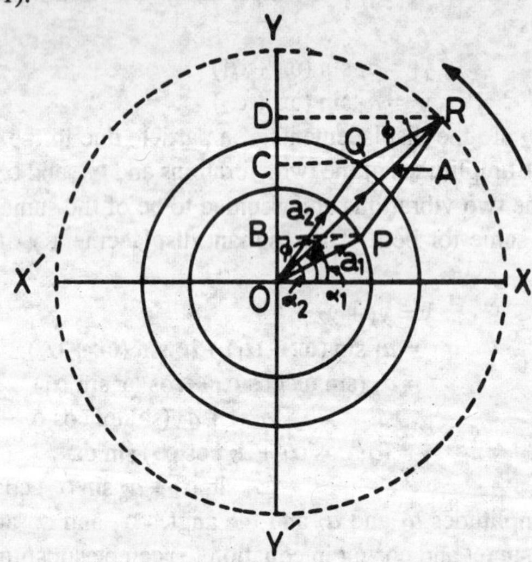

Fig. 2.1

$$\angle POX = \alpha_1 \text{ and } \angle QOX = \alpha_2$$
$$OP = a_1,\ OQ = a_2 \text{ and } OR = A.$$

The displacements of a particle along the line YY' are given by

$$OB = y_1 = a_1 \sin(\omega t + \alpha_1)$$

and $$OC = y_2 = a_2 \sin(\omega t + \alpha_2)$$

where a_1 and a_2 are the amplitudes and α_1 and α_2 are the epoch angles.

The resultant of the vectors OP an OQ is given by OR.

According to the parallelogram law of vectors

$$OR^2 = A^2 = a_1^2 + a_2^2 + 2a_1 a_2 \cos \angle QOP$$
$$OR^2 = a_1^2 + a_2^2 + 2a_1 a_2 \cos(\alpha_2 - \alpha_1) \quad \ldots (1)$$

From the Δs OPB and OQC

$$y_1 = OB = a_1 \sin \alpha_1$$

and $$y_2 = OC = a_2 \sin \alpha_2$$

But the projection of OQ on the Y-axis is equal to the projection of PR on the y-axis.

$\therefore$ $$y_2 = OC = BD = a_2 \sin \alpha_2$$

Resultant displacement

$$y = OB + BD = y_1 + y_2$$
$$= a_1 \sin \alpha_1 + a_2 \sin \alpha_2 \quad \ldots (2)$$

Similarly the projections of OP and OQ on the X-axis will be $a_1 \cos \alpha_1$ and $a_2 \cos \alpha_2$.

$\therefore$ $$RD = a_1 \cos \alpha_1 + a_2 \cos \alpha_2$$

In the $\Delta\ ORD$, $\angle ORD = \phi$

and $$\tan \phi = \frac{OD}{RD} = \frac{a_1 \sin \alpha_1 + a_2 \sin \alpha_2}{a_1 \cos \alpha_1 + a_2 \cos \alpha_2} \quad \ldots (3)$$

Thus the diagonal OR represents completely the resultant of two collinear simple harmonic motions. The resultant amplitude A and the epoch angle ϕ are given by the equations (1) and (3). The resultant displacement is represented by the equation

$$y = A \sin(\omega t + \phi) \quad \ldots (4)$$

If $\alpha_1 = \alpha_2$, then $\alpha_1 - \alpha_2 = 0$ and $A = a_1 + a_2$ and $\tan \alpha = \tan \phi$.

Example 2.1. *Two SHMs acting simultaneously on a particle are given by the equations*

$$y_1 = 2 \sin(\omega t + \pi/6)$$
$$y_1 = 3 \sin(\omega t + \pi/3)$$

Calculate (i) amplitude, (ii) phase constant and (iii) time period of the resultant vibration.

Here $$y_1 = 2 \sin(\omega t + \pi/6) \quad \ldots (1)$$

and $$y_2 = 3 \sin(\omega t + \pi/3) \quad \ldots (2)$$

These equations are similar to the equations,

$$y_1 = a_1 \sin(\omega t + \alpha_1) \quad \ldots (3)$$

and

$$y_2 = a_2 \sin(\omega t + \alpha_2) \quad \ldots (4)$$

The resultant vibration is,

$$y = A \sin(\omega t + \phi) \quad \ldots (5)$$

Here

$$A^2 = a_1^2 + a_2^2 + 2a_1a_2 \cos(\alpha_1 - \alpha_2) \quad \ldots (6)$$

and

$$\tan\phi = \frac{a_1 \sin\alpha_1 + a_2 \sin\alpha_2}{a_1 \cos\alpha_1 + a_2 \cos\alpha_2} \quad \ldots (7)$$

Here

$$a_1 = 2, \ a_2 = 3$$

$$\alpha_1 = \frac{\pi}{6}, \ \alpha_2 = \frac{\pi}{3}.$$

(*i*) Resultant amplitude

$$A = \sqrt{a_1^2 + a_2^2 + 2a_1a_2 \cos(\alpha_1 - \alpha_2)}$$

$$A = \sqrt{4 + 9 + 2 \times 2 \times 3 \cos\left(-\frac{\pi}{6}\right)}$$

$$A = \mathbf{4{\cdot}939}.$$

(*ii*)

$$\phi = \tan^{-1}\left[\frac{2 \times 0{\cdot}5 + 3 \times 0{\cdot}866}{2 \times 0{\cdot}866 + 3 \times 0{\cdot}5}\right]$$

$$\phi = \tan^{-1}[1{\cdot}114]$$

$$\phi = 48{\cdot}1^0$$

$$\phi = \frac{48{\cdot}1 \times \pi}{180} = \frac{4\pi}{15} \text{ (app)}$$

$\therefore$ Phase constant $= \left(\omega t + \dfrac{4\pi}{15}\right)$.

(*iii*) The resultant time period is the same as the individual time periods.

Example 2.2. *Two SHMs acting simultaneously on a particle are given by*

$$y_1 = \sin\left(\omega t + \frac{\pi}{3}\right)$$

$$y_2 = 2 \sin\omega t$$

Find the equation of the resultant vibration.

Here

$$y_1 = \sin\left(\omega t + \frac{\pi}{3}\right) \quad \ldots (1)$$

and

$$y_2 = 2 \sin\omega t \quad \ldots (2)$$

These equations are similar to

$$y_1 = a_1 \sin(\omega t + \alpha_1)$$

$$y_2 = a_2 \sin(\omega t + \alpha_2)$$

$$\therefore \qquad a_1 = 1, a_2 = 2, \alpha_1 = \frac{\pi}{3}, \alpha_2 = 0$$

The resultant vibration is given by

$$y = A \sin(\omega t + \phi)$$

Here
$$A = \sqrt{a_1^2 + a_2^2 + 2\, a_1 a_2 \cos(\alpha_1 - \alpha_2)}$$

$$A = \sqrt{1 + 4 + 2 \times 1 \times 2 \cos\left(\frac{\pi}{3}\right)}$$

$$A = \sqrt{7} = 2{\cdot}645$$

$$\tan\phi = \frac{a_1 \sin\alpha_1 + a_2 \sin\alpha_2}{a_1 \cos\alpha_1 + a_2 \cos\alpha_2}$$

$$\tan\phi = \frac{1 \times 0{\cdot}866 + 0}{1 \times 0{\cdot}5 + 2 \times 1} = \frac{0{\cdot}866}{2{\cdot}5}$$

$$\phi = 19{\cdot}1^\circ = \frac{19{\cdot}1 \times \pi}{180} \text{ radian} = 0{\cdot}1061\,\pi$$

$$\therefore \qquad y = 2{\cdot}645 \sin(\omega t + 0{\cdot}1061\,\pi).$$

2.3. Composition of a Number of Simple Harmonic Motions Acting along a Line

When a particle is influenced by a number of collinear simple harmonic vibrations of different amplitudes and different epoch angles, the resultant vibration can be obtained by the polygon method. In Fig. 2.2, *OP*, *PQ* and *QR* are the vectors representing three simple harmonic vibrations. The amplitude of the individual vibrations a_1, a_2 and a_3 and α_1, α_2 and α_3 are the corresponding epoch angles. The vector *OR* represents and resultant vibration. The amplitude and the epoch angle of the resultant vibration are *A* and ϕ respectively. Proceeding in the same way this method can be employed when a number of collinear simple harmonic vibrations influence the same particle of the medium.

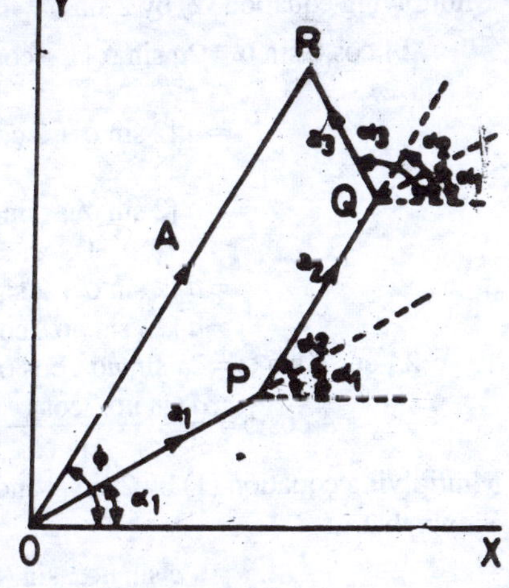

Fig. 2.2.

Let n simple harmonic vibrations of the same amplitude a and epoch angles $0, 2\alpha, 4\alpha \ldots 2(n-1)\alpha$ influence a vibrating particle (Fig. 2.3). If the displacements of the vibrating particle are considered along the y-axis, the individual displacements are given by

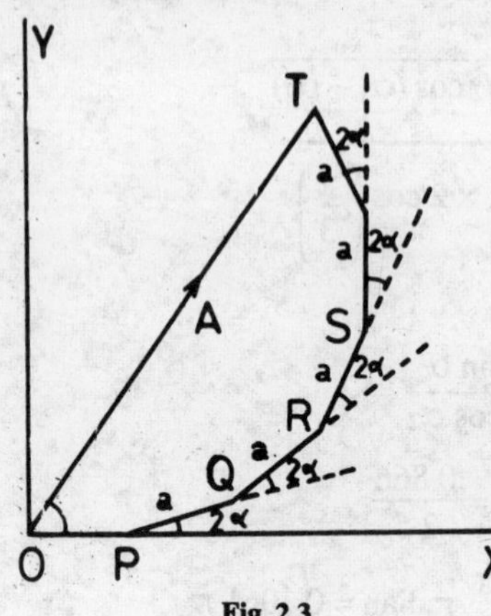

Fig. 2.3.

$$y_1 = a \sin(\omega t - 0)$$

$$y_2 = a \sin(\omega t - 2\alpha)$$

Let A be the amplitude of the resultant. Vibration and ϕ the epoch angle. Then

$$y = A \sin(\omega t - \phi).$$

The projections of the individual vectors OP, PQ, QR etc. on the y-axis are given by

$$0, a \sin 2\alpha, a \sin 4\alpha \text{ etc.}$$

Similarly the projections on the X-axis are given by

$$a, a\cos 2\alpha, a \cos 4\alpha \text{ etc.}$$

If OT represents the resultant vector, then $A \sin\phi$ will give the projection along the Y-axis and $A\cos\phi$ gives the projection along the X-axis.

$$\therefore \quad A\sin\phi = 0 + a\sin 2\alpha + a \sin 4\alpha + \ldots\ldots + a \sin 2(n-1)\alpha$$

$$= a[\sin 2\alpha + \sin 4\alpha + \ldots\ldots + \sin 2(n-1)\alpha] \quad \ldots (1)$$

Similarly,

$$A\cos\phi = a + a\cos 2\alpha + a\cos 4\alpha + \ldots\ldots + a\cos 2(n-1)\alpha$$

$$a[1 + \cos 2\alpha + \cos 4\alpha + \ldots\ldots + \cos 2(n-1)\alpha] \quad \ldots (2)$$

Multiplying equation (2) by $2\sin\alpha$

$$2A\cos\phi\sin\alpha = 2a\sin\alpha\,[1 + \cos 2\alpha + \cos 4\alpha + \ldots\ldots + \cos 2(n-1)\alpha]$$

$$= a[2\sin\alpha + 2\cos 2\alpha . \sin\alpha + 2\cos 4\alpha \sin\alpha + \ldots\ldots + 2\cos 2(n-1)\alpha\sin\alpha]$$

$$= a[2\sin\alpha + (\sin 3\alpha - \sin\alpha) + (\sin 5\alpha - \sin 3\alpha) + \ldots\ldots + \{\sin(2n-1)\alpha - \sin(2n-3)\alpha\}]$$

$$= a[2\sin\alpha + 2\sin(2n-1)\alpha]$$

$$= a[2 . \sin n\alpha . \cos(n-1)\alpha]$$

$$\therefore \quad 2A\cos\phi\sin\alpha = 2a\sin n\alpha . \cos(n-1)\alpha$$

$$\therefore \quad A\cos\phi = \frac{a\sin n\alpha . \cos(n-1)\alpha}{\sin\alpha} \quad \ldots (3)$$

Multiplying equation (1) by $2\sin\alpha$ and proceeding in a similar way, it can be shown that

$$A\sin\phi = \frac{a\sin n\alpha . \sin(n-1)\alpha}{\sin\alpha} \quad \ldots (4)$$

Squaring equations (3) and (4) and adding

$$A^2 (\sin^2 \phi + \cos^2 \phi) = A^2$$

$$= \frac{a^2 \sin^2 n\alpha}{\sin^2 \alpha} [\sin^2 (n-1)\alpha + \cos^2 (n-1)\alpha]$$

$$= \frac{a^2 \sin^2 n\alpha}{\sin^2 \alpha} \quad \ldots (5)$$

$$\therefore \quad A = \frac{a \sin n\alpha}{\sin \alpha} \quad \ldots (6)$$

Dividing equation (4) by (3)

$$\frac{A \sin \phi}{A \cos \phi} = \tan \phi = \frac{a \sin n\alpha \,.\, \sin (n-1)\alpha \,.\, \sin \alpha}{\sin \alpha \,.\, a \sin n\alpha \,.\, \cos (n-1)\alpha}$$

$$= \tan (n-1)\alpha$$

or, $$\phi = (n-1)\alpha \quad \ldots (7)$$

where ϕ represents the epoch angle for the resultant vibration, n is the number of simple harmonic vibrations influencing a particle and α is half the increase in the epoch angle between successive vibrations.

2.4. Composition of Two Simple Harmonic Vibrations of Equal Time Periods Acting at Right Angles

Let $$x = a \sin (\omega t + \alpha) \quad \ldots (1)$$

and $$y = b \sin \omega t \quad \ldots (2)$$

represent the displacements of a particle along the X and Y-axes due to the influence of two simple harmonic vibrations acting simultaneously on a particle in perpendicular directions. Here, the two vibrations are of the same time period but are of different amplitudes and different phase angles.

From equation (2), $$\sin \omega t = \frac{y}{b}$$

$$\therefore \quad \cos \omega t = \sqrt{1 - \frac{y^2}{b^2}}$$

From equation (1), $$\frac{x}{a} = [\sin \omega t \cos \alpha + \cos \omega t \sin \alpha] \quad \ldots (3)$$

Substituting the values of $\sin \omega t$ and $\cos \omega t$ in equation (3)

$$\frac{x}{a} = \left[\frac{y}{b} \cos \alpha + \sqrt{1 - \frac{y^2}{b^2}} \,.\, \sin \alpha\right]$$

or $$\frac{x}{a} - \frac{y}{b} \cos \alpha = \sqrt{1 - \frac{y^2}{b^2}} \sin \alpha$$

Squaring

$$\frac{x^2}{a^2}+\frac{y^2}{b^2}\cos^2\alpha-\frac{2xy}{ab}\cos\alpha=\left(1-\frac{y^2}{b^2}\right)\sin^2\alpha$$

or $$\frac{x^2}{a^2}+\frac{y^2}{b^2}[\sin^2\alpha+\cos^2\alpha]-\frac{2xy}{ab}\cos\alpha=\sin^2\alpha$$

$\therefore$ $$\frac{x^2}{a^2}+\frac{y^2}{b^2}-\frac{2xy}{ab}\cos\alpha=\sin^2\alpha \quad \ldots (4)$$

This represents the general equation of an ellipse. Thus, due to the super imposition of two simple harmonic vibrations, the displacement of the particle will be along a curve (Fig. 2.4) given by equation (4).

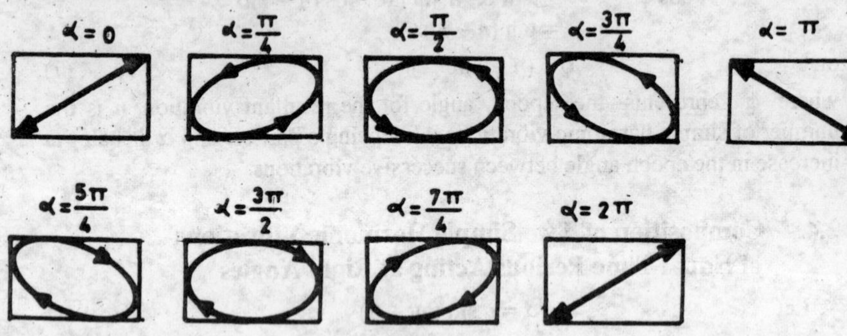

Fig. 2.4.

The resultant vibration of the particle will depend upon the value of α. Figure 2.4 represents the resultant vibration for values of α changing from 0 to 2 π.

Special cases

(*i*) If $\alpha=0$ or 2π ; $\cos\alpha=1$; $\sin\alpha=0$

$$\frac{x^2}{a^2}+\frac{y^2}{b^2}-\frac{2xy}{ab}=0$$

or $$\frac{x}{a}-\frac{y}{b}=0$$

or $$y=\frac{b}{a}x.$$

This represents the equation of the straight line *BD* (Fig. 2.5) *i.e.,* the particle vibrates simple harmonically along the line *DB*.

(*ii*) *If* $\alpha=\pi$; $\sin\alpha=0$;

$$\cos\alpha=-1$$

$$\therefore \quad \frac{x^2}{a^2}+\frac{y^2}{b^2}+\frac{2xy}{ab}=0$$

$$\left(\frac{x}{a}+\frac{y}{b}\right)=0$$

$$y=-\frac{b}{a}x. \qquad \ldots(6)$$

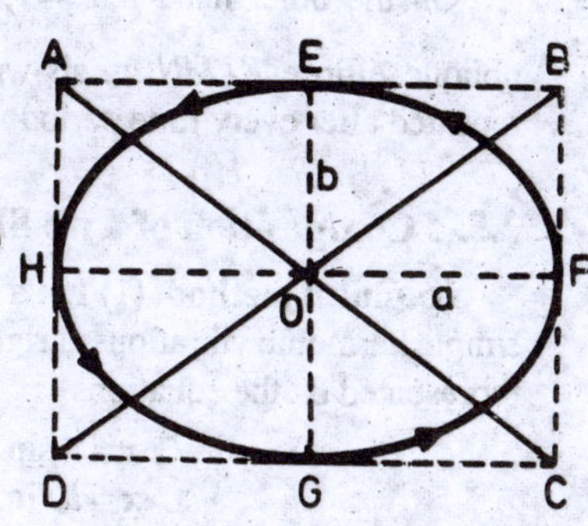

Fig. 2.5.

This represents the equation of the straight line AC (Fig. 2.5).

(*iii*) if $\quad \alpha=\frac{\pi}{2}$ or $\frac{3\pi}{2}$

sin $\quad \alpha=1$; $\cos\alpha=0$

$$\therefore \quad \frac{x^2}{a^2}+\frac{y^2}{b^2}=1$$

This represents the equation of the ellipse $EHGF$ (Fig. 2.5) with a and b as the semi-major and semi-minor axes.

(*iv*) If $\alpha=\frac{\pi}{2}$ or $\frac{3\pi}{2}$

and $\quad a=b$;

then $\quad \frac{x^2}{a^2}+\frac{y^2}{a^2}=1$

or $\quad x^2+y^2=a^2$

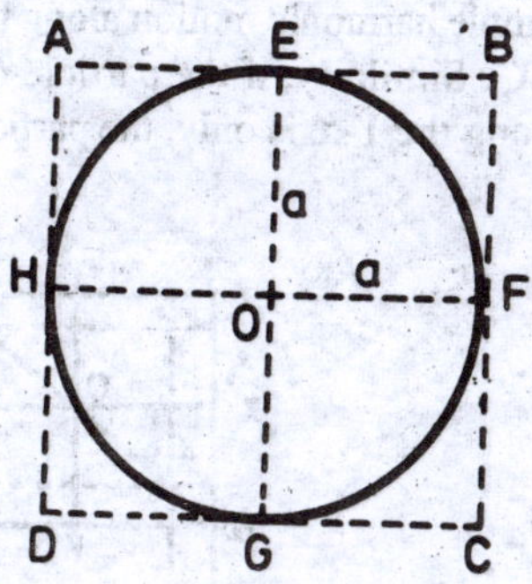

Fig. 2.6.

This represents the equation of a circle of radius a (Fig. 2.6).

(*v*) if $\quad \alpha=\frac{\pi}{4}$ or $\frac{7\pi}{4}$, the resultant vibration is an oblique ellipse $KLMN$ as shown in Fig. 2.7(*i*).

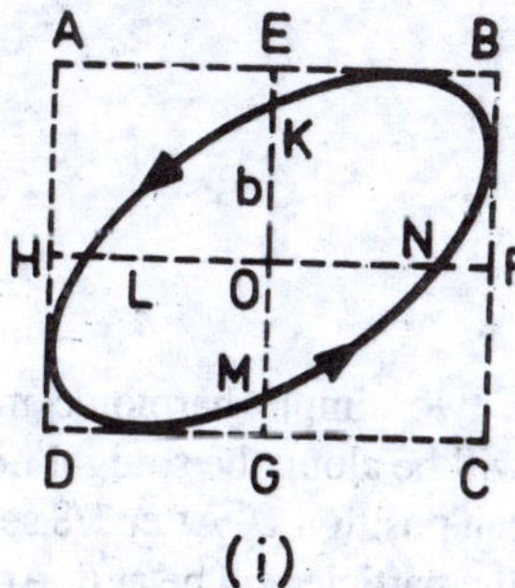

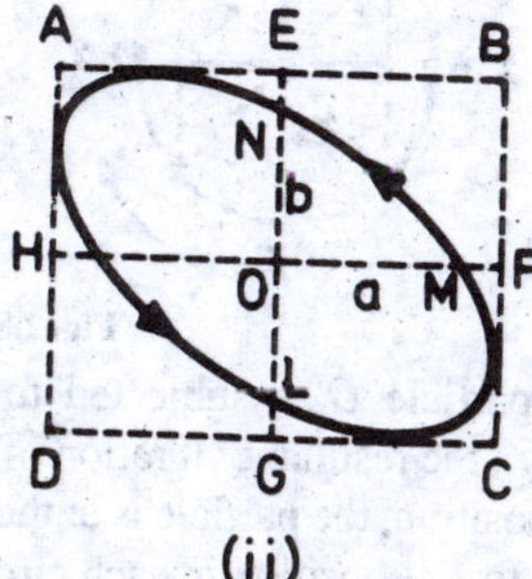

Fig. 2.7.

On the other hand if $\alpha = \frac{3\pi}{4}$ or $\frac{5\pi}{4}$, the resultant vibration is again an oblique ellipse *KLMN* as shown in Fig. 2.7 (*ii*). The cycle of changes is repeated after every time period.

2.5. Composition of Two SHMs at Right Angles of Equal Periods

Graphical method. (1) Let a particle be influenced simultaneously by two simple harmonic vibrations at right angles to each other. The two vibrations are represented by the equations

$$x = a \sin \omega t$$
$$x = b \sin \omega t$$

Here the phase difference between the two vibrations is zero, time periods are equal and the amplitudes are unequal.

Draw two circles of reference with centres C_1 and C_2 and radii a and b respectively. Divide each circle into eight equal parts, marked 0, 1, 2, . . . 7, 8. The angular frequency in each case is ω. If the particle O is subjected to the simple harmonic motion along the ***X*-axis** only, the particle will vibrate along XX'. Similarly, if the particle O is subjected to the simple harmonic motion along the ***Y*-axis** only, the particle will vibrate along YY' (Fig. 2.8).

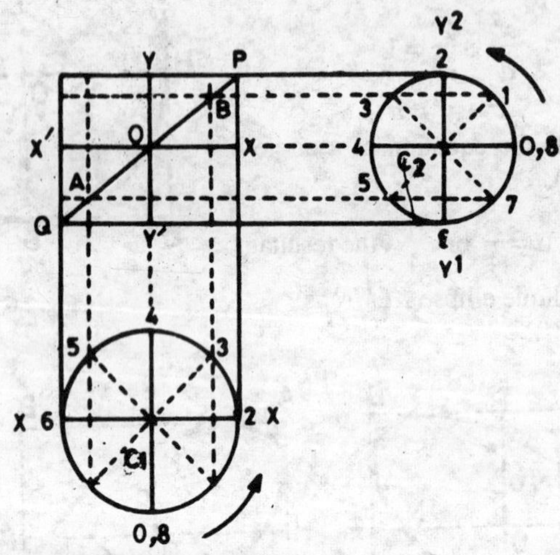

Fig. 2.8.

When the particle O is subjected to the two simple harmonic motions simultaneously ; the resultant vibration of O will be along the straight line PQ. At zero, zero position, the particle is at the mean position O. After $T/8$ seconds, corresponding to 1, 1 position in each circle, the particle will be at B. After $T/4$ seconds, corresponding to 2, 2 positions in each circle, the particle will be at P.

In this way the positions of the resultant displacement of the particle will be, B after $3T/8$ seconds, O after $T/2$ seconds, A after $5T/8$ seconds, Q after $3T/4$ seconds, A after $7T/8$ seconds and again at O after T seconds. In this way the resultant vibration is along POQ. The amplitude of the resultant vibration is OP.

Here $$(OP)^2 = (OX)^2 + (OY)^2$$
$$(OP)^2 = a^2 + b^2$$
$$OP = \sqrt{a^2 + b^2}$$

The angular frequency and time period remain the same as for the two constituent vibrations.

(2) Let a particle be influenced simultaneously by two simple harmonic vibrations at right angles to each other. The two vibrations are represented by the equations

$$x = a \sin (\omega t + \alpha)$$
and $$y = b \sin \omega t$$

Suppose the phase difference $\alpha = \pi/2$ and the time periods are equal. The amplitudes are a and b.

Draw two circles of reference with centres C_1 and C_2 and radii a and b respectively. Divide each circle into eight equal parts, marked 0, 1, 2, . . .7, 8.

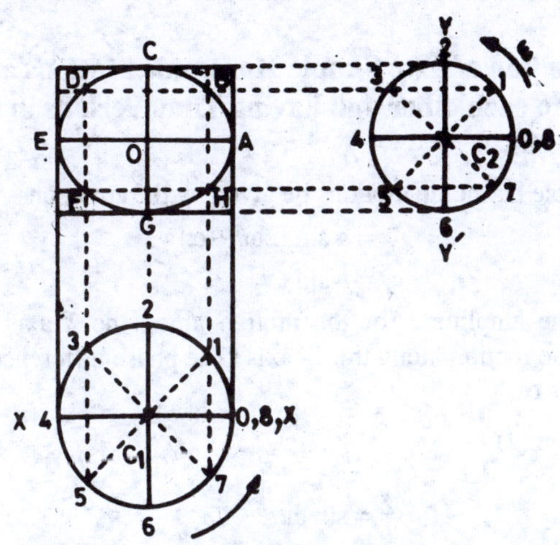

Fig. 2.9.

The angular frequency in each case is ω. If the particle O is subjected to the simple harmonic motion along the X-axis only, the particle will vibrate along XX'. Similarly if the particle O is subjected to the simple harmonic motion along the Y-axis only, the particle will vibrate along YY'. Here, the initial position for vibration along XX' is at the extreme position at $t = 0$. The points on the circle of reference are marked showing that there is a phase difference of $\pi/2$ between the two vibrations (Fig. 2.9).

When the particle O is subjected to the two SHMs simultaneously, the resultant vibration of O will be along an **ellipse** having a and b as semi-major and semi-minor axes. The resultant position after T/8 seconds corresponding to 1, 1 position in each circle of reference will be the point B. Similarly C, D, E, F etc. will be the resultant positions after successive time intervals of $T/8$ seconds. The motion of the particle O will be along the ellipse $ABCDEFGHA$. The angular frequency and time period remain the same as for the two constituent vibrations.

Here at any instant

$$\frac{x^2}{a^2}+\frac{y^2}{b^2}=1$$

If a and b are equal, then the resultant motion is given by the equation

$$x^2+y^2=a^2$$

which represents a circle with centre O and radius a.

[**Note**. If a particle is moving along a circular path of radius a with angular frequency ω, it can be considered to be under the influence of two rectangular SHMs having equal amplitude of value a and angular frequency ω, and a phase difference of $\pi/2$.]

2.6. Composition of Two Simple Harmonic Motions at Right Angles to each other and having Time Periods in the Ratio 1 : 2

Let two simple harmonic motions be given by the equations

$$x = a \sin (2\omega t + \alpha) \quad \ldots (1)$$

and

$$y = b \sin \omega t \quad \ldots (2)$$

Here a is the amplitude for the motion along the X-axis and b is the amplitude for the motion along the Y- axis. The phase difference between the two vibrations is α.

From equation (2)

$$\frac{y}{b} = \sin \omega t$$

and

$$\cos \omega t = \sqrt{1-\sin^2 \omega t} \quad \ldots (3)$$

or

$$\cos \omega t = \sqrt{1-\frac{y^2}{b^2}} \quad \ldots (4)$$

From equation (1)

$$\frac{x}{a} = \sin(2\omega t + \alpha)$$
$$= \sin 2\,\omega t \cos\alpha + \cos 2\omega t \sin\alpha$$
$$= 2\sin\omega t \cos\omega t \cos\alpha + (1 - 2\sin^2\omega t)\sin\alpha$$

Substituting the values of $\sin\omega t$ and $\cos\omega t$,

$$\frac{x}{a} = 2\cdot\frac{y}{b}\cdot\sqrt{1-\frac{y^2}{b^2}}\cos\alpha + \left(1 - 2\frac{y^2}{b^2}\right)\sin\alpha$$

or

$$\left[\frac{x}{a} - \left(1-\frac{2y^2}{b^2}\right)\sin\alpha\right] = \frac{2y}{b}\cos\alpha\sqrt{1-\frac{y^2}{b^2}}$$

$$\left[\left(\frac{x}{a} - \sin\alpha\right) + \frac{2y^2}{b^2}\sin\alpha\right] = \frac{2y\cos\alpha}{b}\sqrt{1-\frac{y^2}{b^2}}$$

Squaring both sides

$$\left(\frac{x}{a} - \sin\alpha\right)^2 + \frac{4y^4}{b^2}\sin^2\alpha + 2\left(\frac{x}{a} - \sin\alpha\right)\frac{2y^2}{b^2}\sin\alpha = \frac{4y^2\cos^2\alpha}{b^2}\left(1-\frac{y^2}{b^2}\right)$$

$$\left(\frac{x}{a} - \sin\alpha\right)^2 + \frac{4y^4}{b^4}(\sin^2\alpha + \cos^2\alpha) - \frac{4y^2}{b^2}(\sin^2\alpha + \cos^2\alpha) + \frac{4y^2}{b^2}\cdot\frac{x}{a}\sin\alpha = 0$$

$$\left(\frac{x}{a} - \sin\alpha\right)^2 + \frac{4y^4}{b^4} - \frac{4y^2}{b^2} + \frac{4y^2}{b^2}\cdot\frac{x}{a}\sin\alpha = 0$$

$$\left(\frac{x}{a} - \sin\alpha\right)^2 + \frac{4y^2}{b^2}\left[\frac{y^2}{b^2} + \frac{x}{a}\sin\alpha - 1\right] = 0 \qquad \ldots (5)$$

Equation (5) represents the general equation of a curve having two loops. The resultant motion of the particle for different values of α is given in

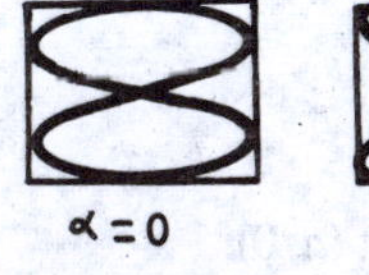

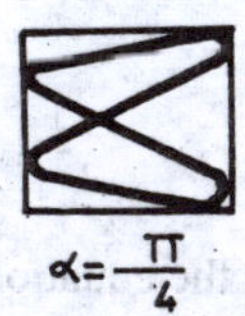

$\alpha = 0$ $\alpha = \frac{\pi}{4}$ $\alpha = \frac{\pi}{2}$ $\alpha = -\frac{\pi}{4}$ $\alpha = \pi$

Fig. 2.10

Fig. 2.10 For a phase difference of 0, π and 2 π, the resultant motion gives the figure of eight.

Special cases

(*i*) When $\alpha = 0, \pi, 2\pi$ etc.

$\sin\alpha = 0$

From equation (5)

$$\frac{x^2}{a^2} + \frac{4y^2}{b^2}\left(\frac{y^2}{b^2} - 1\right) = 0.$$

This equation represents the figure of eight and has two loops.

(*ii*) When $\alpha = \frac{\pi}{2}$

$\sin\alpha = +1$

From equation (5)

$$\left(\frac{x}{a} - 1\right)^2 + \frac{4y^2}{b^2}\left(\frac{y^2}{b^2} + \frac{x}{a} - 1\right) = 0$$

or
$$\left(\frac{x}{a} - 1\right)^2 + \frac{4y^2}{b^2}\left(\frac{x}{a} - 1\right) + \frac{4y^4}{b^4} = 0$$

or
$$\left[\left(\frac{x}{a} - 1\right) + \frac{2y^2}{b^2}\right]^2 = 0$$

or
$$\left(\frac{x}{a} - 1\right) + \frac{2y^2}{b^2} = 0$$

or
$$\frac{2y^2}{b^2} = -\left(\frac{x}{a} - 1\right)$$

$$y^2 = -\frac{b^2}{2}\left(\frac{x}{a} - 1\right)$$

$$y^2 = -\frac{b^2}{2a}(x - a)$$

This represents the equation of a parabola, with vertex at $(a, 0)$.

2.7. Composition of Two SHMs at Right Angles with Time Periods in the Ratio 1 : 2

Graphical Method

Let a particle be influenced simultaneously by two simple harmonic vibrations at right angles to each other. The two vibrations are represented by the equations

$$x = a \sin 2\,\omega t$$
$$y = b \sin\,\omega t$$

Here the phase difference between the two vibrations is zero, amplitudes are unequal and the time periods are in the ratio of 1 : 2.

Draw two circles of reference with centres C_1 and C_2 and radii a and b respectively. Divide the circle with centre C_1 into 4 equal parts and the circle with centre C_2 into 8 equal parts. The angular frequencies are 2ω and ω. If the particle O is subjected to the SHM along the X-axis only, the particle will vibrate along xx' (Fig. 2.11).

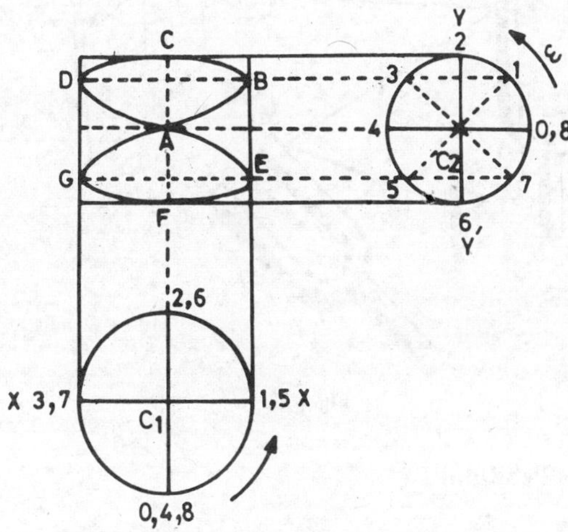

Fig. 2.11.

Similarly, if the particle O is subjected to the SHM along the Y-axis only, the particle will vibrate along yy'.

When the particle O is subjected to the two SHMs simultaneously, the resultant vibration of O will be along a curve $ABCDAEFGA$ which represents the figure of eight. At zero-zero position the particle is at the mean position A. After equal intervals of time the positions B, C, D etc. are obtained in the same order from the two circles of reference (Fig. 2.11). The angular frequency of the resultant vibration is ω.

2.8. Experimental Methods for Obtaining Lissajous' Figures

1. Optical Method

A and *B* are two tuning forks with frequencies in the ratio 2 : 1. The prongs of *A* vibrate in a horizontal plane and the prongs of *B* in a vertical plane. M_1 and M_2 are two small highly polished plane mirrors attached to the prongs of *A* and *B* respectively. It is adjusted that the spot of light after reflection from M_1 and M_2 is obtained at *O*, the centre of the screen. When only the tuning fork *A* vibrates the spot of light moves along XX'. When only the tuning fork *B* vibrates, the spot of light moves along YY'. When *A* and *B* vibrate simultaneously and are in phase, the spot of light traces the figure of eight (Fig. 2.12).

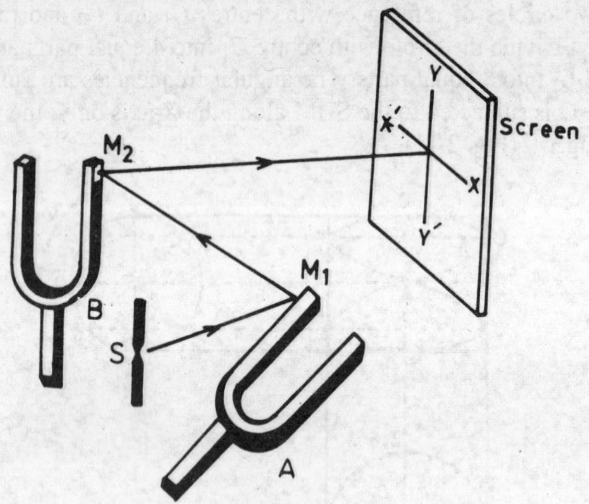

Fig. 2.12.

2. Blackburn's Pendulum

Blackburn's pendulum consists of a funnel having a fine nozzle. It is supported by three strings meeting at *E*. (Fig. 2.13).

EACBDA is a string passing over the Books *C* and *D*. The length *AE* can be changed by moving the knot at *A*. When the pendulum is displaced about the point *A* along XX' it vibrates with a time period t_1.

Here $$t_1 = 2\pi\sqrt{\frac{l_1}{g}}$$

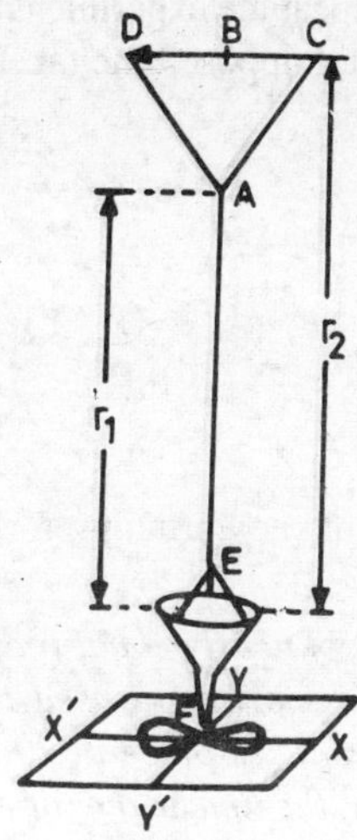

Fig. 2.13.

When the pendulum is displaced about the point B, along YY', it vibrates with a time period t_2.

Here $$t_2 = 2\pi\sqrt{\frac{l_2}{g}}$$

When both the motions are given simultaneously, the nozzle of the funnel traces the Lissajous' figure. If fine sand is put in the funnel, the sand falls through the nozzle and traces a fine curve. The shape and size of the curve will depend upon the relative amplitudes and time periods of the two rectangular SHMs.

These two experiments can be easily demonstrated in the laboratory.

2.9. Uses of Lissajous' Figures

1. To determine the ratio of the time periods. Lissajous' figures help in determining the ratio of the time periods of two constituent vibrations. From the Lissajous' figures, the number of times the curve touches the horizontal and the vertical sides of a rectangle bounding the Lissajous' figure, is found. If the curve touches the horizontal side m times and the vertical side n times, the ratio of the time periods is $m : n$. In the figure of eight (Fig. 2.10) the curve touches the horizontal side once and the vertical side twice. Hence the time periods are in the ratio 1 : 2.

2. Determination of the frequency of a tuning fork. Suppose the given tuning fork A has an unknown frequency n_1. Take a tuning fork B of nearly the same frequency as that of A. Suppose the frequency of $B=n$. Set the two tuning forks to vibrate in perpendicular planes as in the optical method for obtaining Lissajous' figures. Obtain the Lissajous' figure. As the tuning forks differ slightly in frequency, the phase difference between the two, changes with time. The Lissajous' figures obtained are as shown in Fig. 2.4. Due to the continuous change in shape the pattern changes in shape continuously with the phase changing from 0 to 2π. Suppose the complete cycle of change takes place in t seconds. Then, the difference in frequencies of A and $B = \frac{1}{t}$.

Theretore, frequency of $$A = n \pm \left(\frac{1}{t}\right)$$

Now attach a little wax to the tuning fork A. Repeat the experiment and note the time taken to complete one cycle of operation. Suppose time taken is t_1. If t_1 is less than t, the frequency of

$$A = \left(n + \frac{1}{t}\right)$$

If t_1 is greater than t, then the frequency of

$$A = \left(n - \frac{1}{t}\right)$$

Hence, the correct value of the frequency of unknown tuning fork can be determined.

Example 2.3. *Two tuning froks A and B are of nearly equal frequencies. The frequency of A is 288. When the two tuning forks are used to obtain Lissajous' figures, the complete cycle of changes takes place in 20* seconds. *When the tuning fork B is loaded with a little wax, the time taken for one cycle of change is 10* seconds. *Calculate the original frequency of B.*

Before loading

Frequency of A $= 288$

Time for one complete cycle $= 20$ s

Difference in frequencies $= \frac{1}{20} = 0{\cdot}05$

Possible frequencies of B are

either $288 + 0{\cdot}05 = 288{\cdot}05$

or $288 - 0{\cdot}05 = 287{\cdot}95$

After loading, the cycle of changes takes place after 10 seconds *i.e.* time taken for the cycle decreases. Suppose the frequency of B is $288{\cdot}05$. After loading, the frequency of B will be lowered and its difference with the frequency of A becomes less. Therefore, the time taken for each cycle of changes will be more than 20 seconds. Hence it is not possible.

Suppose the frequency of B is 287·95. After loading its difference with the frequency of A is increased and the cycle of operation will take place in less time.

Hence the original frequency of $B = \mathbf{287{\cdot}95}$.

Example 2.4. *Two tuning forks A and B are of nearly equal frequencies. Frequency of A is 256. When the two tuning forks are used to obtain Lissajous' figures the complete cycle of changes takes place in 10* seconds. *When the tuning fork B is loaded with a little wax, the time taken is 20* seconds. *Calculate the frequency of B before loading.*

Frequency of A $= 256$

Time for one complete cycle $= 10$ seconds

$\therefore$ Difference in frequencies $= \frac{1}{10}$

$= 0{\cdot}1$

Possible frequencies of B are

either $256 + 0{\cdot}1 = 256{\cdot}1$

or $256 - 0{\cdot}1 = 255{\cdot}9$

When B is loaded with wax, the difference in frequencies

$= \frac{1}{20} = 0{\cdot}05$

Possible frequencies of B are

either $256 + 0{\cdot}05 = 256{\cdot}05$

or $256 - 0{\cdot}05 = 255{\cdot}95$

Suppose the frequency of B before loading is $256{\cdot}1$. After loading its frequency decreases and can be $256{\cdot}05$ or $255{\cdot}95$.

Therefore the original frequency of B = **256·1**.

If it is supposed that the original frequency of B before loading is 255·9, after loading its frequency should decrease. But it is given that possible frequencies after loading are 256·05 and 255·95 which are higher than 255·9. Therefore, this value is not possible.

Hence the original frequency of B = **256·1.**

Example 2.5. *In an experiment to obtain Lissajous' figures, one tuning fork is of frequency 250 Hz and a circular figure occurs after every five seconds. What deductions may be made about the frequency of the other tuning fork.*

(Delhi, 1976)

Frequency of A = 250 Hz

Time for one complete cycle = 5 s

$\therefore$ Difference in frequencies $= \frac{1}{5}$

= 0·2 Hz

Possible frequencies of B are

either $250 + 0{\cdot}2 =$ **250·2 Hz**

or $250 - 0{\cdot}2 =$ **249·8 Hz.**

Example 2.6. *Two tuning forks produce Lissajous' figures. The figure changes in form from a parabola to a figure of 8 and again a parabola, the whole cycle taking 6* seconds. *If the frequency of one is 100, find the possible frequencies of the other.* (Delhi, 1990)

Frequency of $A = 100$ Hz

Time for one complete cycle $= 6$ s

Difference in frequency $= \frac{1}{6}$ Hz

$= 0{\cdot}167$ Hz

Possible frequencies of B are

either $100 + 0{\cdot}167 = \mathbf{100{\cdot}167}$ **Hz**

or $100 - 0{\cdot}167 = \mathbf{99{\cdot}833}$ **Hz**

EXERCISES

1. Explain how two simple harmonic vibrations acting simultaneously on a particle and in perpendicular directions, can be compounded. Deduce an expression for the resultant vibration.

2. Deduce the resultant vibration of a particle influenced by two mutually perpendicular simple harmonic vibrations, having the same amplitude but differing in phase by (*i*) 0, (*ii*) $\pi/4$, (*iii*) $\pi/2$, and (*iv*) π.

3. Two simple harmonic vibrations acting at right angles to each other have the time periods in the ratio 1 : 2. The phase difference between the two vibrations is $\pi/2$. Show graphically that the resultant curve is a parabola.

4. Two mutually perpendicular simple harmonic vibrations of amplitudes a and b are acting simultaneously on a particle. Their time periods are in the ratio 1 : 2 and their phase difference is ϕ. Derive an expression for the resultant vibration of the particle.

5. What are Lissajous' figures? Show how they can be produced and demonstrated ? *(Calcutta, 1973; Madras, 1974)*

6. What are Lissajous' figures? Discuss briefly their importance in acoustical measurements. *(Delhi, 1974)*

7. Describe an optical method for obtaining Lissajous' figures.

8. What are Lissajous' figures ? Give the analytical treatment of the formation of these figures when the period of vibration of two simple harmonic motions are in the ratio of 1 : 1. Explain how these figures are useful in the laboratory. *(Delhi, 1973)*

9. What are Lissajous' figures ? Discuss the formation of Lissajous' figures when the periods of the two vibrations are equal and the phase difference is $\pi/2$. *(Delhi, 1972)*

10. What are Lissajous' figures? Calculate the resultant of two simple harmonic motions of equal time period when they act at right angles to one another and when the phase difference is (*i*) zero and (*ii*) $\pi/2$. *(Delhi, 1971)*

11. Find the resultant of two simple harmonic motions of equal periods when they act at right angles to each other. Discuss different important cases analytically. How can these be demonstrated.

[Delhi, (Sub), 1976]

12. Find the resultant of two simple harmonic motions of periods in the ratio 1 : 2 and having (*a*) zero phase difference and (*b*) phase difference of 45°. *[Delh (Sub), 1976]*

13. Find the resultant of two simple harmonic motions of equal period when they act at right angles to one another. Discuss the different important cases. *(Delhi, 1976)*

14. What are Lissajous' figures ? Describe in detail two important experimental methods for producing them. *[Delhi (Supp), 1976]*

15. Discuss in detail the resulting motion when two simple harmonic motions having a frequency ratio 1 : 1 and acting at right angles to each other superimpose. *(Kanpur, 1975)*

16. The lower of the two tuning forks has a frequency of 380 hertz and a complete cycle of changes takes place in 10 seconds. Calculate the frequency of the other fork. *(Delhi, 1972)*

[Ans. 380·1]

17. Two tuning forks *A* and *B* are used to produce Lissajous' figures. It is found that the figure completes its cycle in 10 seconds. If the frequency of *A* (slightly larger than that of *B*) is 200, calculate the frequency of *B*.

[Behrampur (Hons), 1972]

[Ans. 199·9 hertz)

18. Define simple harmonic motion. Two simple harmonic oscillators of same frequency and amplitude but in perpendicular directions are compounded. Find the resultant motion. *[Utkal, 1990]*

CHAPTER 3

Free, Forced and Resonant Vibrations

3.1. Free Vibrations

When the bob of a simple pendulum (in vacuum) is displaced from its mean position and left, it executes simple harmonic motion. The time period of oscillation depends only on the length of the pendulum and the acceleration due to gravity at the place. The pendulum will continue to oscillate with the same time period and amplitude for any length of time. In such cases there is no loss of energy by friction or otherwise. In all similar cases, the vibrations will be undamped free vibrations. The amplitude of swing remains constant.

3.2. Undamped Vibrations

For a simple harmonically vibrating particle, the kinetic energy for displacement y, is given by

$$\frac{1}{2} m \left(\frac{dy}{dt}\right)^2$$

At the same instant, the potential energy of the particle is $\frac{1}{2} Ky^2$ where K is the restoring force per unit displacement.

The total energy at any instant,

$$= \frac{1}{2} m \left(\frac{dy}{dt}\right)^2 + \frac{1}{2} Ky^2$$

For an undamped harmonic oscillator, this total energy remains constant.

$$\therefore \quad = \frac{1}{2} m \left(\frac{dy}{dt}\right)^2 + \frac{1}{2} Ky^2 = \text{constant} \qquad \ldots (1)$$

Differentiating equation (1) with respect to time,

$$m\frac{d^2 y}{dt^2} + Ky = 0 \quad \ldots (2)$$

$$\frac{d^2 y}{dt^2} + \left(\frac{K}{m}\right)y = 0 \quad \ldots (3)$$

Equation (3) is similar to the equation

$$\frac{d^2 y}{dt^2} + \omega^2 y = 0 \quad \ldots (4)$$

Here $\omega^2 = \left(\frac{K}{m}\right)$

The solution for equation (4) is

$$y = a \sin(\omega t - \alpha)$$

$$\therefore \quad y = a \sin\left[\sqrt{\frac{K}{m}}\, t - \alpha\right]$$

The frequency of oscillation is

$$n = \frac{\omega}{2\pi} = \frac{1}{2\pi}\sqrt{\frac{K}{m}}$$

Thus, in the case of undamped free vibrations, the differential equation is

$$\frac{d^2y}{dt^2} + \left(\frac{K}{m}\right) y = 0 \quad \ldots (5)$$

This is only an ideal case. In the first chapter, for the motion of a pendulum, loaded spring, LC circuit etc., it has been assumed that the vibrations are free and undamped.

3.3. Damped Vibrations

In actual practice, when the pendulum vibrates in air medium, there are frictional forces and consequently energy is dissipated in each vibration. The amplitude of swing decreases continuously with time and finally the oscillations die out. Such vibrations are called **free damped** vibrations. The dissipated energy appears as heat either within the system itself or in the surrounding medium.The dissipative force due to friction etc. (resistance in LCR circuit) is proportional to the velocity of the particle at that instant. Let $\mu \frac{dy}{dt}$ be the dissipative force due to friction etc.

This term is to be introduced in equation (2).

Therefore, the differential equation in the case of Free-damped vibrations is,

$$m\frac{d^2 y}{dt^2} + Ky + \mu\frac{dy}{dt} = 0 \quad \ldots (6)$$

or $$\frac{d^2 y}{dt^2}+\left(\frac{\mu}{m}\right)\frac{dy}{dt}+\left(\frac{K}{m}\right)y=0 \qquad \ldots(7)$$

This equation is similar to a general differential equation,

$$\frac{d^2y}{d^2}+2b\frac{dy}{dt}+k^2 y=0 \qquad \ldots(8)$$

The solution of this equation is

$$y=ae^{-bt}\sin(\omega t-\alpha) \qquad \ldots(9)$$

The general solution of equation (7) is also given by

$$y=A\ e^{(-b+\sqrt{b^2+k^2})\ t}+B\ e^{(-b-\sqrt{a^2-k^2})\ t}$$

Here $$b=\frac{\mu}{2m} \text{ and } k^2=\frac{K}{m}$$

and $$\omega=\sqrt{k^2-b^2}$$

or $$\omega=\sqrt{\frac{K}{m}-\frac{\mu^2}{4m^2}}$$

or $$n=\frac{\omega}{2\pi}=\frac{1}{2\pi}\sqrt{k^2-b^2}$$

3.4. Damped SHM in an Electrical Circuit

In the case of an electrical circuit, the force equation is replaced by the voltage equation. The circuit consists a condenser C, inductance L and resistance R (Fig. 3.1).

When the condenser C is charged by pressing the morse key, it gets discharged through an inductance L and resistance R when the key is released (Fig. 3.1).

Suppose, during discharge, at any instant, the charge on the condenser = Q, current flowing = I and rate of fall of current = $\frac{dI}{dt}$.

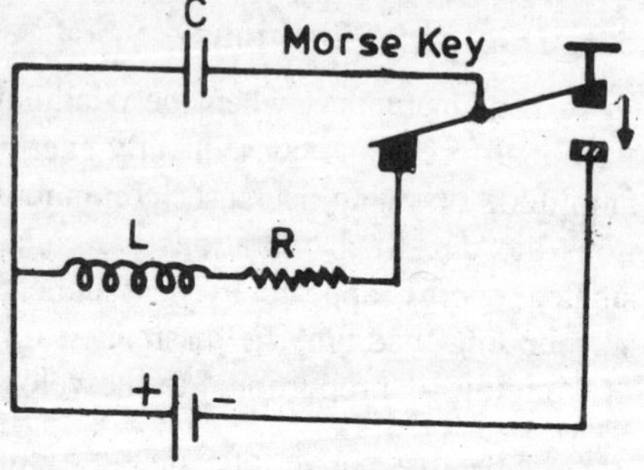

Fig. 3.1:

In this case, $$\frac{Q}{C}+RI+L\frac{dI}{dt}=0$$

But. $$I=\frac{dQ}{dt} \text{ and } \frac{dI}{dt}=\frac{d^2Q}{dt^2}$$

$$\therefore \qquad L\frac{d^2 Q}{dt^2} + R\frac{dQ}{dt} + \frac{Q}{C} = 0$$

$$\frac{d^2 Q}{dt^2} + \frac{R}{L}\frac{dQ}{dt} + \frac{Q}{LC} = 0 \qquad \ldots (1)$$

Taking $\frac{R}{L} = 2b$ and $\frac{1}{LC} = k^2$

we get $$\frac{d^2 Q}{dt^2} + 2b\frac{dQ}{dt} + k^2 Q = 0$$

The general solution of this equation is

$$Q = A\,e^{(-b+\sqrt{b^2-k^2})\,t} + B\,e^{(-b-\sqrt{b^2-k^2})\,t} \qquad \ldots (2)$$

When $t = 0$, $Q = Q_0$ and from equation (2)

$$A + B = Q_0 \qquad \ldots (3)$$

Differentiating equation (2),

$$\frac{dQ}{dt} = A\,(-b+\sqrt{b^2-k^2})\;e^{(-b+\sqrt{b^2-k^2})\,t} + B\,(-b-\sqrt{b^2-k^2})\;e^{(-b-\sqrt{b^2-k^2})\,t}$$

When, $t = 0, \ \frac{dQ}{dt} = 0$

$$\therefore \qquad A\,(-b+\sqrt{b^2-k^2}) + B\,(-b-\sqrt{b^2-k^2}) = 0$$

$$-b\,(A+B) + \sqrt{b^2-k^2}\,(A-B) = 0$$

$$-bQ_0 + \sqrt{b^2-k^2}\,(A-B) = 0$$

$$\therefore \qquad A - B = \frac{bQ_0}{\sqrt{b^2-k^2}} \qquad \ldots (4)$$

Adding (3) and (4),

$$2A = Q_0\left(1 + \frac{b}{\sqrt{b^2-k^2}}\right)$$

$$A = \frac{Q_0}{2}\left(1 + \frac{b}{\sqrt{b^2-k^2}}\right) \qquad \ldots (5)$$

Subtracting (4) from (3),

$$2B = Q_0\left(1 - \frac{b}{\sqrt{b^2-k^2}}\right)$$

$$B = \frac{Q_0}{2}\left(1 - \frac{b}{\sqrt{b^2-k^2}}\right) \qquad \ldots (6)$$

Substituting the values of A and B in equation (2),

$$Q = \frac{Q_0}{2}\left(1 + \frac{b}{\sqrt{b^2 - k^2}}\right) e^{(-b + \sqrt{b^2 - k^2})\, t}$$

$$+ \frac{Q_0}{2}\left(1 - \frac{b}{\sqrt{b^2 - k^2}}\right) e^{(-b - \sqrt{b^2 - k^2})\, t}$$

Substituting the values of b and k

$$Q = \frac{Q_0}{2}\left[\left(1 + \frac{R/2L}{\sqrt{\frac{R^2}{4L^2} - \frac{1}{LC}}}\right) e^{(-R/2L + \sqrt{R^2/4L^2 - 1/LC})\, t}\right.$$

$$\left. + \left(1 - \frac{R/2L}{\sqrt{\frac{R^2}{4L^2} - \frac{1}{LC}}}\right) e^{(-R/2L - \sqrt{(R^2/4L^2 - 1/LC)})\, t}\right] \quad \ldots (7)$$

Special Cases

(1) When $\frac{R^2}{4L^2} = \frac{1}{LC}$.

From equation (7)

$$Q = Q_0\, e^{-Rt/2L}$$

which shows that the discharge is *aperiodic,* and *critically damped.* The curve is an exponential decay curve a (Fig. 3.2).

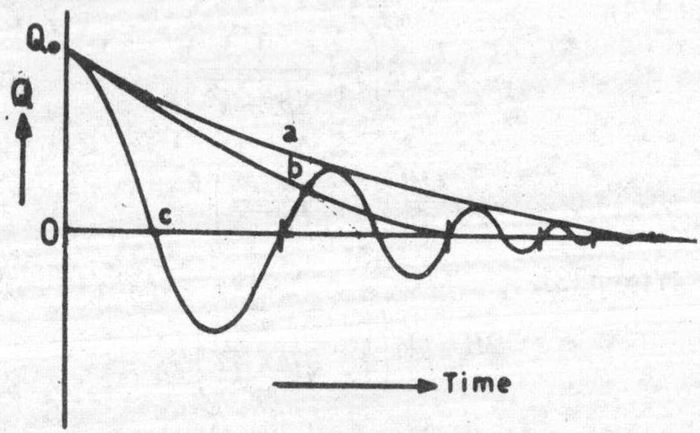

Fig. 3.2.

(2) When $\frac{R^2}{4L^2} > \frac{1}{LC}$, the square roots are real and the discharge is *non-oscillatory* and is said to be dead beat as represented by curve b in Fig. 3.2. In this case $R^2 > \frac{4L}{C}$.

(3) When $\frac{R^2}{4L^2} < \frac{1}{LC}$, the square roots are imaginary. From equation (7), taking,

$$\frac{R^2}{4L^2} = b^2, \quad \frac{R^2}{4L^2} - \frac{1}{LC} = b^2 - k^2$$

and
$$j = \sqrt{-1},$$

$$Q = \frac{Q_0}{2}\left[\left(1 + \frac{b}{j\sqrt{k^2 - b^2}}\right) e^{(-b + j\sqrt{j^2 - b^2})\cdot t} + \left(1 - \frac{b}{j\sqrt{k^2 - b^2}}\right) e^{(-b - j\sqrt{j^2 - b^2})\, t}\right]$$

$$Q = Q_0\, e^{-bt}\left[\left(\frac{e^{j\sqrt{k^2 - b^2}\, t} + e^{-j\sqrt{k^2 - b^2}\, t}}{2}\right) + \frac{b}{\sqrt{(k^2 - b^2)}}\left(\frac{e^{j\sqrt{k^2 - b^2}\, t} + e^{-j\sqrt{k^2 - b^2}\, t}}{2j}\right)\right]$$

$$Q = Q_0\, e^{-bt}\left[\cos\sqrt{k^2 - b^2}\, t + \frac{b}{\sqrt{k^2 - b^2}} \cdot \sin\sqrt{k^2 - b^2}\, t\right]$$

$$Q = \frac{Q_0\, e^{-bt}\, k}{\sqrt{k^2 - b^2}}\left[\frac{\sqrt{k^2 - b^2}}{k} \cos\sqrt{k^2 - b^2}\, t + \frac{b}{k} \sin\sqrt{k^2 - b^2}\, t\right]$$

Put $\frac{\sqrt{k^2 - b^2}}{k} = \sin\theta$, and $\frac{b}{k} = \cos\theta$

Then, $\tan\theta = \frac{\sqrt{k^2 - b^2}}{b}$

$$\therefore\ Q = \frac{Q_0\, e^{-bt}\, k}{\sqrt{k^2 - b^2}}\left[\sin(\sqrt{k^2 - b^2}\, t)\cos\theta + \cos(\sqrt{k^2 - b^2}\, t)\sin\theta\right]$$

$$Q = \frac{Q_0\, e^{-bt}\, k}{\sqrt{k^2 - b^2}} \sin\left(\sqrt{k^2 - b^2}\, t + \theta\right)$$

Substituting the values of b^2 and k^2

$$Q = \frac{Q_0 e^{-\frac{R}{2L}t}}{\sqrt{\left(\frac{l}{LC} - \frac{R^2}{4L^2}\right) LC}} \sin\left(\sqrt{\left(\frac{l}{LC} - \frac{r^2}{4L^2}\right)} \quad t + \theta\right) \qquad \ldots (8)$$

In this case, the discharge is oscillatory as represented by the curve c (Fig. 3.2). This dischage is of simple harmonic type and the natural frequency of the circuit,

$$f = \frac{1}{2\pi}\sqrt{\frac{1}{LC} - \frac{R^2}{4L^2}} \qquad \ldots (9)$$

When $$R = 0,\ f = \frac{1}{2\pi\sqrt{LC}} \qquad \ldots (10)$$

Results. (1) When $R^2 > \frac{4L}{C}$, the discharge is non-oscillatory and dead beat.

(2) When $R^2 = \frac{4L}{C}$, the discharge is 'aperiodic' and critically damped.

(3) When $R^2 < \frac{4L}{C}$, the discharge is oscillatory and the natural frequency of the circuit.

$$f = \frac{1}{2\pi}\sqrt{\frac{1}{LC} - \frac{R^2}{4L^2}}.$$

Example 3.1. *A condenser of capacity 1 uF, an inductance of 0·2* henry *and a resistance of 800* ohms *are joined in series. Is the circuit oscillatory?*

Here $L = 1\ \mu F = 10^{-6} F,\ R = 800$ ohms, $L = 0{\cdot}2$ henry.

$$\therefore \quad R^2 = (800)^2 = 640000 = 6{\cdot}4 \times 10^5 \qquad \ldots (1)$$

$$\frac{4L}{C} = \frac{4 \times 0{\cdot}2}{10^{-6}} = 8 \times 10^5 \qquad \ldots (2)$$

As $R^2 <= \frac{4L}{C}$, the *circuit is oscillatory.*

Example 3.2. *In an oscillatory circuit, $L = 0{\cdot}2$* henry, *$C = 0{\cdot}0012\ \mu F$. What is the maximum value of resistance for the circuit to be oscillatory?*

Here, $L = 0{\cdot}2$ henry, $C = 0{\cdot}0012\ \mu\text{F} = 12 \times 10^{-10}\ \text{F}$

$$R^2 = \frac{4L}{C}$$

$$R = \sqrt{\frac{4L}{C}} = \sqrt{\frac{4 \times 0{\cdot}2}{12 \times 10^{-10}}} = \mathbf{2{\cdot}582 \times 10^4 \ ohms.}$$

A maximum resistance of $R = 2{\cdot}582 \times 10^4$ ohms can be connected in series so that the discharge remains oscillatory. For the discharge to be oscillatory R should be less than $2{\cdot}582 \times 10^4$ ohms.

Example 3.3. *Find whether the discharge of a condenser through the following inductive circuit is oscillatory :*

$C = 0{\cdot}1$ μF, $L = 10$ millihenry, $R = 200$ ohms .

If the circuit is oscillatory, calculate its frequency.

Here $C = 0{\cdot}1 \times 10^{-6}\, F, L = 10 \times 10^{-3}\, H, R = 200$ ohms

$$R^2 = (200)^2 = 40000 = 4 \times 10^4$$

$$\frac{4L}{C} = \frac{4 \times 10 \times 10^{-3}}{0{\cdot}1 \times 10^{-6}} = 4 \times 10^5$$

As $R^2 < \frac{4L}{C}$, the circuit is oscillatory

$$f = \frac{1}{2\pi}\sqrt{\frac{1}{LC} - \frac{R^2}{4L^2}} = \frac{1}{2\pi}\sqrt{\frac{1}{10 \times 10^{-3} \times 0{\cdot}1 \times 10^{-6}} - \frac{40000}{4 \times (10 \times 10^{-3})^2}}$$

$$= \frac{1}{2\pi}\sqrt{10^9 - 10^8} = \frac{1}{2\pi}\sqrt{9 \times 10^8} = \frac{3}{2\pi} \times 10^4 = \mathbf{4772{\cdot}7 \ Hz.}$$

3.5. Forced Vibrations

The time period of a body executing simple harmonic motion depends on the dimensions of the body and its elastic properties. The vibrations of such a body die out with time due to dissipation of energy. If some external periodic force is constantly applied on the body, it continues to oscillate under the influence of such external forces. Such vibrations of the body are called *forced vibrations.*

Initially, the amplitude of the swing increases, then decreases with time, becomes minimum and again increases. This will be repeated if the external periodic force is constantly applied on the system. In such cases the body will finally be forced to vibrate with the same frequency as that of the applied force. The frequency of the forced vibration is different from the natural frequency of vibration of the body. The amplitude of the forced vibration of the body depends on the difference between the natural frequency and the frequency of the applied force. The amplitude will be large if difference in frequencies is small and *vice versa.*

For forced vibrations, equation (6) Art. 3.3 is modified in the form,

$$m\frac{d^2y}{dt^2}+Ky+\mu\frac{dy}{dt}=F\sin pt \qquad \ldots(1)$$

Here p is the angular frequency of the applied periodic force.

The particular solution of equation (1) representing the forced vibrations is

$$y=a\sin(pt-\alpha) \qquad \ldots(2)$$

$$\therefore \qquad \frac{dy}{dt}=ap\cos(pt-\alpha) \qquad \ldots(3)$$

$$\frac{d^2y}{dt^2}=-ap^2\sin(pt-\alpha)=-p^2y \qquad \ldots(4)$$

Substituting these values in equation (1)

$$-mp^2a\sin(pt-\alpha)+Ka\sin(pt-\alpha)+\mu ap\cos(pt-\alpha)=F\sin pt$$

$$-mp^2a[\sin pt\cos\alpha-\cos pt\sin\alpha]+Ka[\sin pt\cos\alpha-\cos pt\sin\alpha]$$
$$+\mu ap[\cos pt\cos\alpha+\sin pt\sin\alpha]-F\sin pt=0 \qquad \ldots(5)$$

When $\sin pt=1$; $\cos pt=0$

$$\therefore \qquad -mp^2a\cos\alpha+Ka\cos\alpha+\mu ap\sin\alpha-F=0 \qquad \ldots(6)$$

When $\cos pt=1$; $\sin pt=0$

$$\therefore \qquad +mp^2a\cos\alpha-Ka\sin\alpha+\mu ap\cos\alpha=0 \qquad \ldots(7)$$

Dividing equation (7) by $\cos\alpha$ and simlifying

$$\tan\alpha=\frac{\mu p}{K-mp^2}=\frac{A}{B} \qquad \ldots(8)$$

From equation (8)

$$\sin\alpha=\frac{A}{\sqrt{A^2+B^2}} \qquad \ldots(9)$$

$$\cos\alpha=\frac{B}{\sqrt{A^2+B^2}} \qquad \ldots(10)$$

Dividing equation (6) by $\cos\alpha$

$$-mp^2a+Ka+\mu ap\tan\alpha-\frac{F}{\cos\alpha}=0$$

or $$a[(K-mp^2)+\mu p\tan\alpha]=\frac{F}{\cos\alpha}$$

But $(K-mp^2)=B$, and $\mu p=A$

Substituting the values of tan α and cos α

$$a\left[B+\frac{A^2}{b}\right]=\frac{F\sqrt{A^2+B^2}}{B}$$

$$a=\frac{F}{\sqrt{A^2+B^2}}$$

Substituting the values of A and B

$$a=\frac{F}{\sqrt{\mu^2 p^2+(K-mp^2)^2}} \qquad \ldots (11)$$

$$\therefore \qquad y = a \sin(pt-\alpha)$$

or

$$y=\frac{F}{\sqrt{\mu^2 p^2+(K-mp^2)^2}}\sin(pt-\alpha) \qquad \ldots (12)$$

Applying the boundary conditions, another solution is obtained when $F = 0$. This corresponds to free vibrations. In the case of free vibrations the solution is

$$y = a\,e^{-bt}\sin(\omega t-\alpha) \qquad \ldots (13)$$

The general solution will include both the particular solutions for free and forced vibrations.

$$\therefore \qquad y=a\,e^{-bt}\sin(\omega t-\alpha)+\frac{F}{\sqrt{\mu^2 p^2+(K-mp^2)^2}}\sin(pt-\alpha) \qquad \ldots (14)$$

Here $b=\dfrac{\mu}{2m}$.

Example 3.4. *The equation for displacement of a point on a damped oscillator is given by*

$$x = 5\,e^{-0.25\,t}\,\sin\left(\frac{\pi}{2}\right)t \text{ metre}$$

Find the velocity of the oscillating point at

$t=\dfrac{T}{4}$ and T, where T is the time period of the oscillator.

(IAS, 1987)

Here,

$$x = 5\,e^{-0.25\,t}\,\sin\left(\frac{\pi}{2}\right)t \qquad \ldots (i)$$

This equation is similar to the equation

$$x = a\,e^{-bt}\sin\omega t \qquad \ldots (ii)$$

Comparing (*i*) and (*ii*)

$$\omega=\frac{\pi}{2}$$

Time period

$$T=\frac{2\pi}{\omega}=\frac{2\pi}{\pi/2}=\mathbf{4\ s}$$

Differentiating equation *(i)* with respect to time *t*,

$$\text{velocity, } \frac{dx}{dt} = 5(-0{\cdot}25)\, e^{-0{\cdot}25t} \sin\left(\frac{\pi}{2}\right)t$$

$$+ 5 \times \left(\frac{\pi}{2}\right) e^{-0{\cdot}25t} \cos\left(\frac{\pi}{2}\right)t \qquad \ldots (iii)$$

(*i*) When $t = \frac{T}{4}$,

$$t = \frac{4}{4} = 1 \text{ s}$$

$$\frac{dx}{dt} = -1{\cdot}25\, e^{-0{\cdot}25} \sin\left(\frac{\pi}{2}\right) + \left(\frac{5\pi}{2}\right) e^{-0{\cdot}25} \cos\left(\frac{\pi}{2}\right)$$

$$\frac{dx}{dt} = -1{\cdot}25\, e^{-0.25} = (-1{\cdot}25) \times (0{\cdot}368)^{0{\cdot}25}$$

$$= -1{\cdot}25 \times 0{\cdot}779$$

$$= \mathbf{-0{\cdot}974 \text{ m/s.}}$$

–ve sign shows that velocity is in opposite direction.

(*ii*) When $t = T = 4$ s

$$\frac{dx}{dt} = -1{\cdot}25\, e^{-0{\cdot}25 \times 4} \sin\left(\frac{\pi}{2}\right)4 + \left(\frac{5\pi}{2}\right) e^{-0{\cdot}25 \times 4} \cos\left(\frac{\pi}{2}\right)4$$

$$= \left(\frac{5\pi}{2}\right) e^{-1}$$

$$= \left(\frac{5\pi}{2}\right) \times 0{\cdot}368$$

$$= \mathbf{2{\cdot}89 \text{ m/s.}}$$

3.6. Resonance and Sharpness of Resonance

In the case of forced vibrations, the general solution for the displacement at any instant is given by

$$y = a\, e^{-bt} \sin(\omega t - \alpha) + \frac{F}{\sqrt{\mu^2 p^2 + (K - mp^2)^2}} \sin(pt - \alpha)$$

If the effect of viscocity of the medium is small, the amplitude,

$$\frac{F}{\sqrt{\mu^2 p^2 + (K - mp^2)^2}}$$

under the action of the driving force is maximum when the denominator is minimum. It is possible if $K - mp^2 = 0$ or $K = mp^2$

or $$p = \sqrt{\frac{K}{m}}$$

Further, the amplitude will be infinite if μ is also zero. The oscillations will have maximum amplitude and this state of vibration of a system is called *resonance.* It means that, when the forced frequency is equal to the natural frequency of vibration of the body, resonance takes place. If friction is present, the amplitude at resonance

$$= \frac{F}{\mu p} = \frac{F}{\mu\sqrt{K/m}}$$

or amplitude at resonance $$= \frac{F}{\mu}\sqrt{\frac{m}{K}}.$$

In the case of sound, the study of sharpness of resonance is of great importance. Sharpness of resonance refers to the fall in amplitude with change in frequency on each side of the maximum amplitude.

The particular solution for displacement in the case of forced vibrations is,

$$y = \frac{F}{\sqrt{\mu^2 p^2 + (K - mp^2)^2}} \sin(pt - \alpha) \qquad \ldots (1)$$

Differentiating equation (1) with respect to time

$$\frac{dy}{dt} = \frac{Fp}{\sqrt{\mu^2 p^2 + (K - mp^2)^2}} \cos(pt - \alpha) \qquad \ldots (2)$$

The velocity *(dy/dt)* is maximum when cos ($pt - \alpha$) is maximum *i.e.* the instant at which the particle crosses the mean position.

$$\left(\frac{dy}{dt}\right)_{\max} = \frac{Fp}{\sqrt{\mu^2 p^2 + (K - mp^2)^2}} \qquad \ldots (3)$$

Kinetic energy of the vibrating particle at the instant of crossing the mean position is given by

$$\text{K.E.} = \frac{1}{2} m \left(\frac{dy}{dt}\right)^2_{\max}$$

$$\text{K.E.} = \frac{\frac{1}{2} mF^2 p^2}{\mu^2 p^2 + (K - mp^2)^2} \qquad \ldots (4)$$

The mean square of the driving force per unit mass

$$= \frac{\left(\frac{O + F^2}{2}\right)}{m} = \frac{F^2}{2m} \qquad \ldots (5)$$

Dividing equation (4) by $\frac{F^2}{2m}$ we get *kinetic energy per unit force* which is called the response R.

$$R = \frac{\frac{1}{2}mF^2 p^2}{\mu^2 p^2 + (K - mp^2)^2} \div \frac{F^2}{2m}$$

$$R = \frac{m^2 p^2}{\mu^2 p^2 + (K - mp^2)^2}$$

$$R = \frac{p^2}{\frac{\mu^2 p^2}{m^2} + \left(\frac{K}{m} - p^2\right)^2} \qquad \ldots (6)$$

The natural frequency of the system in the absence of damping is $\sqrt{\frac{K}{m}}$

Therefore, the term $\left(\frac{K}{m} - p^2\right)$ in equation (6) represents the extent to which the natural frequency of the system deviates from the forced frequency.

When $$\frac{K}{m} = p^2$$

the natural frequency coincides with the forced frequency, and the value of R will be maximum. From equation (6)

$$R = \frac{p^2}{\frac{\mu^2 p^2}{m^2}} = \frac{m^2}{\mu^2} = \left(\frac{m}{\mu}\right)^2 \qquad \ldots (7)$$

The response $$R \propto \frac{1}{\mu}$$

It means that the response R is inversely proportional to the frictional force. In the absence of friction, the response is maximum.

The term $\left(\frac{K}{m} - p^2\right)$ in equation (6), refers to mistuning. The larger is its value, the greater is the system away from resonance.

The graph between p/ω along the X-axis and the response R along the Y-axis is shown in Fig. 3.3.

(*i*) When p/ω is equal to 1 the response is maximum. For curve A, μ is large and for curve C, μ is less. The response decreases for values of p/ω greater than 1 or less than 1.

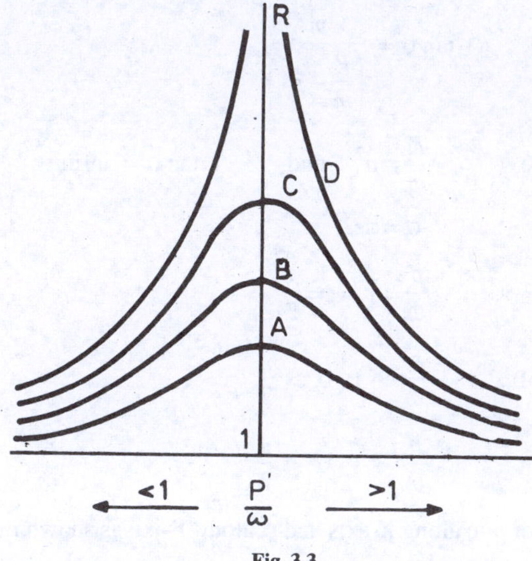

Fig. 3.3.

(*ii*) When the frictional forces are absent, *i.e.*, $\mu = 0$, R is infinite and the sharpness of resonance is maximum.

(*iii*) The sharpness of resonance decreases with increase in the value of μ.

(*iv*) The sharpness of resonance dies rapidly even for a very small change in the value of p/ω from 1, in the case, where μ is minimum.

In the case of the resonance tube, the damping force is large and the graph will be similar to the curve A in Fig. 3.3. The resonance persists over a wide range and it is difficult to exactly locate the position of maximum sharpness of resonance. Hence the results obtained with the resonance tube apparatus are not very accurate.

In the case of the sonometer wire, the damping forces are small and the graph will be similar to curve C in Fig. 3.3. In this case the sharpness of resonance is maximum in a very narrow region. Even a slight variation in length or tension reduces the sharpness considerably. The vibrations die out rapidly. Thus, the results obtained with a sonometer are accurate.

3.7. Phase of Resonance

Considering the phase lead of the forced vibrations with reference to the driving force, in equation (8) of Art. 3.5,

$$\tan \alpha = \frac{\mu p}{K - mp^2}$$

$$\tan \alpha = \frac{\frac{\mu p}{m}}{\left(\frac{K}{m} - p^2\right)} \qquad \ldots (8)$$

At resonance $\frac{K}{m} = p^2$ and $\tan \alpha =$ infinity

i.e. $\alpha = \pi/2$

It means, for $\frac{p}{\omega} = 1, \ \alpha = \frac{\pi}{2}$

For values of $\frac{p}{\omega} > 1, \ \alpha > \frac{\pi}{2}$

For values of $\frac{p}{\omega} < 1, \ \alpha < \frac{\pi}{2}$

The graph for p/ω along X-axis and α along Y-axis is shown in Fig. 3.4.

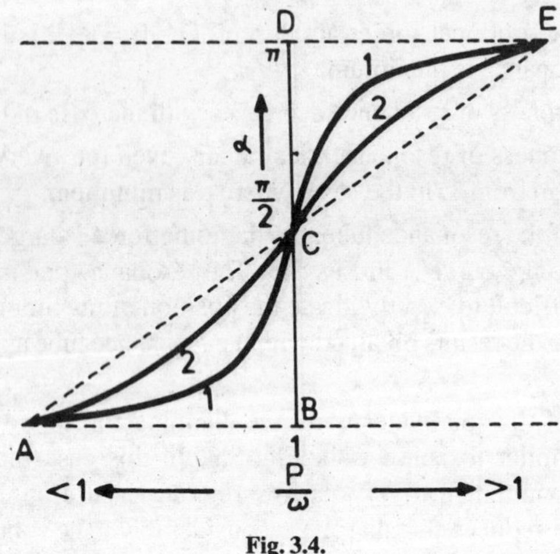

Fig. 3.4.

The shape of the curve will also depend on the value of μ, *i.e.* the external frictional forces.

(1) For values of $\mu = 0$, the curve $ABCDE$ represents the graph.

(2) For large values of μ, the curve will be similar to curve 2.

(3) For small values of μ, the curve will be similar to curve 1.

3.8. Quality Factor

The sharpness of resonance is inversely proportional to μ and can be represented in terms of quality factor. Suppose p_1 and p_2 are the values of the angular frequencies of the driving force on the two sides of the resonance angular frequency p_o at which the power in the response is one half.

The power at any instant is the product of the force and the velocity at that instant.

$$\therefore \text{ Average value of power} = \frac{\mu F^2 p^2}{2\left[\mu^2 p^2 + (K - mp^2)^2\right]} \qquad \ldots (1)$$

The power will be maximum when $(K - mp^2) = 0$

$$\therefore \qquad \text{Maximum power} = \frac{\mu F^2 p^2}{2\mu^2 p^2} = \frac{F^2}{2\mu} \qquad \ldots (2)$$

The average power will be half the maximum power when

$$\mu^2 p^2 = (K - mp^2)^2$$

or
$$K - mp^2 = \pm\ \mu p$$

or
$$mp^2 \pm \mu p - K = 0$$

or
$$p = \frac{\mu \pm \sqrt{\mu^2 + 4Km}}{2m}$$

One set of values of p_1 and p_2 are

$$p_1 = \frac{+\mu + \sqrt{\mu^2 + 4Km}}{2m}$$

$$p_2 = \frac{-\mu + \sqrt{\mu^2 + 4Km}}{2m}$$

$$\therefore \qquad p_1 - p_2 = \frac{2\mu}{2m} = \frac{\mu}{m}$$

The quality factor
$$Q = \frac{p_0}{p_1 - p_2} = \frac{\sqrt{K/m}}{\mu/m} = \frac{\sqrt{K/m}}{\mu} \qquad \ldots (3)$$

3.9. Examples of Forced and Resonant Vibrations

1. An interesting example of forced vibrations and resonance is noticed in the case of troops marching over a suspension bridge. The marching troops exert a certain *periodic force* on the bridge and the bridge is set into forced vibrations. If all the soldiers march in step, each will exert a periodic force in phase and this result in the increased swing of the bridge. If, however the natural frequency of the swing coincides with the forced frequency of the marching force, resonance occurs and the amplitude of the swing becomes very

high. Due to this reason it is dangerous and the bridge may give way. Consequently the troops while crossing the bridge are usually ordered to march out of step.

2. A child exerts a periodic force to set the swing to move to and fro. The child also knows to give the push to the swing at periodic intervals. The amplitude of the swing gradually increases. When the natural frequency of the swing coincides with the forced frequency of the force exerted by the child, resonance occurs and the swing oscillates with a large amplitude.

3. When the prong of a tuning fork is struck against a rubber pad, the prongs vibrate with the natural frequency of the tuning fork. In this case the energy is dissipated only by the prongs of the tuning fork and the intensity of sound is low. If the handle of the vibrating tuning fork is pressed against the table, the particles of the table are set into forced vibrations. A loud sound is heard because energy is dissipated through a large area.

Take two tuning forks A and B of the same frequency. Mount each tuning fork on a hollow wooden box open at one end (Fig. 3.5). The length of the box

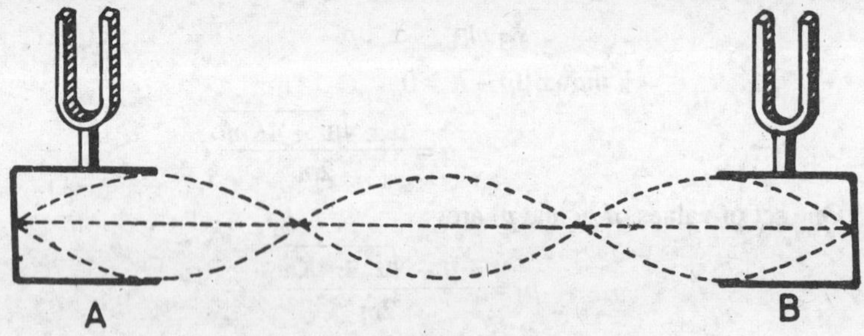

Fig. 3.5.

is such that the natural frequency of vibration of the air column in the box is equal to the natural frequency of the tuning fork.

When the prong of the tuning fork A is struck against a rubber pad, it is set into vibration. The air column in the box A resonates and maximum sound is heard. After some time it is found that the tuning fork on the box B also begins to vibrate although it has not been struck with the hammer. Here the tuning fork on the box B is forced to vibrate with its natural frequency due to resonance of the air column in B.

EXERCISES

1. Explain clearly the terms free vibration, forced vibration and resonance. Give examples.
2. Discuss the effect of a periodic force in producing vibrations in an elastic body. Give the necessary conditions for the production of forced vibrations.

3. Show that in the case of forced vibrations,

$$m\frac{d^2 y}{dt^2}+Ky+\mu\frac{dy}{dt}=F\sin pt.$$

Hence explain the phenomenon of resonance.

4. Discuss sharpness of resonance and explain clearly the factors on which the sharpness depends.
5. What are forced vibrations ? Discuss the phenomenon of resonance.
6. A system executing damped simple harmonic motion is subjected to an external periodic force. Derive the expression for the forced vibration and obtain the condition of resonance. Illustrate your answer with examples.
7. Distinguish between free and forced vibrations. Give the necessary conditions for the production of forced vibrations. *(Lucknow, 1974; Osmania, 1971)*
8. Give the theory of forced vibrations and discuss sharpness of resonance.
9. Discuss analytically the phenomenon of forced vibrations and resonance. Give two examples of resonance. *(Delhi, 1971)*
10. Derive the equation of forced vibrations and discuss the conditions of resonance. Explain sharpness of resonance. *(Delhi, 1971)*
11. Calculate the frequency of amplitude resonance in the case of forced vibrations. *[Delhi (Hons), 1976]*
12. Discuss the theory of forced vibrations in the presence of a damping force proportional to the velocity of the vibrating particle. Deduce the condition of resonance for energy and displacement of the particle and discuss the dependence of the response on the damping factor. *[Behrampur (Hons), 1972]*
13. What are free and forced vibrations? Give the theory of forced vibrations and explain what do you mean by sharpenss of resonance. *(Delhi, 1972)*
14. Distinguish between free and forced vibrations. Give the theory of forced vibrations and discuss the conditions of resonance. *(Delhi, 1974)*
15. Give the theory of slightly damped forced vibrations. Prove that vibrating systems with very small damping and large damping have respectively sharp and flat resonance responses. Give two examples of each. *(Panjab, 1973)*
16. Write the differential equation for the motion of a particle executing forced vibrations in a resisting medium. Explain the physical meaning of each term and each constant in the equation. Solve the above equation and obtain the condition for amplitude resonance. Explain what is meant by sharpness of resonance. *(Nagpur, 1974)*

17. Discuss the phenomenon of sharpness of resonance and show how it depends on the damping factor. *(Kanpur, 1975)*
18. What are free, damped and forced vibrations? Give the theory of forced vibrations and discuss the condition of resonance. *[Delhi (Sub.), 1976]*
19. What are damped vibrations? Obtain an expression for the displacement in the case of a damped oscillatory motion. Discuss the effect of damping on the natural frequency. *[Delhi (Sub.) Supp., 1976]*
20. What is a forced vibration? Discuss mathematically, the vibration of a system executing damped simple harmonic motion when subjected to an external periodic force. What is sharpness of resonance? *(Delhi, 1976)*
21. Define quality factor and bandwidth of the sharpness of resonance. Obtain quality factor for a driven harmonic oscillator at resonance. *(Bhagalpur, 1990)*
22. Explain, in brief, (*i*) free oscillations, (*ii*) forced oscillation and (*iii*) phenomena of resonance. *(Bhagalpur, 1990)*
23. What do you understand by damped vibrations? Obtain an expression for displacement as a function of time for a damped oscillator. What is the effect of damping on the natural frequency of the oscillator? *(Delhi, 1991)*
24. Discuss the phenomenon of sharpness of resonance and show how it depends on the damping factor? *(Delhi, 1992)*
25. (*a*) Derive the differential equation of damped oscillatory motion and give its general solution.

    (*b*) What type of motion do you get when the damping is small? *(Delhi, 1991)*

CHAPTER 4

Wave Motion

4.1. Wave Motion

Wave motion is a form of disturbance which travels through the medium due to the repeated periodic motion of the particles of the medium about their mean positions, the disturbance being handed over from one particle to the next. When a stone is dropped into a pond containing water, waves are produced at the point where the stone strikes the water in the pond. The waves travel outward, the particles of water vibrate only up and down about their mean positions. Water particles do not travel along with the wave. Similarly when a tuning fork is set into vibration, it produces waves in air. The wave travels from one particle to the next but the particles of air vibrate about their mean positions.

It is essential to understand the concept of wave motion in the study of various branches in Physics. Wave motion, in general, refers to the transfer of energy from one point to another point of the medium. Transference of various forms of energy like sound, heat, light, X-rays, γ-rays, radio-waves etc. takes place in the form of wave motion. For the transference of energy through a medium, the medium must possess the properties of *elasticity, inertia and negligible frictional resistance.*

4.2. What Propagates in Wave Motion?

Before studying the characteristics of the different forms of wave motion, it is essential to clearly understand–*what is propagated in a wave motion* ? The answer to this question is that the physical condition due to a disturbance generated at some point in the medium is propagated to other points in the medium. In all the waves, the particles of the medium vibrate about their mean positions. Hence, in the case of wave motion, it is not matter that is propagated but it is only state of motion of the matter that is propagated. It is a form of *dynamic condition* that is propagated from one point to the other point in the medium.

According to the laws of Physics, any dynamic condition is related to momentum and energy. To conclude, it may be said that in wave motion *momentum and energy* are transferred or propagated. It is not a case of propagation of matter as a whole.

4.3. Characteristics of Wave Motion

1. Wave motion is a disturbance produced in the medium by the repeated periodic motion of the particles of the medium.

2. Only the wave travels forward whereas the particles of the medium vibrate about their mean positions.

3. There is a regular phase change between the various particles of the medium. The particle ahead starts vibrating a little later than a particle just preceding it.

4. The velocity of the wave is different from the velocity with which the particles of the medium are vibrating about their mean positions. The wave travels with a uniform velocity whereas the velocity of the particles is different at different positions. It is maximum at the mean position and zero at the extreme position of the particles.

There are two types of wave motions :

(*i*) Transverse and (*ii*) Longitudinal.

Sound waves are longitudinal waves and light waves are transverse waves.

4.4. Transverse Wave Motion

In this type of wave motion, the particles of the medium vibrate at right angles to the direction of propagation of the wave.

To understand the propagation of transverse waves in a medium consider nine particles of the medium and the circle of reference (Fig. 4.1). The particles are vibrating about their mean positions up and down and the wave is travelling from left to right. The disturbance takes $T/8$ seconds to travel from one particle to the next.

(1) At $t = 0$, all the particles are at their mean position.

(2) After $T/8$ seconds, particle 1 travels a certain distance upward and the disturbance reaches particle 2.

(3) After $2T/8$ seconds, particle 1 has reached its extreme position and the disturbance has reached particle 3.

(4) After $3T/8$ seconds, particle 1 has completed 3/8 of its vibration and the disturbance has reached particle 4. The positions of particles 2 and 3 are also shown in Fig. 4.1.

(5) In this way after $T/2$ seconds, particle 1 has come back to its mean position and the particles 2, 3 and 4 are at the positions shown in the diagram.

The disturbance has reached particle 5.

In this way the process continues and the positions of the particles after *5T/8, 6T/8, 7T/8* and *T* seconds are shown in the diagram.

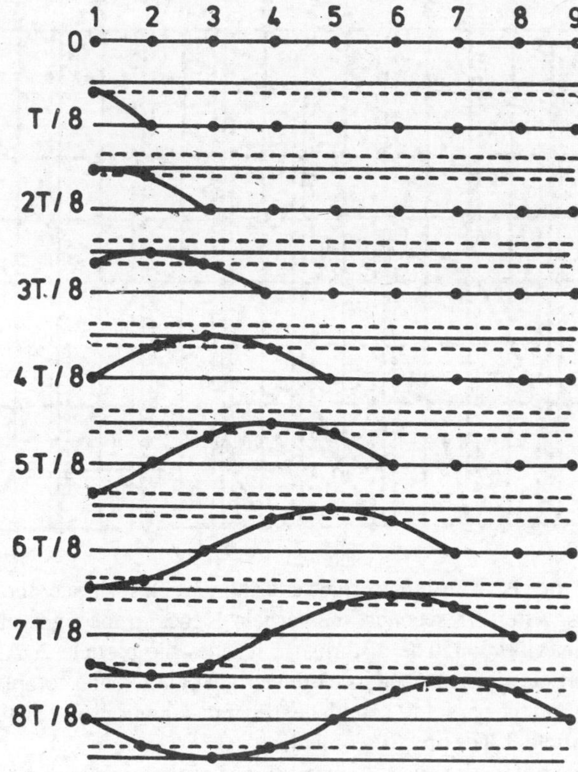

Fig. 4.1.

After *T* seconds, the particles 1, 5 and 9 are at their mean positions. The wave has reached particle 9. Particles 1 and 9 are in the same phase. The wave has travelled a distance between particles 1 and 9 in the time in which the particle 1 has completed one vibration.

The top point on the wave at the maximum distance from the mean position is called *crest*, while the point at the maximum distance below the mean position is called trough. Thus in a transverse wave, crests and troughs are alternately formed. The contour of the displaced particles of the medium represents the wave. In the case of transverse (or longitudinal) progressive waves, this contour continuously changes position in space and the wave seems to advance in the direction of propagation.

4.5. Longitudinal Wave Motion

In this type of wave motion, particles of the medium vibrate along the direction of propagation of the wave.

Consider nine particles of the medium and the circle of reference (Fig. 4.2).

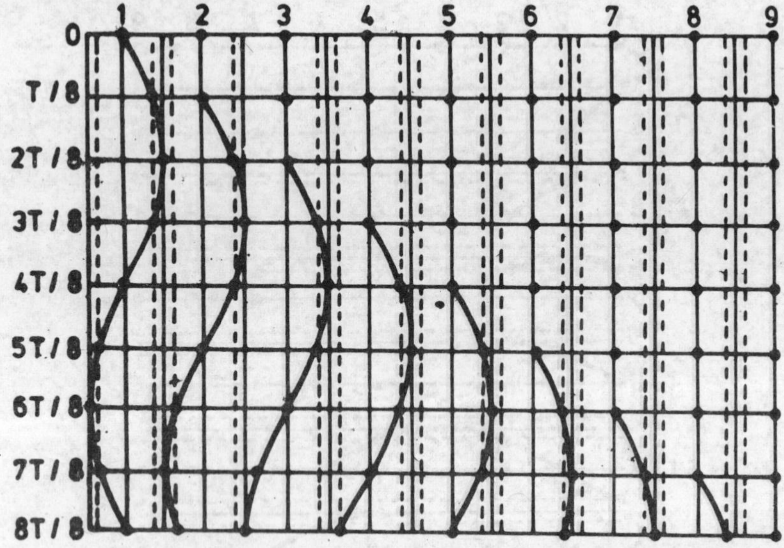

Fig. 4.2.

The wave travels from left to right and the particles vibrate about their mean positions. After *T*/8 seconds, the particle 1 goes to the right and completes 1/8 of its vibration. The disturbance reaches the particle 2. After *T*/4 seconds the particle 1 has reached its extreme right position and completes 1/4 of its vibration and the particle 2 completes 1/8 of its vibration. The disturbance reaches the particle 3. The process continues.

After one complete time period, the positions of the various particles is as shown in the diagram. The wave has reached particle 9. Here 1 and 9 are again in the same phase. Here particles 1, 5 and 9 are at their mean positions. The particles 1 and 3 are close to the particle 2. This is the position of condensation. Similarly particles 9 and 8 are close to the particle 7. This is also the position of condensation or compression. On the other hand, particles 4 and 6 are far away from the particle 5. This is the position of rarefaction. Hence in a longitudinal wave motion, condensations and rarefactions are alternately formed.

4.6. Definitions

Wavelength. It is the distance travelled by the wave in the time in which the particle of the medium completes one vibration. It is also defined as the distance between two nearest particles in the same phase.

The distance AB (Fig. 4.3) is equal to the wavelength λ.

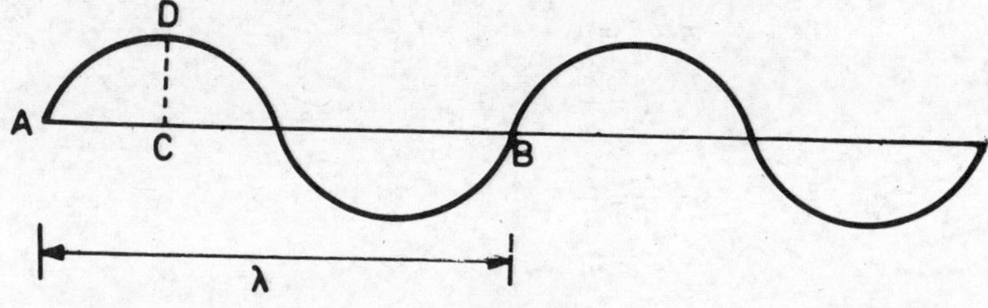

Fig. 4.3.

Frequency. It is the number of vibrations made by a particle in one second.

Amplitude. It is the maximum displacement of the particle from its mean position of rest. In the diagram CD is the amplitude.

Time period. It is the time taken by a particle to complete one vibration.

Suppose frequency = n

Time taken to complete n vibrations = 1 second.

Time taken to complete 1 vibration = $\frac{1}{n}$ second.

From the definition of time period, time taken to complete one vibration is the time period (T)

$$\therefore \qquad T = \frac{1}{n} \qquad \text{or} \qquad nT = 1$$

$$\therefore \qquad \text{Frequency} \times \text{Time period} = 1$$

Vibration. It is the to and fro motion of a particle from one extreme position to the other and back again. It is also equal to the motion of a particle from the mean position to one extreme position, then to the other extreme position and finally back to the mean position.

Phase. It is defined as the ratio of the displacement of the vibrating particle at any instant to the amplitude of the vibrating particle or it is defined as the fraction of the time interval that has elapsed since the particle crossed the mean position of rest in the positive direction or it is also equal to the angle swept by the radius vector since the vibrating particle last crossed its mean position of rest.

4.7. Relation between Frequency and Wavelength

Velocity of the wave is the distance travelled by the wave in one second.

$$\text{Velocity} = \frac{\text{Distance}}{\text{Time}}$$

Wavelength (λ) is the distance travelled by the wave in one time period (T).

$$\text{Velocity} = \frac{\text{Wavelength}}{\text{Time period}} = \frac{\lambda}{T}$$

But, frequency × time period = 1

$$n \times T = 1$$

$$T = \frac{1}{n}$$

$$v = \frac{\lambda}{T} = \frac{\lambda}{\frac{1}{n}}$$

$$v = n\lambda$$

Example 4.1. *If the frequency of a tuning fork is 400 and the velocity of sound in air is 320* metres/s, *find how far sound travels while the fork completes 30 vibrations.*

Here, $n = 400$, $v = 320$ metres/second, $\lambda = ?$

$$v = n\lambda$$

or $$\lambda = \frac{v}{n} = \frac{320}{400} = \textbf{0·8 metre}$$

∴ Distance travelled by the wave when the fork completes 1 vibration
= 0·8 metre

Distance travelled by the wave when the fork completes 30 vibrations
= 0·8 × 30 = **24 metres.**

4.8. Properties of Longitudinal Progressive Waves

1. The particles of the medium vibrate simple harmonically along the direction of propagation of the wave.

2. All the particles have the same amplitude, frequency and time period.

3. There is a gradual phase difference between the successive particles.

4. All the particles vibrating in phase will be at a distance equal to $n\lambda$. Here n =1, 2, 3 etc. It means the minimum distance between two particles vibrating in phase is equal to the wavelength.

5. The velocity of the particle is maximum at their mean position and it is zero at their extreme positions.

6. When the particle moves in the same direction as the propagation of the wave, it is in a region of compression.

7. When the particle moves in a direction opposite to the direction of propagation of the wave, it is in a region of rarefaction.

8. When the particle is at the mean position, it is a region of maximum compression or rarefaction.

9. When the particle is at the extreme position, the medium around the particles has its normal density, with compression on one side and rarefaction on the other.

10. Due to the repeated periodic motion of the particles, compressions and rarefactions are produced continuously. These compressions and rarefactions travel forward along the wave and transfer energy in the direction of propagation of the wave.

4.9. Demonstration of Transverse Waves

(*a*) Wave Apparatus

The formation of transverse waves can be demonstrated in the laboratory with the help of the wave motion apparatus. The apparatus consists of a vertical rectangular frame fixed on a horizontal base (Fig. 4.4). The axle passes escentrically through a number of circular discs with grooved edges. The circular discs are equidistant. Vertical rods carrying spherical balls at their upper ends rest on the circumference of their respective discs. When the axle is rotated, the rods are displaced vertically through different distances. It is adjusted that there is a gradual phase difference between the successive rods. Each ball will execute simple harmonic motion and will complete one vibration in one rotation of the discs. When the axle is continuously rotated with a uniform speed, the balls show a wave pattern and the transverse wave appears to progress in the forward direction with the formation of alternate crests and troughs. It is observed that the balls move in the vertical direction and the wave advances in the horizontal direction. Therefore, transverse waves are produced.

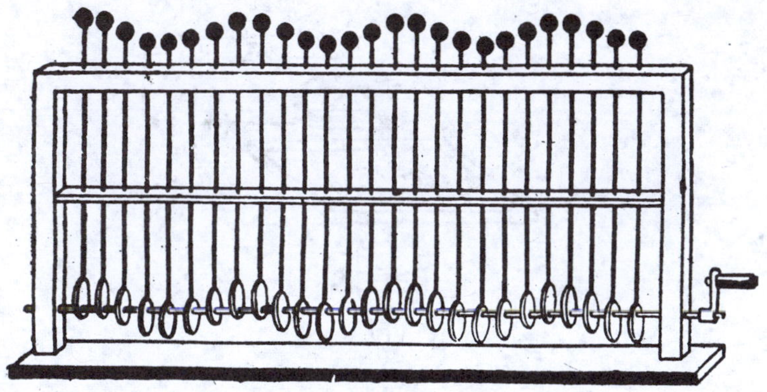

Fig. 4.4.

(*b*) Ripple Tank

The formation of transverse waves on the surface of water can be demonstrated in the laboratory with the help of ripple tank apparatus.

The ripple tank apparatus consists of a shallow rectangular dish with a glass bottom. The edges of the dish are sloping outwards so as to avoid interference between the direct and the reflected waves. An electrically maintained tuning fork having a fine style fixed to one of its prongs is adjusted so that the tip of the style just dips in water. The field of view is illuminated from below with the help of a strong source of light *S* and a condensing lens system *C* (Fig. 4.5).

The tuning fork is set into vibration. The style oscillates vertically up and down and transverse waves are produced on the surface of water. These waves originate at the tip of the style and travel radially outwards. The image of the water surface is obtained on the screen by reflection from the mirror *M*. The wave pattern is observed on the screen.

If a stroboscope is used so that the dish is illuminated intermittently with the same frequency as that of the tuning fork, a stationary pattern is obtained on the screen.

Using the ripple tank apparatus, interference phenomenon can be demonstrated fixing two styles to the same prong of the tuning fork.

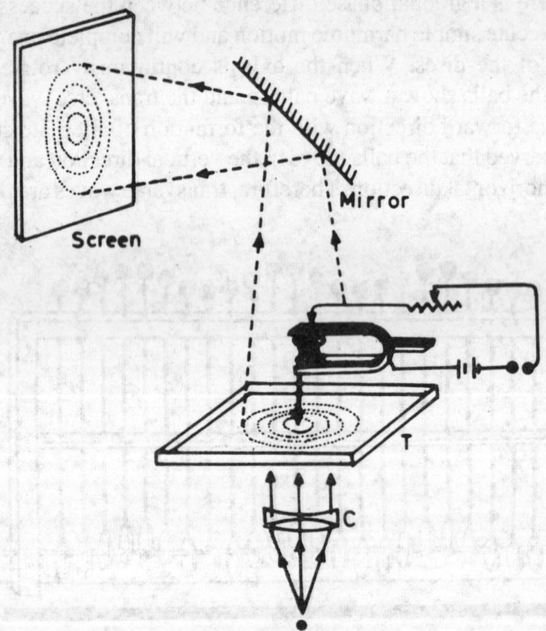

Fig. 4.5.

In this case, amplitude and frequency of vibrations produced by the two styles will be the same. The interference pattern can be observed on the screen.

If instead of a style, a thin edged blade is used, plane progressive transverse waves are produced on the surface of water.

4.10. Demonstration of Longitudinal Waves

The formation of longitudinal waves can be demonstrated with the help of a spring. One end of the spring is fixed to a handle H and the other end is free. A small push is given to the handle. The first three turns are compressed, the rest of the spring is in the relaxed position (Fig. 4.6). When the handle is brought back to its original position, the compression travels forward and there is rarefaction between the handle and the compression. If the handle is kept fixed at the initial position it is seen that with time, the compression and rarefaction travel forward as shown in Fig. 4.6. This demonstrates the formation of longitudinal waves in which any particle on the spring vibrates simple harmonically along the direction of propagation of the wave.

If the handle is vibrating continuously, continuous compressions and rarefactions are produced alternately all along the spring.

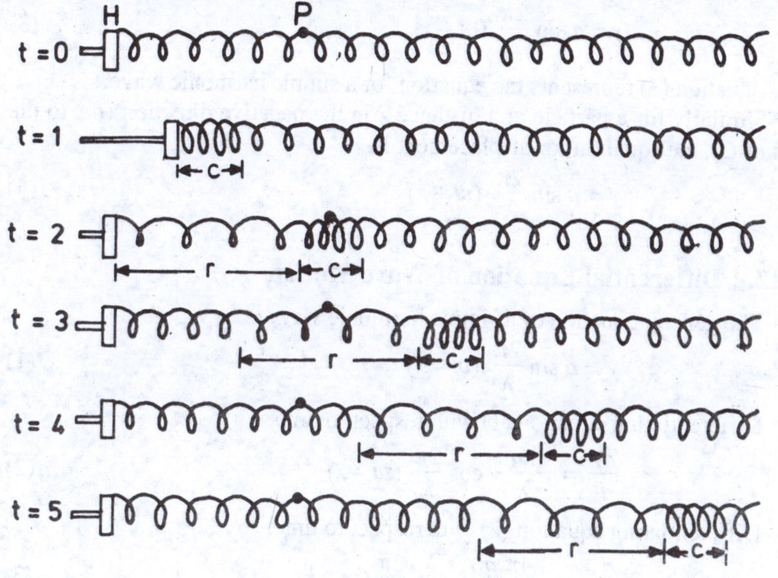

Fig. 4.6.

4.11. Equation of a Simple Harmonic Wave

Consider a particle O in a medium. Let the displacement at any instant of time be given by

$$y = a \sin \omega t \qquad \ldots (1)$$

Consider another particle A at a distance x from the particle O to its right. Here it is assumed that the wave is travelling with a velocity v from left to right *i.e.* from particle O towards A. The displacement at A is given by

$$y = a \sin (\omega t - \alpha) \qquad \ldots (2)$$

where α is the phase difference between the particles O and A. For a phase difference of 2π, the path difference is λ. Suppose for a phase difference of α, the path difference is x.

$$\therefore \quad \frac{\alpha}{x} = \frac{2\pi}{\lambda}$$

$$\therefore \quad \alpha = \frac{2\pi x}{\lambda}$$

Also $$\omega = \frac{2\pi}{T} = \frac{2\pi v}{\lambda}$$

Substituting the values of α and ω in equation (2),

$$y = a \sin\left(\frac{2\pi vt}{\lambda} - \frac{2\pi x}{\lambda}\right)$$

$$y = a \sin \frac{2\pi}{\lambda} (vt - x) \quad \ldots (3)$$

Equation (3) represents the equation for a simple harmonic wave.

Similarly for a particle at a distance x in the negative direction (*i.e.* to the left of O), the equation for displacement is,

$$y = a \sin \frac{2\pi}{\lambda} (vt + x) \quad \ldots (4)$$

4.12. Differential Equation of Wave Motion

The general equation of a simple harmonic wave is,

$$y = a \sin \frac{2\pi}{\lambda} (vt - x) \quad \ldots (1)$$

Differentiating equation (1) with respect to time,

$$\frac{dy}{dt} = \frac{2\pi a v}{\lambda} \cos \frac{2\pi}{\lambda} (vt - x) \quad \ldots (2)$$

Differentiating equation (2) with respect to time,

$$\frac{d^2y}{dt^2} = -\frac{4\pi^2 a v^2}{\lambda^2} \sin \frac{2\pi}{\lambda} (vt - x) \quad \ldots (3)$$

To find the value of compression, differentiate equation (1) with respect to x,

$$\frac{dy}{dx} = -\frac{2\pi a}{\lambda} \cos \frac{2\pi}{\lambda} (vt - x) \quad \ldots (4)$$

To find the rate of change of compression with respect to distance, differentiate equation (4) with respect to x,

$$\frac{d^2y}{dx^2} = -\frac{4\pi^2 a}{\lambda^2} \sin \frac{2\pi}{\lambda} (vt - x) \quad \ldots (5)$$

From equations (2) and (4)

$$\frac{dy}{dt} = -v\ \frac{dy}{dx} \qquad \ldots (6)$$

From equations (3) and (5)

$$\frac{d^2 y}{dt^2} = v^2\ \frac{d^2 y}{dx^2} \qquad \ldots (7)$$

Equation (7) represents the differential equation of wave motion.

The general differential equation of wave motion can be written as

$$\frac{d^2 y}{dt^2} = K\frac{d^2 y}{dx^2} \qquad \ldots (8)$$

Here $\quad K = v^2$

or $\quad v = \sqrt{K}$

Thus, knowing the value of K, the value of the wave velocity can be calculated.

4.13. Particle Velocity and Wave Velocity

The equation for a simple harmonic wave is given by

$$y = a \sin \frac{2\pi}{\lambda}(vt - x) \qquad \ldots (1)$$

Here v is the velocity of the wave and y is the displacement of the particle. The velocity of the particle $U = dy/dt$.

∴ Differentiating equation (1) with respect to time t,

$$U = \frac{dy}{dt} = \frac{2\pi av}{\lambda} \cos \frac{2\pi}{\lambda}(vt - x) \qquad \ldots (2)$$

The maximum value of the particle velocity is

$$U_{max} = \frac{2\pi av}{\lambda} \qquad \ldots (3)$$

$$\therefore \text{[Maximum Particle Velocity]} = \frac{2\pi a}{\lambda}\ \text{[Wave Velocity]}$$

To find the particle acceleration, differentiate equation (2) with respect to time

$$f = \frac{d^2 y}{dt^2} = -\frac{4\pi^2 av^2}{\lambda^2} \sin \frac{2\pi}{\lambda}(vt - x) \qquad \ldots (4)$$

$$f = -\frac{4\pi^2 v^2}{\lambda^2}\left[a \sin \frac{2\pi}{\lambda}(vt - x)\right]$$

$$\therefore \qquad f = -\left(\frac{4\pi^2 v^2}{\lambda^2}\right) y$$

The acceleration is maximum when $y = a$

$$\therefore \qquad f_{max} = -\left(\frac{4\pi^2 v^2}{\lambda^2}\right) a \qquad \ldots (5)$$

The negative sign shows that the acceleration of the particle is directed towards its mean position.

Differentiating equation (1) with respect to x

$$\frac{dy}{dx} = -\frac{2\pi a}{\lambda} \cos \frac{2\pi}{\lambda}(vt - x) \qquad \ldots (6)$$

dy/dx represents the slope of the displacement curve.

From equations (2) and (6)

$$U = \frac{dy}{dt} = -v\left(\frac{dy}{dx}\right) \qquad \ldots (7)$$

$$\therefore \left[\begin{array}{c}\text{Particle velocity}\\ \text{at any instant}\end{array}\right] = \left[\text{Wave velocity}\right]\left[\begin{array}{c}\text{Slope of the displacement}\\ \text{curve at that instant}\end{array}\right]$$

4.14. Distribution of Velocity and Pressure in a Plane Progressive Wave

For a plane progressive wave

$$y = a \sin \frac{2\pi}{\lambda}(vt - x) \qquad \ldots (1)$$

The particle velocity,

$$U = \frac{dy}{dt} = \frac{2\pi a v}{\lambda} \cos \frac{2\pi}{\lambda}(vt - x) \qquad \ldots (2)$$

The strain in the medium is dy/dx. If dy/dx is positive, it represents a region of rarefaction. If dy/dx is negative, it represents a region of compression. The bulk modulus of elasticity of the medium.

$$K = \frac{\text{Change in pressure}}{\text{Volume strain}}$$

$$\therefore \qquad K = \frac{-dP}{(dy/dx)}$$

$$\therefore \qquad dP = -K \cdot \left(\frac{dy}{dx}\right)$$

$$\therefore \qquad dP = K\left(-\frac{dy}{dx}\right) \qquad \ldots (3)$$

It means if dy/dx is negative, dP is + ve *i.e.* it is a region of compression. If dy/dx is positive, dP is negative, *i.e.* it is a region of rarefaction.

Differentiating equation (1) with respect to x

$$\frac{dy}{dx} = -\frac{2\pi a}{\lambda} \cos \frac{2\pi}{\lambda}(vt - x)$$

or

$$dP = K\left(-\frac{dy}{dx}\right)$$

$$dP = \frac{2\pi Ka}{\lambda} \cos \frac{2\pi}{\lambda}(vt - x) \qquad \ldots (4)$$

The graphs for displacement, velocity and change in pressure are shown in Fig. 4.7.

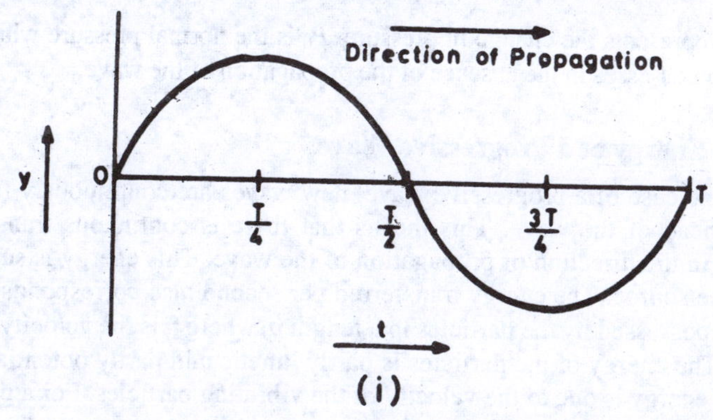

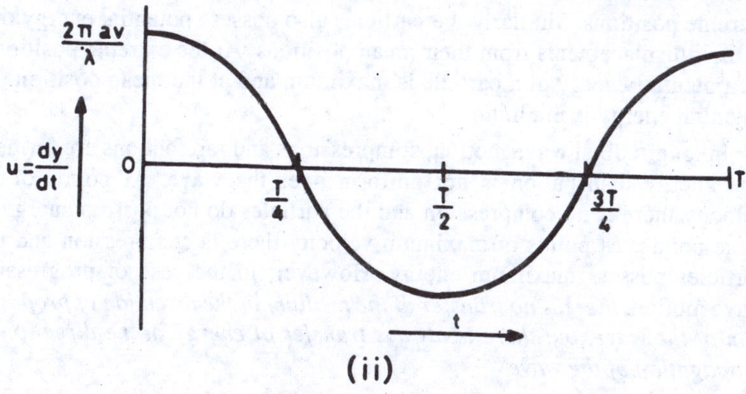

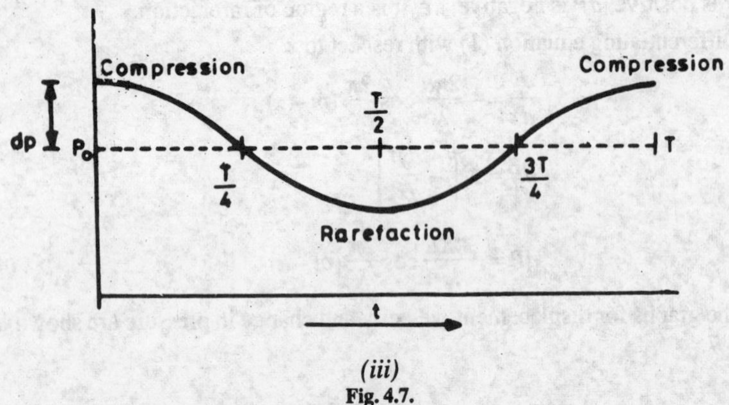

(iii)
Fig. 4.7.

dP represents the change in pressure. P_0 is the normal pressure which the medium possesses in the absence of the propagation of the wave.

4.15. Energy of a Progressive Wave

In the case of a progressive wave, new waves are continuously formed at the head of the wave. This means that there is continuous transfer of energy in the direction of propagation of the wave. This energy is supplied from the source. The energy transferred per second also corresponds to the energy possessed by the particles in a length v, where v is the velocity of the wave. The energy of the particles is partly kinetic and partly potential. The kinetic energy is due to the velocity of the vibrating particles. For a particle executing simple harmonic motion, the velocity is maximum at the mean position and it is zero at the extreme positions. Consequently the kinetic energy of the particle at the mean position is maximum and zero at the extreme positions. Similarly the particles also possess potential energy due to their displacements from their mean positions. At the extreme positions, the potential energy of a particle is maximum and at the mean position the potential energy is minimum.

In longitudinal wave motion, compressions and rarefactions are formed. The energy distribution is not uniform over the wave. At points of no velocity there is no compression and the particles do not possess energy at these points. At points of maximum velocity there is compression and the particles possess maximum energy. However, in the case of progressive wave motion, *there is no transfer of the medium in the direction of propagation of the wave, but there is always transfer of energy in the direction of propagation of the wave.*

Analytical Treatment

The equation of a simple harmonic wave is

$$y = a \sin \frac{2\pi}{\lambda}(vt - x) \qquad \ldots (1)$$

The particle velocity U at any instant can be obtained by differentiating equation (1) with respect to time

$$\therefore \qquad U = \frac{dy}{dt} = \frac{2\pi av}{\lambda} \cos \frac{2\pi}{\lambda}(vt - x) \qquad \ldots (2)$$

The acceleration of the particle at that instant,

$$f = \frac{dU}{dt}$$

$$\therefore \qquad f = \frac{d^2 y}{dt^2}$$

Differentiating equation (2) with respect to time,

$$f = \frac{d^2 y}{dt^2} = -\frac{4\pi^2 av^2}{\lambda^2} \sin \frac{2\pi}{\lambda}(vt - x) \qquad \ldots (3)$$

The –ve sign shows that acceleration is directed towards the mean position.

Potential Energy

To move the particle from its mean position to a distance y, work has to be done against acceleration.

Work done for a displacement dy

$$= F dy$$

Let ρ be the density of the medium

Work done per unit volume for a displacement dy

$$= \rho \left(+ \frac{4\pi^2 av^2}{\lambda^2} \sin \frac{2\pi}{\lambda}(vt - x) \right) dy$$

Total work done for a dispacement y

$$= \int_0^y \rho \left(+ \frac{4\pi^2 av^2}{\lambda^2} \sin \frac{2\pi}{\lambda}(vt - x) \right) dy$$

But $\qquad y = a \sin \frac{2\pi}{\lambda}(vt - x)$

$\therefore$ Potential energy per unit volume

$$= \left(\frac{4\pi^2 \rho v^2}{\lambda^2} \right) \int_0^y y\,dy$$

$$= \frac{4\pi^2 \rho v^2 y^2}{2\lambda^2}$$

$$\text{P.E.} = \frac{2\pi^2 \rho v^2 y^2}{\lambda^2}$$

$$\text{P.E.} = \frac{2\pi^2 \rho v^2 . a^2}{\lambda^2} \sin^2 \left[\frac{2\pi}{\lambda}(vt - x) \right] \quad \ldots (4)$$

K.E. per unit volume $= \frac{1}{2} \rho U^2$

$$\text{K.E.} = \frac{1}{2}\rho \left[\frac{2\pi a v}{\lambda} \cos \frac{2\pi}{\lambda}(vt - x) \right]^2$$

$$\text{K.E.} = \frac{2\pi^2 \rho v^2 a^2}{\lambda^2} \cos^2 \left[\frac{2\pi}{\lambda}(vt - x) \right] \quad \ldots (5)$$

Total energy per unit volume,

$E = \text{P. E.} + \text{K.E.}$

$$= \frac{2\pi^2 \rho v^2 a^2}{\lambda^2} \left[\sin^2 \frac{2\pi}{\lambda}(vt - x) + \cos^2 \frac{2\pi}{\lambda}(vt - x) \right]$$

$$E = \frac{2\pi^2 \rho v^2 a^2}{\lambda^2} \quad \ldots (6)$$

But $v = n\lambda$

$\therefore$ $E = 2\pi^2 \rho n^2 a^2$ $\quad \ldots (7)$

The average kinetic energy per unit volume and the average potential energy per unit volume are equal and each is equal to half the total energy per unit volume.

Average K.E. per unit volume $= \pi^2 \rho n^2 a^2$ $\quad \ldots (8)$

Average P.E. per unit volume $= \pi^2 \rho n^2 a^2$ $\quad \ldots (9)$

Suppose that the area of cross section of a parallel beam of radiation is 1 unit and the velocity of the wave is v.

Volume $= 1 \times v = v$

$\therefore$ Energy transfer per unit area per second.

$= E \times v$

$= 2\pi^2 \rho n^2 a^2 v$ $\quad \ldots (10)$

Energy transfer per second is also called energy current per unit area of cross section.

It is to be remembered that the potential and kinetic energies of every particle will change with time, but the average kinetic energy per unit volume and the average potential energy per unit volume remain constant. The current density per unit area of cross-section and the total energy per unit volume also remain constant.

Example 4.2. *When a simple harmonic wave is propagated through a medium, the displacement of a particle (in* cm*) at any instant of time is given by*

$$y = 10 \sin \frac{2\pi}{100} (36000\, t - 20)$$

Calculate, the amplitude of the vibrating particle, wave velocity, wavelength, frequency and time period.

The general equation of a simple harmonic wave is,

$$y = a \sin \frac{2\pi}{\lambda} (vt - x) \qquad \ldots (1)$$

Here $$y = 10 \sin \frac{2\pi}{100} (36000\, t - 20) \qquad \ldots (2)$$

Comparing equations (1) and (2)

$$\mathbf{a = 10\ cm}$$

$$\boldsymbol{\lambda} \mathbf{= 100\ cm}$$

Wave velocity $$\mathbf{v = 36000\ cm/s}$$

Frequency $$n = \frac{v}{\lambda} = \frac{36000}{100}$$

or $$\mathbf{n = 360\ hertz}$$

Time period $$T = \frac{1}{n} = \frac{1}{360} \textbf{ second.}$$

Example 4.3. *A simple harmonic wave of amplitude 8 units tranverses a line of particles in the direction of the positive X-axis. At any given instant of time, for a particle at a distance of 10* cm *from the origin, the displacement is +6 units, and for a particle at a distance of 25* cm *from the origin, the displacement is +4 units. Calculate the wavelength.*

$$y = a \sin \frac{2\pi}{\lambda} (vt - x)$$

or $$\frac{y}{a} = \sin 2\pi \left(\frac{t}{T} - \frac{x}{\lambda} \right)$$

(1) In the first case

$$\frac{y_1}{a} = \sin 2\pi \left(\frac{t}{T} - \frac{x_1}{\lambda} \right)$$

Here $y_1 = +6, \quad a = 8, \quad x_1 = 10$ cm.

$$\therefore \quad \frac{6}{8} = \sin 2\pi\left(\frac{t}{T} - \frac{10}{\lambda}\right) \quad \ldots (1)$$

(2) In the second case,

$$\frac{y_2}{a} = \sin 2\pi\left(\frac{t}{T} - \frac{x_2}{\lambda}\right)$$

Here $y_2 = +4, \quad a = 8, \quad x_2 = 25$

$$\therefore \quad \frac{4}{8} = \sin 2\pi\left(\frac{t}{T} - \frac{25}{\lambda}\right) \quad \ldots (2)$$

From equation (1)

$$0{\cdot}75 = \sin 2\pi\left(\frac{t}{T} - \frac{10}{\lambda}\right)$$

But $$\sin\left(\frac{48{\cdot}6\pi}{180}\right) = 0{\cdot}75$$

$$\therefore \quad 2\pi\left(\frac{t}{T} - \frac{10}{\lambda}\right) = \frac{48{\cdot}6\,\pi}{180} \quad \ldots (3)$$

or $$\frac{t}{T} - \frac{10}{\lambda} = \frac{48{\cdot}6}{360} \quad \ldots (4)$$

From equation (2)

$$0{\cdot}5 = \sin 2\pi\left(\frac{t}{T} - \frac{25}{\lambda}\right)$$

But $$\sin\frac{\pi}{6} = 0{\cdot}5$$

$$\therefore \quad 2\pi\left(\frac{t}{T} - \frac{25}{\lambda}\right) = \frac{\pi}{6} \quad \ldots (5)$$

$$\frac{t}{T} - \frac{25}{\lambda} = \frac{1}{12} \quad \ldots (6)$$

Subtracting (6) from (4)

$$\frac{25}{\lambda} - \frac{10}{\lambda} = \frac{48{\cdot}6}{360} - \frac{1}{12}$$

or $$\boldsymbol{\lambda = 290{\cdot}8 \text{ cm.}}$$

Example 4.4. *The velocity of a simple harmonic wave is 30 cm/s. At a time t=0 the displacement of a particle is given by*

$$y = 4 \sin 2\pi \left(\frac{x}{100} \right)$$

Find the equation for the displacement at a time t = 2 s.

The general equation of a simple harmonic wave is

$$y = a \sin \frac{2\pi}{\lambda} (vt - x)$$

or $$y = a \sin \left(\frac{2\pi t}{T} - \frac{2\pi x}{\lambda} \right]$$

When $$t = 0$$

$$y = a \sin \left(- \frac{2\pi x}{\lambda} \right)$$

or $$y = - a \sin \frac{2\pi x}{\lambda} \quad \ldots (1)$$

At $t = 0$, the given equation is

$$y = 4 \sin 2\pi \left(\frac{x}{100} \right) \quad \ldots (2)$$

Comparing equations (1) and (2)

$$a = - 4 \text{ and } \lambda = 100 \text{ cm}$$

At time $$t = 2 \text{ s}$$

$$y = a \sin \frac{2\pi}{\lambda} (vt - x)$$

Here $$a = - 4 \text{ cm}, \ \lambda = 100 \text{ cm}, t = 2 \text{ s}$$

$$y = - 4 \sin \frac{2\pi}{100} (30 \times 2 - x)$$

$$y = - 4 \sin \left[\frac{6\pi}{5} - 2\pi \left(\frac{x}{100} \right) \right]$$

or $$y = 4 \sin \left[2\pi \left(\frac{x}{100} \right) - \frac{6\pi}{5} \right]$$

Example 4.5. *A simple harmonic wave is represented by*

$$y = 10 \sin \left(\frac{2\pi t}{T} + \alpha \right)$$

The time period is 30 s. *At time t=0, the displacement is 5* cm. *Calculate (i) phase angle at t=7·5* s, *and (ii) phase difference between two positions at a time interval of 6* s.

Here $y = 10 \sin\left(\frac{2\pi t}{T} + \alpha\right)$

$T = 30$ s

(*i*) At $t = 0$

$y = 5$ cm

$\therefore$ $5 = 10 \sin \alpha$

$\sin \alpha = 0{\cdot}5$

or $\alpha = \frac{\pi}{6}$ **radian**

(*ii*) At $t = 7{\cdot}5$ s,

the phase angle $\phi = \left(\frac{2\pi t}{T} + \alpha\right)$

$$\phi = \frac{2\pi \times 7{\cdot}5}{30} + \frac{\pi}{6}$$

$$\phi = \frac{2\pi}{3} \ \textbf{radian}$$

The phase difference for a time interval of 6 s,

$$\theta = \frac{2\pi t}{T}$$

Here $t = 6$ s

$$\theta = \frac{2\pi \times 6}{30} = \frac{2\pi}{5} \ \textbf{radians.}$$

Example 4.6. *A source of sound has a frequency of 512* Hz *and an amplitude of 0·25* cm. *What is the flow of energy across a* square cm per second, *if the velocity of sound in air is 340 m/s and the density of air is 0·00129* g/cm^3?

(Delhi, 1976)

Here $n = 512$ Hz

$a = 0{\cdot}25$ cm

$\rho = 0{\cdot}00129$ g/cm^3

$v = 340$ m/s $= 34000$ cm/s

Total energy per unit volume

$$= 2\pi^2 \rho n^2 a^2$$

Energy transferred across a sq-cm per second

$$= 2\pi^2 \rho n^2 a^2 v$$

$$= 2 \times (3{\cdot}14)^2 \times 0{\cdot}00129 \times (512)^2 \times (0{\cdot}25)^2 \times 34000$$

$$= \mathbf{1{\cdot}417 \times 10^7 \ ergs/cm^2 - s.}$$

EXERCISES

1. Explain the term wave motion. Distinguish clearly between transverse and longitudinal wave motion.
2. Describe briefly the properties of longitudinal progressive waves.
3. Describe experiments to demonstrate the formation of (*i*) transverse waves and (*ii*) longitudinal waves.
4. Obtain the equation for a plane simple harmonic wave.
5. Derive the relation

$$\frac{d^2y}{dt^2} = v^2 \cdot \frac{d^2y}{dx^2}.$$

6. Distinguish between particle velocity and wave velocity and obtain the relation between the two.
7. Show that for a plane progressive wave, on the average, half the energy is kinetic and half potential. Hence derive the expression for intensity of sound in terms of pressure amplitude. *(Delhi, 1992)*
8. Show that in the case of progressive longitudinal waves

   Particle velocity=Wave velocity × Compression. *(Punjab, 1973)*
9. Show that

$$U = -v\frac{dy}{dx}$$

   where U is the particle velocity and v is the wave velocity. *(Punjab, 1993)*
10. Discuss the variation of velocity and pressure in a plane progressive wave.
11. Show that the energy of a plane progressive wave is given by

$$E = 2\pi^2 \rho n^2 a^2.$$

12. Discuss the distribution of energy in a plane progressive wave.
13. Show how a transverse wave may be represented by a curve and explain how the velocity of any particle can be determined from the curve.
14. Derive the equation of wave motion in the form

$$y = a \sin\frac{2\pi}{\lambda}(vt-x).$$

15. Deduce the equation of a simple harmonic wave travelling in the positive x direction in the form

$$y = a \sin\frac{2\pi}{\lambda}(vt-x)$$

   and explain how energy is distributed in such a progressive wave. *(Delhi, 1991)*

CHAPTER 5

Velocity of Sound

5.1. Origin of Sound

Sound is produced by a vibrating body. When a gong of a bell is struck with a hammer, sound is produced. The bell is set into vibration and sound is propagated through air. These vibrations reach the ear and the ear drum is set into vibration. These vibrations are communicated to the brain. By touching the gong with the hand, one can feel the vibration of the gong. Similarly the cycle bell produces sound due to the vibrations produced by the gong. When the cycle bell is touched with hand, the vibrations are stopped and the bell does not produce sound. If a pith ball pendulum is held in contact with the edges of a vibrating gong, the pith ball moves to and fro. This shows that the gong vibrates as long as the sound is produced.

A tuning fork is set into vibration by striking one of its prongs against a rubber pad. The vibration of the prongs of the tuning fork can be seen. Similarly, vibrating strings, air columns, vibrating plates etc. produce sound.

5.2. Material Medium is a Necessity

It can be proved by means of an experiment that a material medium is a necessity for the propagation of sound waves. In the absence of a medium, no sound waves can travel.

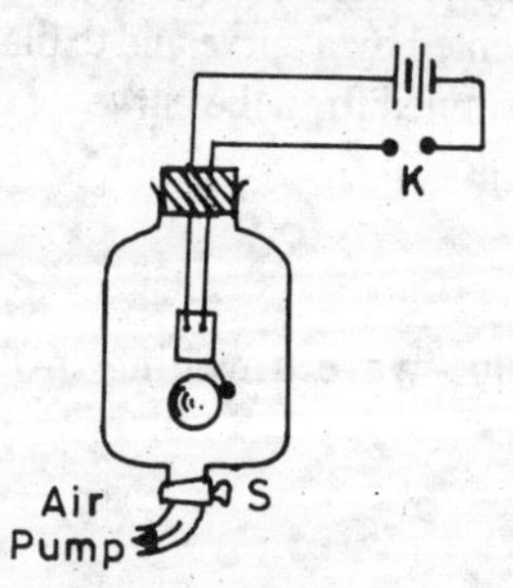

Fig. 5.1

Experiment. Take a jar and fix a bell inside it as shown. The connecting wire passes through an air-tight rubber cork fixed to the neck of the jar. The electric bell is connected to a battery and a key outside the jar (Fig. 5.1). The jar is placed on the platform of an air pump. Insert the key. The bell rings and the sound is heard. When the air in the jar is removed gradually, the sound becomes fainter although the same current is passing through the bell. After some time, when the pressure of air inside the jar is extremely low, only a very feeble sound is heard. It is not possible to create perfect vacuum with an air pump. In vacuum, no sound travels. Hence material medium is a necessity for the propagation of sound waves.

5.3. Velocity of Longitudinal Waves in Gases

Consider a long tube of area of cross-section a. Let A and B be two cross-sections of the tube. Suppose, in the region of normal density, the medium moves from right to left along the length of the tube with a velocity U. The sound waves travel from left to right with a velocity U. In such a case, the

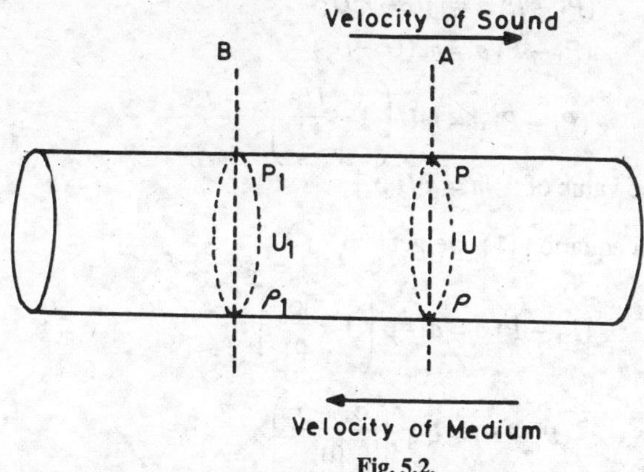

Fig. 5.2.

positions of condensations and rarefactions in the tube with respect to the ground will remain fixed. For any small region, the pressure, velocity and density of the medium remain constant. The values of pressure, velocity and density of the medium will be different cross-sections of the tube.

Let A correspond to the region of normal density and B correspond to the region of condensation. At A, the pressure is P, density is ρ and velocity of the medium is U. At B, the pressure is P_1, density is ρ_1 and velocity of the medium is U_1.

As there is no change in the average density of the medium, the masses of the medium crossing the sections A and B in one second are equal.

Mass of the medium entering the cross-section at A in one second

$$= a \times U \times \rho$$

Mass of the medium leaving the cross-section at B in one second

$$= a \times U_1 \times \rho_1$$

$$\therefore \quad m = aU\rho = aU_1 \rho_1 \qquad \ldots (1)$$

If B is at a region of condensation, its density will be higher than the normal density ρ at A.

$$\therefore \quad \rho_1 > \rho$$

and $\quad U_1 < U$

The momentum per second of the medium entering at $A = mU$

The momentum per second of the medium leaving at $B = mU_1$.

Change in momentum per second $= mU - mU_1$. This change in momentum per second is due to the difference in pressure $(P_1 - P)$ between A and B. Since the tube has an area of cross section a, the force applied,

$$F = (P_1 - P)\, a.$$

Also, according to Newton's second law of motion, rate of change of momentum is equal to the applied force.

$$\therefore \quad (P_1 - P)\, a = mU - mU_1$$

$$(P_1 - P)\, a = m(U - U_1)$$

$$(P_1 - P)\, a = mU\left(1 - \frac{U_1}{U}\right) \quad \ldots (2)$$

Substituting the value of $\quad m = a\,U\rho$,

and $\dfrac{U_1}{U} = \dfrac{\rho}{\rho_1}$, from equation (1), we get

$$(P_1 - P)\, a = aU^2\rho\left(1 - \frac{\rho}{\rho_1}\right)$$

$$(P_1 - P) = U^2\rho\left(\frac{\rho_1 - \rho}{\rho_1}\right)$$

$$\therefore \quad U^2\rho = \frac{(P_1 - P)}{\left(\dfrac{\rho_1 - \rho}{\rho_1}\right)} \quad \ldots (3)$$

The bulk modulus of elasticity of the medium,

$$E = \frac{(P_1 - P)}{\left(\dfrac{V - V_1}{V}\right)}$$

But $\quad V \propto \dfrac{1}{\rho}$

$$\therefore \quad \frac{V - V_1}{V} = \frac{\rho_1 - \rho}{\rho_1}$$

$$\therefore \quad E = \frac{P_1 - P}{\left(\dfrac{\rho_1 - \rho}{\rho_1}\right)} \quad \ldots (4)$$

From equations (3) and (4)

$$U^2\rho = E$$

$$U^2 = \frac{E}{\rho}$$

or $$U = \sqrt{\frac{E}{\rho}} \qquad \ldots (5)$$

The formula is true only in the case of *plane waves* where the original disturbance is of very small amplitude. This requirement is fulfilled by sound waves of ordinary intensity. The relation also holds good for the velocity of simple harmonic plane waves through homogeneous isotropic media *viz.*, solids, liquids and gases.

This relation is also applicable to torsional waves in solids and transverse vibrations in strings. However, in the case of transverse vibrations of bars and plates, the formula needs modification.

5.4. Newton's Formula for Velocity of Sound

The velocity of sound in a medium–solid, liquid or gas, depends on the elasticity and density of the medium.

$$U = \sqrt{\frac{E}{\rho}}$$

where U is the velocity of sound in the medium, E the elasticity and ρ the density of the medium.

Newton also found that the elasticity of air is equal to its pressure. He assumed that the temperature of air remains constant when sound waves travel through air. The process is isothermal and Boyle's law can be applied. At the point of condensation, pressure increases and volume decreases.

Suppose, initial pressure $= P$

Initial volume $= V$

Increase in pressure $= p$

Decrease in volume $= v$

Final pressure $= (P + p)$

Final volume $= (V - v)$

$$\therefore \qquad (P + p)(V - v) = PV$$

$$PV + pV - Pv - pv = PV$$

(pv is negligibly small)

$$\therefore \qquad pV = Pv \quad \text{or} \quad P = \frac{pV}{v} \qquad \ldots (1)$$

From the definition of elasticity of the medium

$$E = \frac{\text{Change in pressure}}{\dfrac{\text{Change in volume}}{\text{Original volume}}}$$

$$E = \frac{p}{\dfrac{v}{V}} = \frac{pV}{v} \qquad \ldots (2)$$

Equating (1) and (2)

$$E = P$$

$\therefore$ Substituting $E = P$ in the original formula

$$U = \sqrt{\frac{P}{\rho}} \qquad \ldots (3)$$

Taking normal pressure $P = 76$ cm of Hg

$= 76 \times 13{\cdot}6 \times 981$ dynes/sq cm

and density of air at NTP = 0·001293 g/cc

$$U = \sqrt{\frac{76 \times 13{\cdot}6 \times 981}{0{\cdot}001293}} = 28000 \text{ cm/s}$$

The experimental value determined from various experiments gives the velocity of sound at NTP

= 33200 cm/s.

Therefore, there is a difference of about 5200 cm/s between the theoretical and the experimental values. The large difference cannot be attributed to the experimental errors. Newton was unable to explain the error in his formula and the correction was explained by a French Scientist Laplace.

Laplace correction. According to Laplace, when sound waves travel through air, there is condensation and rarefaction in the particles of the medium. Where there is condensation, particles come near each other and are heated up. Where there is rarefaction, particles go apart and there is fall of temperature. Therefore, the temperature does not remain constant when sound waves travel through air or any other gas. The process is not isothermal but it is adiabatic. The total quantity of heat of the system as a whole remains constant. It neither gains nor loses any heat to the outside. During an adiabatic process

$$PV^{\gamma} = \text{constant}$$

Suppose, initial pressure $= P$

Initial volume $= V$

Change in pressure $= p$

Change in volume $= v$

$$PV^{\gamma} = (P+p)\ (V-v)^{\gamma}$$

$$PV^{\gamma} = P\left[1+\frac{p}{P}\right]V^{\gamma}\left[1-\frac{v}{V}\right]^{\gamma}$$

$$1 = \left[1+\frac{p}{P}\right]\left[1-\frac{v}{V}\right]^{\gamma}$$

$$1 = \left[1+\frac{p}{P}\right]\left[1-\frac{\gamma v}{V}\right]$$

$$1 = 1+\frac{p}{P}+\frac{\gamma v}{V}$$

Neglecting $\frac{\gamma p v}{PV}$, $\frac{p}{P} = \frac{\gamma v}{V}$

or $$\gamma P = \frac{pV}{v} \qquad \ldots(1)$$

From the definition of elasticity

$$E = \frac{pV}{v} \qquad \ldots(2)$$

Equating (1) and (2)

$$E = \gamma P$$

Substituting this value of $E = \gamma P$ in the equation

$$U = \sqrt{\frac{E}{\rho}}, \qquad \text{we get}$$

$$U = \sqrt{\frac{\gamma P}{\rho}} \qquad \ldots(3)$$

Here γ is the ratio between the two specific heats of air or a gas

For air $\gamma = 1.42$

The velocity of sound in air at NTP

$$U = \sqrt{\frac{1{\cdot}42 \times 76 \times 13{\cdot}6 \times 981}{0{\cdot}001293}}$$

$$= 33280 \text{ cm/s}$$

This value is in agreement with the experimental value

Hence the correct formula for the velocity of sound in air or a gas is

$$U = \sqrt{\frac{\gamma P}{\rho}}$$

5.5. Effect of Temperature

The density of air or a gas changes with the change in temperature.

At 0°C, $$U_0 = \sqrt{\frac{\gamma P}{\rho_0}} \qquad \ldots (1)$$

At t^0C, $$U_t = \sqrt{\frac{\gamma P}{\rho_t}} \qquad \ldots (2)$$

Dividing (2) by (1)

$$\frac{U_t}{U_0} = \sqrt{\frac{\rho_0}{\rho_t}} = \sqrt{1 + \frac{1}{273}} = \sqrt{\frac{T}{T_0}}$$

$$U \propto \sqrt{T}$$

Therefore, velocity of sound in air or a gas is directly proportional to the square root of the absolute temperature.

Example 5.1. *Find the temperature at which sound travels in hydrogen with the same velocity as in oxygen at 1000° C. Assume γ to be the same for the two gases.* (Delhi, 1976)

Here, for oxygen $T_1 = 273 + 1000$

$= 1273$ K

Density $= \rho_1$

For hydrogen

$T_2 = ?$

Density $= \rho_2$

$$\frac{\rho_1}{\rho_2} = 16$$

Also $$\frac{T_1}{\rho_1} = \frac{T_2}{\rho_2}$$

$$T_2 = \frac{T_1 \rho_2}{\rho_1}$$

or $$T_2 = \frac{1273 \times 1}{16}$$

$= 79{\cdot}56$ K

$= \mathbf{-193{\cdot}44° \, C.}$

Example 5.2. *The velocity of sound in air at 14°C is 340* metres/second. *What will it be when the pressure of the gas is doubled and its temperature is raised to 157·5 °C ?*

Change of pressure has no effect on the velocity of sound.

Here, $t_1 = 14^\circ$ C, $U_1 = 340$ metres/second.

$t_2 = 157{\cdot}5^\circ$ C, $U_2 = ?$

$$\frac{U_2}{U_1} = \sqrt{\frac{273 + t_2}{273 + t_1}}$$

$$\frac{U_2}{340} = \sqrt{\frac{273 + 157{\cdot}5}{273 + 14}} = \sqrt{\frac{430{\cdot}5}{287}}$$

$$= \sqrt{\frac{3}{2}} = \sqrt{1{\cdot}5}$$

$$U_2 = 340\sqrt{1{\cdot}5} \cong \mathbf{416{\cdot}14 \text{ metres/second.}}$$

Example 5.3. *Calculate the increase in the velocity of sound in air per degree celsius rise in temperature.*

Or

Prove that the velocity of sound in air increases by 61 cm/s *for each degree celsius rise in temperature.*

$$\frac{U_t}{U_0} = \sqrt{\frac{T}{T_0}} = \sqrt{\frac{273 + t}{273}} = \left(1 + \frac{t}{273}\right)^{\frac{1}{2}}$$

$$U_t = U_0\left(1 + \frac{1}{2} \times \frac{t}{273}\right) = U_0\left(1 + \frac{t}{546}\right)$$

(*i*) $U_0 = 332$ metres/second, $t = 1$

$$U_t = 332\left(1 + \frac{1}{546}\right)$$

$$U_t = 332 + 0{\cdot}61 = 332{\cdot}61 \text{ m/s}$$

$\therefore$ $U_t - U_0 = \mathbf{0{\cdot}61}$ **m/s or 61 cm/s.**

Example 5.4. *The velocity of sound in air at 16°*C *is 340* metres/second. *Find the wavelengths in air of a note of frequency 680*Hz *at 16°*C *and 51°*C.

(1) Suppose the wavelength $= \lambda$ at 16°C

$U_1 = 340$ metres/second, n=680Hz

$\therefore$ $$U_1 = n\lambda_1$$

or $$\lambda_1 = \frac{U_1}{n} = \frac{340}{680}$$

$$= \mathbf{0{\cdot}50 \text{ metre.}}$$

(2) Suppose the wavelength = λ_2 at 510° C

$$t_1 = 16° \text{ C}, \qquad U_1 = 340 \text{ metres/second}$$

$$t_2 = 51° \text{ C}, \qquad U_2 = ?$$

$$\frac{U_2}{U_1} = \sqrt{\frac{273 + t_2}{273 + t_1}}$$

$$\frac{U_2}{340} = \sqrt{\frac{273 + 51}{273 + 16}} = \sqrt{\frac{324}{289}} = \frac{18}{17}.$$

$$U_2 = \frac{340 \times 18}{17} = 360 \text{ metres / second}$$

$$\lambda_2 = \frac{U_2}{n} = \frac{360}{680} = \mathbf{0{\cdot}53 \ metre.}$$

5.6. Effect of Pressure

Velocity of sound in air is independent of change of pressure.

$$\text{Volume} = \frac{\text{Mass}}{\text{Density}} = \frac{m}{\rho}$$

$$\therefore \qquad PV = \frac{Pm}{\rho} = \text{constant, if the temperature is constant.}$$

Since, mass remains constant, $\frac{P}{\rho}$ is constant.

In the formula

$$U = \sqrt{\frac{\gamma P}{\rho}}$$

$\sqrt{\frac{P}{\rho}}$ is constant.

Therefore, the velocity of sound in air or a gas is independent of changes in pressure, provided the temperature remains constant.

5.7. Effect of Density of the Medium

Consider two media of densities ρ_1 and ρ_2

$$U_1 = \sqrt{\frac{\gamma P}{\rho_1}} \qquad \ldots (1)$$

and

$$U_2 = \sqrt{\frac{\gamma P}{\rho_2}} \qquad \ldots (2)$$

Dividing (2) by (1)

$$\frac{U_2}{U_1} = \sqrt{\frac{\rho_1}{\rho_2}}$$

or $$U \propto \frac{1}{\sqrt{\rho}} \qquad \ldots (3)$$

Velocity of sound in air or a gas is inversely proportional to the square root of the density of the medium.

Example 5.5. *Prove that the velocity of sound in hydrogen is four times the velocity of sound in oxygen.*

Density of oxygen $= D_1$

Velocity of sound in oxygen $= U_1$

Density of hydrogen $= D_2$

Velocity of sound in hydrogen $= U_2$

$$\frac{D_1}{D_2} = \frac{16}{1} \text{ and } \frac{U_2}{U_1} = \sqrt{\frac{D_1}{D_2}} = \sqrt{\frac{16}{1}} = 4$$

$$U_2 = 4 \times U_1.$$

5.8. Effect of Humidity

The density of water vapour at NTP $= \frac{18}{22400} = 00{\cdot}0008$ g/cc whereas the density of dry air at NTP = 0·001293 g/cc. Therefore, water vapour has a density less than the density of dry air. Consequently the density of moist air is less than the density of dry air. As the velocity of sound is more in a medium of lesser density, velocity of sound in moist air is more than the velocity of sound in dry air.

5.9. Effect of Wind

If the wind is blowing in the direction of propagation of the wave, it will increase the velocity of sound. On the other hand, if the wind is blowing in a direction opposite to the direction of propagation of the sound waves, it will decrease the velocity of sound in air.

5.10. Velocity of Sound in Water

According to Newton's formula, the velocity of sound in water is given by

$$U = \sqrt{\frac{E}{\rho}}$$

Volume elasticity of water $= 2{\cdot}23 \times 10^{10}$ dynes/sq cm and density of water =1 g/cc

$$U=\sqrt{\frac{2\cdot 23\times 10^{10}}{1}}=1493\times 10^{2}\text{ cm/s}$$

$$=1493\text{ metres/s.}$$

The velocity of sound in water is approximately four times the velocity of sound in air.

Experiment. This experiment is due to Colladon and Strum who performed the experiment in a lake at Geneva in 1826.

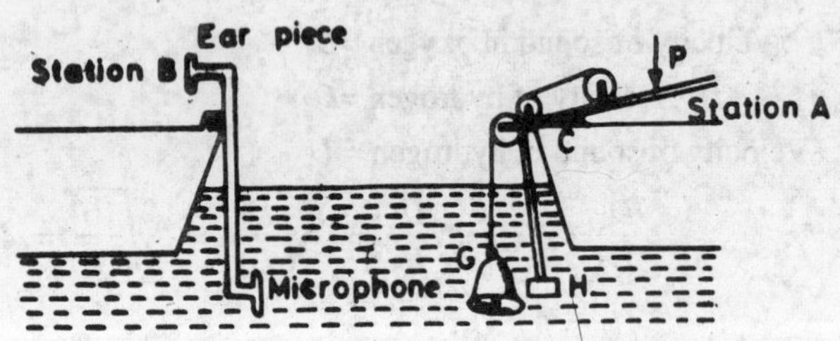

Fig. 5.3.

Two boats A and B were stationed at a distance of about 13,487 metres. A large bell G was suspended in water and a hammer H was arranged as shown in Fig. 5.3. At the station B, another boat B was arranged with a microphone and an ear piece. The person sitting in A presses the handle at the point P. The hammer H strikes against the gong and also a flash of light is produced at C on the boat A. The observer at B, observes the flash at C at the time the hammer H strikes against the bell G. After some time the sound reaches the microphone M and is picked up by the ear piece. The observer at B finds the time interval between the instant he sees the flash of light and the instant the sound reaches him through water. Knowing the distance and time, the velocity of sound in water can be determined.

The value of the velocity of sound in water was found to be 1435 metres per second which agrees with the theoretical result within the limits of experimental error.

5.11. Velocity of Sound in Air

In this case, two microphones M_1 and M_2 are fixed at two stations about 10 kilometres apart (Fig. 5.4.). Microphone M_1 is connected to a primary coil P_1 of a transformer and a battery in the circuit. Microphone M_2 is connected to a primary coil P_2 of a transformer and a battery in the circuit. The secondary coil of the transformer (common for P_1 and P_2) is connected to a sensitive galvanometer. Light from the bulb after reflection from the mirror of the galvanometer falls on a photographic film wrapped on a rotating drum. The drum rotates with a uniform speed.

When a gun is fired at the station A, the microphone M_1 receives the sound. A change in current in the primary coil P_1 is produced and an induced current is produced in the secondary coil S. The galvanometer coil is deflected and the spot on the drum moves up as shown in Fig. 5.4. After some time, sound

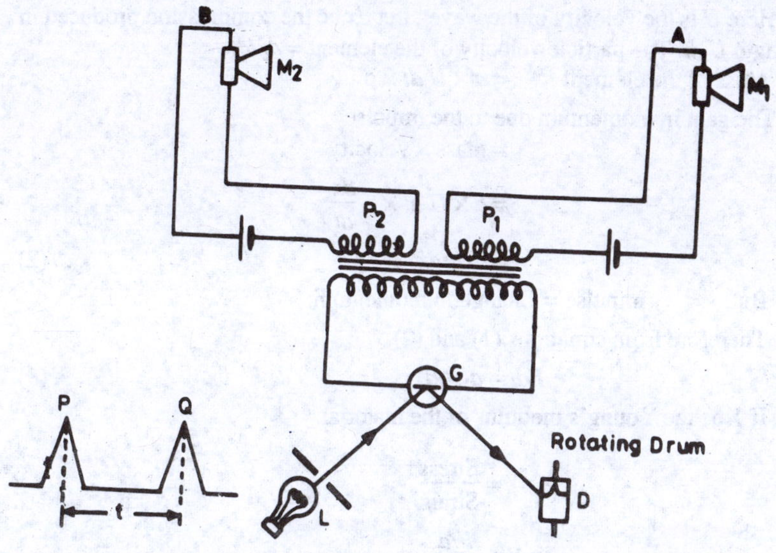

Fig. 5.4.

reaches the microphone M_2 and the change in current in the primary coil P_2 produces an induced current in the secondary coil S. Again deflection, in the galvanometer coil moves the spot of light on the photographic film on the rotating drum.

As the speed of rotation of the drum is known, the time between the two deflections in the galvanometer is calculated. Suppose the distance between the two stations is d and the time taken by sound is t.

Then $$V = \frac{d}{t}.$$

This gives the velocity of sound in air at room temperature.

5.12. Velocity of Sound in Isotropic Solids

In the case of solids, different types of strain are possible and hence different kinds of waves are possible in them. If one end of a thin long rod is struck with a hammer, longitudinal compressional waves are set up. In this case, there is increase in length in the region of rarefaction and decrease in length in the region of condensation. As there is change in length only, it is a case of lougitudinal strain involving only the Young's modulus of elasticity.

Consider a thin long rod of uniform area of cross-section a. Let one end of the rod be struck with a hammer with a force F for a small interval of time dt.

$$\therefore \qquad \text{Impulse} = Fdt \qquad \ldots (1)$$

This impulse produces a compressional wave. The distance travelled by the wave in time $dt = Udt$.

Here U is the velocity of the wave . Let dx be the compression produced in a length $U\,dt$. The particle velocity of the element $= dx/dt$.

Mass of the element $= a \times U\,dt \times \rho$

The gain in momentum due to the impulse

$$= \text{mass} \times \text{velocity}$$

$$= a \times U\,dt \times \rho \frac{dx}{dt}$$

$$= aU\rho dx \qquad \ldots (2)$$

But, Impulse = Change in momentum

Therefore from equations (1) and (2)

$$Fdt = aU\rho dx \qquad \ldots (3)$$

If Y be the Young's modulus of the material,

$$Y = \frac{\text{Stress}}{\text{Strain}}$$

$$Y = \frac{F/a}{\dfrac{adx}{aUdt}}$$

or

$$Fdt = Ya\,\frac{dx}{U} \qquad \ldots (4)$$

Comparing (3) and (4)

$$aU\rho dx = Ya\,\frac{dx}{U}$$

$$U^2 = \frac{Y}{\rho}$$

$$U = \sqrt{\frac{Y}{\rho}}$$

[**Note.** In the case of extended solids, the contraction or expansion in length is not possible. In such a case, the *elongational elasticity*,

$$E = K + \frac{4}{3}\eta$$

is to be used

$$\therefore \qquad U = \sqrt{\frac{K + \frac{4}{3}\eta}{\rho}}$$

Here K is the bulk modulus of elasticity and η is the modulus of rigidity].

Example 5.6. *Calculate the velocity of propagation of longitudinal sound waves through a steel rod. Y for steel* = 2×10^{12} dynes/cm^2.

ρ *for steel* = 7·6 g/cm^3

Here $Y = 2 \times 10^{12}$ dynes/cm^2

ρ = 7·6 g/cm^3

$$\therefore \qquad U = \sqrt{\frac{Y}{\rho}}$$

$$\therefore \qquad U = \sqrt{\frac{2 \times 10^{12}}{7{\cdot}6}}$$

$$U = 51{\cdot}3 \times 10^4 \text{ cm/s}$$

$$= \mathbf{5130 \text{ m/s.}}$$

5.13. Wave Velocity and Molecular Velocity

The velocity of propagation of sound waves through a gas is,

$$U = \sqrt{\frac{\gamma P}{\rho}} \qquad \ldots (1)$$

According to this equation the velocity of sound will depend on the pressure of the gas. Also according to kinetic theory of gases, the pressure of the gas is,

$$P = \frac{1}{3}\rho c^2$$

Here c is the root means square velocity of the molecules at a particular temperature.

$$\therefore \qquad \frac{P}{\rho} = \frac{c^2}{3} \qquad \ldots (2)$$

Substituting the value of P/ρ in equation (1)

$$U = \sqrt{\frac{\gamma c^2}{3}}$$

or $$U = c\sqrt{\frac{\gamma}{3}} \qquad \ldots (3)$$

Also $$c = U\sqrt{\frac{3}{\gamma}} \qquad \ldots (4)$$

From equation (3), if the value of velocity of sound and the ratio of the two specific heats of the gas are known, the root mean square velocity of the molecules can be calculated.

5.14. Velocity of Sound and Frequency

The velocity of sound propagation through a medium is practically constant. It is independent of the change in frequency of sound. It has been found that the velocity of sound in a medium remains constant upto a frequency of about hundred million hertz. The velocity of sound in different media is given in Table 5.1.

Table 5.1

SOLIDS AT 20°C

Substances	*Velocity in* m/s
Lead	1230
Copper	3750
Aluminium	5100
Iron	5130
Granite	6000

LIQUIDS AT 25°C

Liquid	*Velocity in* m/s
Kerosene oil	1315
Mercury	1450
Water	1493
Sea water	1533

GASES AT 0°C

Gas	*Velocity in* m/s
Oxygen	317
Air	332
Nitrogen	339
Steam(100°C)	405
Hydrogen	1270

EXERCISES

1. Obtain an expression for the velocity of a plane progressive wave through a gaseous medium.
2. State and explain Newton's formula for the velocity of sound through air. Discuss Laplace's correction.
3. Discuss the effect of temperature, pressure and humidity on the velocity of sound through air.
4. Describe briefly an experiment to determine the velocity of sound in (*i*) water and (*ii*) air.
5. Show that the velocity of sound in isotropic solids is given by
$$U = \sqrt{\frac{Y}{\rho}}.$$
6. Show that the velocity of sound in the case of extended solids is given by
$$U = \sqrt{\frac{3K + 4\eta}{3\rho}}.$$
7. Distinguish between wave velocity and molecular velocity and show that
$$U = c\sqrt{\frac{\gamma}{3}}.$$
8. Derive from elementary principles an expression for the velocity of sound in air (Newton's formula). Explain Laplace correction.
What is the effect of temperature on the velocity of sound in air. *(Delhi, 1971, 1973)*
9. Derive the formula for the velocity of a plane progressive wave in a fluid. *(Delhi, 1971)*
10. Obtain an expression for the velocity of sound in a gas. Show how it is effected by temperature and pressure. *(Delhi, 1971)*
11. Derive the formula for the velocity of sound waves in a gas. What is Laplace's correction ? Discuss the effects of temperature and pressure on the velocity of sound propagation. *(Delhi, 1972)*
12. Derive an expression for the velocity of longitudinal waves in a metallic rod and hence derive an expression for the fundamental frequency emitted by a rod of finite length. *(Delhi, 1973)*
13. Obtain an expression for the velocity of longitudinal waves in a gas. Discuss Laplace's correction. *(Delhi, 1974)*
14. Show that for every 1°C change in temperature the velocity of sound in air changes by about 0·61 m/s. *(Delhi, 1974)*
15. Derive Newton's equation for the velocity of sound waves in a material medium. Discuss its application to solids and explain what corrections need be made in the case of gases. Explain the effect of various factors that influence the velocity of sound in free atmosphere. *[Delhi (Suppl.), 1976]*

16. Obtain an expression for the velocity of sound in a gas discussing in detail Newton's formula and Laplace's correction. What is the effect of temperature variation on the velocity of sound in a gas. *(Delhi, 1976)*

17. Obtain an expression for the velocity of sound in air. How does the velocity depend on humidity, temperature and pressure? What are the other factors which influence the velocity of sound? *[Delhi (Suppl.), 1976]*

18. At what temperature is the velocity of sound in nitrogen gas is equal to its velocity in oxygen at 20°C. The atomic weights of oxygen and nitrogen are in the ratio 16 :14. *(Delhi, 1971)*
[**Hint.** $T_2 \rho_1 = T_1 \rho_2$] [**Ans.** 16.7°]

19. Derive an expression for the excess pressure at a point in compression waves in a fluid and hence obtain the velocity of propagation of waves. *(Bhagalpur, 1990)*

20. Find an expression for the velocity of longitudinal waves through a homogeneous, elastic medium. *(Guahati, 1992)*

CHAPTER 6

Stationary Waves, Interference and Beats

6.1. Stationary Waves

When two simple harmonic waves of the same amplitude, frequency and time period travel in *opposite directions* in a straight line, the resultant wave obtained is called a stationary or a standing wave. The formation of stationary waves is due to the superposition of the two waves on the particles of the medium.

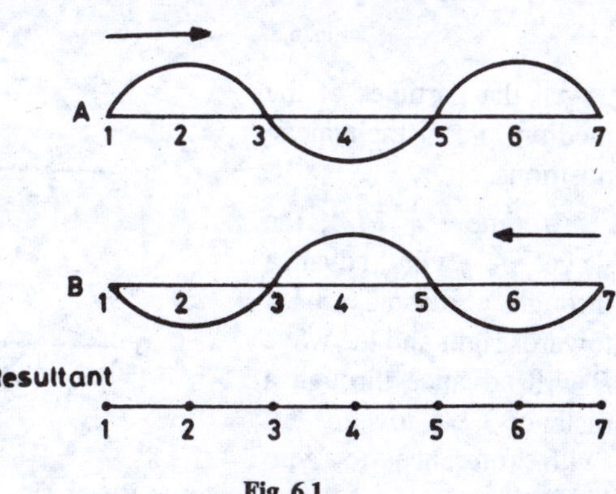

Fig. 6.1.

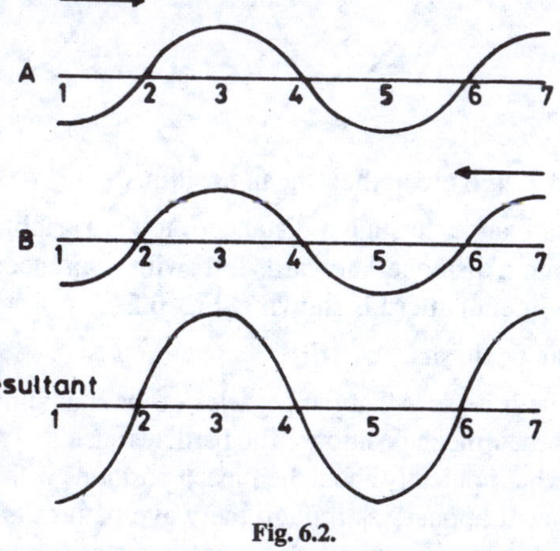

Fig. 6.2.

The formation of stationary waves can be represented graphically as follows :

Consider two wave trains *A* and *B* of the same amplitude, frequency and wavelength travelling in opposite directions. At an instant of time $t = 0$, the waves are as shown in Fig. 6.1. The resultant displacement curve is a straight line. All the particles of the medium are at their mean positions.

At time $t = T/4$, the wave A will advance through a distance $\lambda/4$ towards right, and the wave B will advance through a distance $\lambda/4$ towards left. The resultant displacement pattern is shown in Fig. 6.2.

The particles 1, 3, 5, and 7 are at their extreme positions and particles 2, 4 and 6 are at their mean positions.

At time $t=T/2$, the wave A will advance through a distance $\lambda/2$ towards right and the wave B will advance through a distance $\lambda/2$ towards left (with reference to zero time).

The resultant displacement pattern is shown in Fig. 6.3.

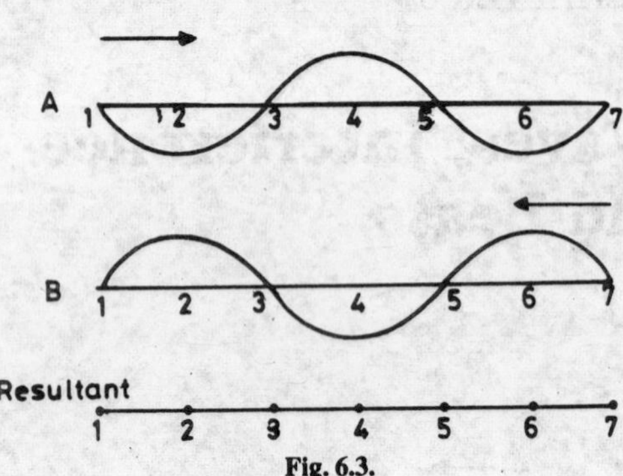

Fig. 6.3.

All the particles of the medium are at their mean positions.

At time $t = 3T/4$, the wave A will advance through a distance $3\lambda/4$ towards right and the wave B will advance through a distance $3\lambda/4$ towards left (with reference to zero time).

The resultant displacement pattern is shown in Fig. 6.4.

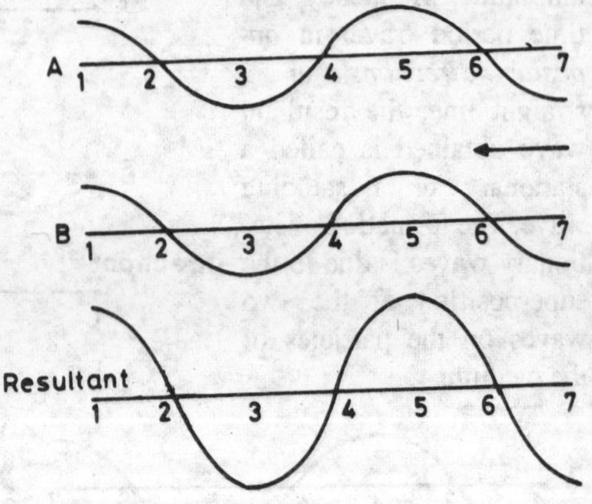

Fig. 6.4.

The particles 1, 3, 5 and 7 are at their extreme positions and 2, 4, 6 are at their mean positions.

At time $t = T$, the wave A will advance through a distance λ towards right and the wave B will advance through a distance λ towards left (with reference to zero time). The resultant displacement pattern is shown in Fig. 6.5.

All the particles are at their mean positions.

From the patterns discussed above it is clear that the particles of the medium such as 2, 4, 6 etc. always remain at their mean positions. The particles such as 1, 3, 5, 7 etc. continue to vibrate simple harmonically about their mean positions with double the amplitude of each wave. It appears as though the wave pattern is stationary in space. The resultant displacement patterns at intervals of time,

$0, \frac{T}{4}, \frac{T}{2}, \frac{3T}{4}, T$

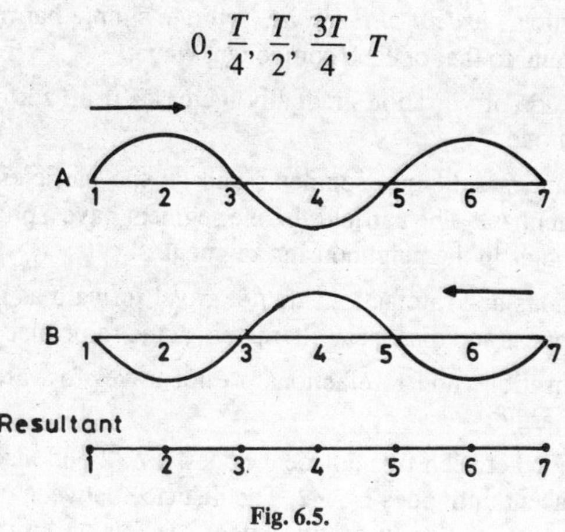

Fig. 6.5.

are shown in Fig. 6.6.

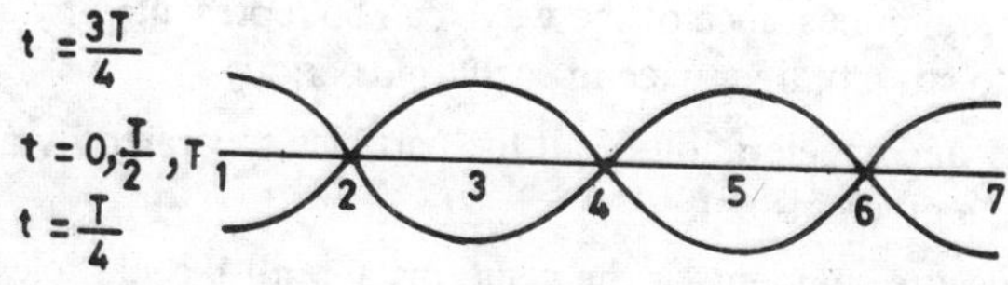

Fig. 6.6.

The positions of the particles 2, 4, 6 etc. which always remain at their mean positions are called *nodes*. Node is a position of zero displacement and maximum strain.

The positions of the particles 1, 3, 5, 7 etc. which vibrate simple harmonically with maximum amplitude (twice the amplitude of each wave) are called antinodes. At the antinodes, the strain is minimum. The distance between any two consecutive nodes or antinodes is equal to $\lambda/2$. Between a node and an antinode, the amplitude gradually increases from zero to maximum.

6.2. Properties of Stationary Longitudinal Waves

The stationary waves are formed due to the superposition of two simple harmonic longitudinal progressive waves of the same amplitude and periodic time and travelling in opposite directions. The important properties of these waves are :

(1) In these waves, nodes and antinodes are formed alternately. Nodes are the positions where the particles are at their mean positions having maximum strain. Antinodes are the positions where the particles vibrate with maximum amplitude having minimum strain.

(2) All the particles *except at the nodes,* vibrate simple harmonically with the time period equal to that of each component wave.

(3) The amplitude of vibration gradually increases from zero to maximum from node to antinode.

(4) The medium is split into segments and all the particles of the same segment vibrate in phase. The particles in one segment have a phase difference of π with the particles in the neighbouring segment.

(5) Condensations and rarefactions do not travel forward as in progressive waves but they appear and disappear alternately at the same place.

(6) As condensations and rarefactions do not travel forward, there is no transference of energy.

(7) The distance between two adjacent nodes is $\lambda/2$ and also the distance between two adjacent antinodes is $\lambda/2$. The distance between a node and the adjacent antinode is $\lambda/4$. Between two nodes there is an antinode and *vice versa.*

(8) The general appearance of the wave can be represented by a sine curve but it reduces to straight line twice in each time period.

(9) Velocity and acceleration of all the particles separated by a distance λ are the same at a given instant.

(10) In the same segment, at the same instant, all the particles will be in phase and their velocities and accelerations will be maximum or minimum at the same instant.

6.3. Tuning Fork

XY is a rod which vibrates transversely with nodes at *N* and *N*. (Fig. 6.7). When this rod is bent gradually as shown in Fig. 6.7 *(ii), (iii), (iv)* the nodes come closer. As the curvature is gradually increased, the nodes shift towards each other more and more. When the rod becomes U-shaped, the positions of nodes and antinodes are as shown in Fig. 6.7 (*iv*). When this U-shaped rod is attached to a metal stem, they come still closer and this arrangement is called a tuning fork. The tuning fork vibrates in three portions as shown in Fig. 6.7 (*v*). The free ends of the tuning fork behave as antinodes. The position where the stem is attached, also behaves as an antinode.

A tuning fork is of great use as a source of standard frequency. It is made of an alloy of nickel and steel. A tuning fork maintains its pitch for a number of years and it can be set into vibration when one of the prongs is struck against a hard rubber pad. Tuning forks of frequencies 256, 288, 320, 341.33, 384, 426.66, 480 and 512 are commonly manufactured.

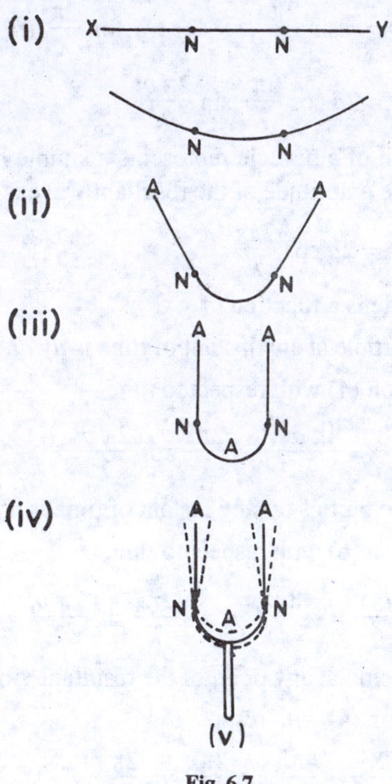

Fig. 6.7

6.4. Analytical Treatment

Stationary waves are formed in an open end organ pipe or a closed end organ pipe. Stationary waves are also formed with a stretched string fixed at one end and free at the other end or fixed at the other end.

(a) Open end organ pipe or string free at the other end

Consider a simple harmonic wave given by the equation,

$$y_1 = a \sin \frac{2\pi}{\lambda}(vt - x) \qquad \textbf{Incident wave} \ldots (1)$$

The displacement of the same particle at the same instant due to the reflected wave is given by

$$y_2 = a \sin \frac{2\pi}{\lambda}(vt + x) \qquad \textbf{Reflected Wave} \ldots. (2)$$

The resultant displacement of the wave,

$$y = y_1 + y_2$$

or $$y = a \sin \frac{2\pi}{\lambda}(vt - x) + a \sin \frac{2\pi}{\lambda}(vt + x) \quad \ldots (3)$$

$$y = 2a \cos \frac{2\pi x}{\lambda} \sin \frac{2\pi vt}{\lambda} \quad \ldots (4)$$

The resultant vibration of a particle represents a simple harmonic wave of the same time period. The amplitude of the resultant wave is given by

$$A = 2a \cos \frac{2\pi x}{\lambda} \quad \ldots (5)$$

The resultant amplitude is a function of *x*.

The velocity of the particle at any instant of time is dy/dt.

Differentiating equation (4) with respect to time

$$\frac{dy}{dt} = \frac{4\pi av}{\lambda} \cos \frac{2\pi x}{\lambda} \cos \frac{2\pi vt}{\lambda} \quad \ldots (6)$$

The acceleration of the particle at any instant of time is d^2y/dt^2.

Differentiating equation (6) with respect to time

$$\frac{d^2y}{dt^2} = -\frac{8\pi^2 av^2}{\lambda^2} \cos \frac{2\pi x}{\lambda} \sin \frac{2\pi vt}{\lambda} \quad \ldots (7)$$

The strain or compression at any point of the resultant vibration is dy/dx.

Differentiating equation (4) with respect to *x*

$$\frac{dy}{dx} = -\frac{4\pi a}{\lambda} \sin \frac{2\pi x}{\lambda} \sin \frac{2\pi vt}{\lambda} \quad \ldots (8)$$

These equations show that the amplitude, velocity, acceleration and strain or compression vary with position and time.

Changes with respect to position

(*i*) Consider the positions, where

$$\sin \frac{2\pi x}{\lambda} = 0 \text{ and } \cos \frac{2\pi x}{\lambda} = \pm 1$$

From equations (4) to (8)

Displacement $$y = \pm 2a \sin \frac{2\pi vt}{\lambda} \quad \ldots (9)$$

Amplitude $$A = \pm 2a \quad \ldots (10)$$

Velocity $$\frac{dy}{dt} = \pm \frac{4\pi av}{\lambda} \cos \frac{2\pi vt}{\lambda} \quad \ldots (11)$$

Acceleration $$\frac{d^2y}{dt^2} = \mp \frac{8\pi^2 av^2}{\lambda^2} \sin \frac{2\pi vt}{\lambda} \quad \ldots (12)$$

Strain $\frac{dy}{dx} = 0$ ... (13)

As the strain is zero, these positions correspond to **antinodes.**

$\because \quad \sin \frac{2\pi x}{\lambda} = 0.$

But $\sin m\pi = 0$

Where $m = 0, 1, 2, 3 \ldots$ etc.

$\therefore \quad \frac{2\pi x}{\lambda} = m\pi,$

or $x = \frac{m\lambda}{2}$

or $x = 0, \frac{\lambda}{2}, \lambda, \frac{3\lambda}{2} \ldots$ etc.

Thus the antinodes are equidistant and separated by $\lambda/2$. At $x = 0$, the position of interface is an antinode. Therefore, the position of the open end of the open end organ pipe or free end of a stretched string corresponds to an **antinode**.

(*ii*) Consider the positions, where

$$\sin \frac{2\pi x}{\lambda} = \pm 1, \text{ and } \cos \frac{2\pi x}{\lambda} = 0$$

From equations (4) to (8)

Displacement $y = 0$ ... (14)

Amplitude $A = 0$ ... (15)

Velocity $\frac{dy}{dt} = 0$ ... (16)

Acceleration, $\frac{d^2 y}{dt^2} = 0$ ... (17)

Strain $\frac{dy}{dx} = \mp \frac{4\pi a}{\lambda} \sin \frac{2\pi vt}{\lambda}$ ... (18)

These positions correspond to nodes.

$\because \quad \cos \frac{2\pi x}{\lambda} = 0,$

But $\cos (2m + 1) \frac{\pi}{2} = 0$

Where $m = 0, 1, 2, 3 \ldots$ etc.

$$\therefore \qquad \frac{2\pi x}{\lambda} = (2m+1)\frac{\pi}{2}$$

or
$$x = \frac{(2m+1)\lambda}{4}$$

$$\therefore \qquad x = \frac{\lambda}{4}, \frac{3\lambda}{4}, \frac{5\lambda}{4}, \ldots \text{ etc.}$$

Thus the antinodes are also equidistant and separated by a distance of $\lambda/2$.

It is clear that between two consecutive antinodes there is a node and *vice versa*. Distance between a node and an antinode is $\lambda/4$.

Changes with respect to time

(*i*) Consider the instant of time, when

$$\sin\frac{2\pi vt}{\lambda} = 0 \text{ and } \cos\frac{2\pi vt}{\lambda} = \pm 1.$$

From equations (4) to (8)

Displacement $\qquad y = 0$

Amplitude $\qquad A = 2a\cos\frac{2\pi x}{\lambda} \qquad$ (independent of time)

Velocity $\qquad \frac{dy}{dt} = \pm\frac{4\pi av}{\lambda}\cos\frac{2\pi x}{\lambda}$

Acceleration $\qquad \frac{d^2y}{dt^2} = 0$

Strain $\qquad \frac{dy}{dx} = 0$

$$\because \qquad \sin\frac{2\pi vt}{\lambda} = 0,$$

$$\sin m\pi = 0,$$

where $\qquad m = 0, 1, 2, 3, \ldots$ etc.

$$\therefore \qquad \frac{2\pi vt}{\lambda} = m\pi$$

or
$$t = \frac{m\lambda}{2v}$$

But $\qquad \frac{v}{\lambda} = n \text{ (frequency) } = \frac{1}{T}$

$$\therefore \qquad t = \frac{mT}{2}$$

or
$$t = 0, \frac{T}{2}, T, \frac{3T}{2} \ldots \text{ etc.}$$

It follows that twice in each time period the particles pass through their mean positions.

(*ii*) Consider the instant of time,

When $\sin \frac{2\pi vt}{\lambda} = \pm 1$ and $\cos \frac{2\pi vt}{\lambda} = 0.$

From equations (4) to (8)

Displacement $y = \pm\ 2a \cos \frac{2\pi x}{\lambda}$

Amplitude $A = 2a \cos \frac{2\pi x}{\lambda}$ (independent of time)

Velocity $\frac{dy}{dt} = 0$

Acceleration $\frac{d^2 y}{dt^2} = \mp \frac{8\pi^2 a v^2}{\lambda^2} \cos \frac{2\pi x}{\lambda}$

The strain or compression,

$$\frac{dy}{dx} = \mp \frac{4\pi a}{\lambda} \sin \frac{2\pi x}{\lambda}$$

$$\because \quad \sin \frac{2\pi vt}{\lambda} = \pm 1,$$

and
$$\cos \frac{2\pi vt}{\lambda} = 0$$

But
$$\sin (2m+1)\frac{\pi}{2} = \pm 1$$

Where
$$m = 0, 1, 2 \ldots \text{etc.}$$

$$\therefore \quad \frac{2\pi vt}{\lambda} = (2m+1)\frac{\pi}{2}$$

or
$$t = \frac{(2m+1)\lambda}{4v}$$

But
$$\frac{v}{\lambda} = n = \frac{1}{T}$$

$$\therefore \quad t = \frac{(2m+1)T}{4} = \frac{mT}{2} + \frac{T}{4}$$

or
$$t = \frac{T}{4}, \frac{3T}{4}, \frac{5T}{4} \ldots \text{etc.}$$

At these instants, the displacement, acceleration and strain are maximum at all positions and the velocity of the particles is zero. At these instants, each particle is at its extreme position and the pattern is stationary at that instant. This instant is called the stationary instant.

(*b*) Closed end organ pipe or string fixed at the other end

Consider a simple harmonic wave given by the equation

$$y_1 = a \sin \frac{2\pi}{\lambda}(vt - x) \qquad \textbf{Incident wave} \quad \ldots (1)$$

The displacement of the same particle at the same instant due to the reflected wave is given by

$$y_2 = -a \sin \frac{2\pi}{\lambda}(vt + x) \qquad \textbf{Reflected wave} \quad \ldots (2)$$

The resultant displacement

$$y = y_1 + y_2$$

$$y = a \sin \frac{2\pi}{\lambda}(vt - x) - a \sin \frac{2\pi}{\lambda}(vt + x) \quad \ldots (3)$$

$$y = -2a \sin \frac{2\pi x}{\lambda} \cos \frac{2\pi\, vt}{\lambda} \quad \ldots (4)$$

The resultant vibration of the particle represents a simple harmonic wave of the same time period. The amplitude of the resultant wave is given by

$$A = 2a \sin \frac{2\pi x}{\lambda} \quad \ldots (5)$$

The resultant amplitude is a function of x.

The velocity of the particle at any instant of time is dy/dt

Differentiating equation (4) with respect to time,

$$\frac{dy}{dt} = \frac{4\pi\, av}{\lambda} \sin \frac{2\pi x}{\lambda} \cos \frac{2\pi\, vt}{\lambda} \quad \ldots (6)$$

The acceleration of the particle at any instant of time is d^2y/dt^2.

Differentiating equation (6) with respect to time,

$$\frac{d^2 y}{dt^2} = \frac{8\pi^2 av^2}{\lambda^2} \sin \frac{2\pi x}{\lambda} \cos \frac{2\pi\, vt}{\lambda} \quad \ldots (7)$$

The strain or compression at any point of the resultant vibrations is dy/dx.

Differentiating equation (4) with respect to x

$$\frac{dy}{dt} = -\frac{4\pi a}{\lambda} \cos \frac{2\pi x}{\lambda} \cos \frac{2\pi\, vt}{\lambda} \quad \ldots (8)$$

These equations show that the amplitude, velocity, acceleration and strain or compression vary with position and time.

Changes with respect to position

(1) Consider the positions, where

$$\sin\frac{2\pi x}{\lambda}=0 \quad \text{and} \quad \cos\frac{2\pi x}{\lambda}=\pm 1.$$

From equations (4) to (8),

Displacement $\quad y=0 \quad$ ... (9)

Amplitude $\quad A=0 \quad$ ... (10)

Velocity $\quad \frac{dy}{dt}=0 \quad$ ... (11)

Acceleration $\quad \frac{d^2 y}{dt^2}=0 \quad$ ... (12)

Strain $\quad \frac{dy}{dx}=\mp\frac{4\pi a}{\lambda}\cos\frac{2\pi vt}{\lambda} \quad$ ... (13)

From equations (9) to (13), it is clear that at these positions, displacement, amplitude, velocity and acceleration are zero but the strain is maximum. These positions correspond to **nodes.**

$\because \quad \sin\frac{2\pi x}{\lambda}=0$

But $\quad \sin m\pi=0$

where $\quad m=0, 1, 2 \ldots$ etc.

$\therefore \quad \frac{2\pi x}{\lambda}=m\pi$

or $\quad x=\frac{m\lambda}{2}$

or $\quad x=0, \frac{\lambda}{2}, \lambda, \frac{3\lambda}{2}, \ldots$ etc.

Thus, the nodes are equidistant and separated by $\lambda/2$. At $x=0$, the position of interface is a node. Therefore, the position of the closed end organ pipe or fixed end of a stretched string corresponds to a **node.**

(ii) Consider the positions, where

$$\sin\frac{2\pi x}{\lambda}=\pm 1 \quad \text{and} \quad \cos\frac{2\pi x}{\lambda}=0$$

From equations (4) to (8),

Displacement $\quad y=\mp 2a\cos\frac{2\pi vt}{\lambda} \quad$ ... (14)

Amplitude $\quad A=\pm 2a \quad$ ... (15)

Velocity $\quad \frac{dy}{dt}=\pm\frac{4\pi av}{\lambda}\sin\frac{2\pi vt}{\lambda} \quad$ ... (16)

Acceleration $\frac{d^2 y}{dt^2} = \pm \frac{8\pi^2 a v^2}{\lambda^2} \cos \frac{2\pi\, vt}{\lambda}$ ... (17)

Strain $\frac{dy}{dx} = 0$ ... (18)

As the strain in zero, these positions correspond to antinodes.

$$\therefore \quad \sin \frac{2\pi x}{\lambda} = \pm 1$$

Also $\sin \frac{(2m+1)\pi}{2} = \pm 1$

where $m = 0, 1, 2, \ldots$ etc.

$$\therefore \quad \frac{2\pi x}{\lambda} = \frac{(2m+1)\pi}{2}$$

or $x = \frac{(2m+1)\lambda}{4}$

or $x = \frac{\lambda}{4}, \frac{3\lambda}{4}, \frac{5\lambda}{4} \ldots$ etc.

Thus, the antinodes are equidistant and separated by $\lambda/2$. The first antinode is at a distance of $\lambda/4$ from the closed end to the closed end pipe or fixed end of a stretched string.

Changes with respect to time

(*i*) At the instant of time, when

$$\sin \frac{2\pi\, vt}{\lambda} = 0 \quad \text{and} \quad \cos \frac{2\pi\, vt}{\lambda} = \pm 1$$

From equations (6) to (8),

Displacement $y = \mp 2a \sin \frac{2\pi x}{\lambda}$

Amplitude $A = 2a \sin \frac{2\pi x}{\lambda}$ (independent of time)

Velocity $\frac{dy}{dt} = 0$

Acceleration $\frac{d^2 y}{dt^2} = \pm \frac{8\pi^2 a v^2}{\lambda^2} \sin \frac{2\pi x}{\lambda}$

Strain $\frac{dy}{dx} = \pm \frac{4\pi a}{\lambda} \cos \frac{2\pi x}{\lambda}$

These equations show that at these instants, the displacement, acceleration and strain are maximum at all positions and the velocity of the particles is zero.

Here $\cos \frac{2\pi\, vt}{\lambda} = \pm 1$

Also $\cos m\pi = \pm 1$

$\therefore$ $$\frac{2\pi vt}{\lambda} = m\pi$$

But $$\frac{v}{\lambda} = \frac{1}{T}$$

or $$t = \frac{mT}{2}$$

$$t = 0, \frac{T}{2}, T, \frac{3T}{2} \ldots \text{etc.}$$

At these instants, each particle is at its extreme position and the pattern is stationary at that instant. This instant is called the stationary instant.

(*ii*) At the instant of time, when

$$\sin\frac{2\pi vt}{\lambda} = \pm 1 \text{ and } \cos\frac{2\pi vt}{\lambda} = 0$$

From equations (4) to (8),

Displacement $y = 0$

Amplitude $A = 2a\sin\frac{2\pi x}{\lambda}$ (independent of time)

Velocity $\frac{dy}{dt} = 0$

Acceleration $\frac{d^2y}{dt^2} = 0$

Strain $\frac{dy}{dx} = 0$

These equations show that displacement, velocity, acceleration and strain are zero.

$\because$ Here $\sin\frac{2\pi vt}{\lambda} = \pm 1$

Also $\sin\frac{(2m+1)\pi}{2} = \pm 1$

$\therefore$ $$\frac{2\pi vt}{\lambda} = \frac{(2m+1)\pi}{2}$$

But $$\frac{v}{\lambda} = \frac{1}{T}$$

$\therefore$ $$t = \frac{(2m+1)T}{4}$$

or $$t = \frac{T}{4}, \frac{3T}{4}, \frac{5T}{2} \ldots \text{etc.}$$

At these instants of time, all the particles are at their mean positions. It follows that twice in each time period the particles pass through their mean positions.

6.5. Energy of a Stationary Wave

When a wave is propagated through a fluid,

$$E = {}^{-p} \Big/ \frac{dy}{dx}$$

where E is the bulk modulus of elasticity, p is the excess of pressure and $\frac{dy}{dx}$ is the volumetric strain.

$$\therefore \qquad p = -E\frac{dy}{dx} \qquad \ldots (1)$$

In the case of an open end organ pipe, strain is given by

$$\frac{dy}{dx} = -\frac{4\pi a}{\lambda}\sin\frac{2\pi x}{\lambda}\sin\frac{2\pi\, vt}{\lambda}$$

Also

$$v = \sqrt{\frac{E}{\rho}}$$

or

$$v^2 = \frac{E}{\rho}$$

or

$$E = v^2\rho$$

Substituting the values of $\frac{dy}{dx}$ and E in equation (1)

$$\therefore \qquad p = v^2\rho\,\frac{4\pi a}{\lambda}\sin\frac{2\pi x}{\lambda}\sin\frac{2\pi\, vt}{\lambda} \qquad \ldots (2)$$

When $\sin\frac{2\pi x}{\lambda} = 1$ and $\sin\frac{2\pi\, vt}{\lambda} = 1$,

the excess of pressure is maximum

$$p = p_0 = v^2\rho\cdot\frac{4\pi a}{\lambda} \qquad \ldots (3)$$

From equations (2) and (3),

$$\therefore \qquad p = p_0\sin\frac{2\pi x}{\lambda}\sin\frac{2\pi\, vt}{\lambda} \qquad \ldots (4)$$

Taking $p_0\sin\frac{2\pi x}{\lambda} = p_x$

$$p = p_x \sin \frac{2\pi\, vt}{\lambda} \qquad \ldots (5)$$

The velocity of the particle,

$$U = \frac{dy}{dt} = \frac{4\pi\, av}{\lambda} \cos \frac{2\pi x}{\lambda} \cos \frac{2\pi\, vt}{\lambda} \qquad \ldots (6)$$

Taking $\frac{4\pi\, av}{\lambda} \cos \frac{2\pi x}{\lambda} = U_x$

$$U = U_x \cos \frac{2\pi\, vt}{\lambda} \qquad \ldots (7)$$

Work done or the energy transfer per unit area in a small interval of time dt

$$= pU \,.\, dt$$

The total energy transfer in time T

$$= \int_0^T pU \,.\, dt$$

Rate of energy transfer

$$= \frac{\int_0^T pU\, dt}{T}$$

$$= \frac{1}{T} \int_0^T \left(p_x \sin \frac{2\pi\, vt}{\lambda} \right) \times \left(U_x \cos \frac{2\pi\, vt}{\lambda} \right) dt$$

$$= \frac{p_x U_x}{T} \int_0^T \sin \frac{2\pi\, vt}{\lambda} \cdot \cos \frac{2\pi\, vt}{\lambda} \cdot dt$$

$$= \frac{p_x U_x}{2T} \int_0^T \sin \frac{4\pi\, vt}{\lambda} \cdot dt$$

But $\int_0^T \sin \frac{4\pi vt}{\lambda} \,.\, dt = 0$

$\therefore$ Rate of energy transfer = 0

Thus in the case of stationary waves no **energy** is *transferred.*

Example 6.1. *Two transverse sin waves, each of amplitude 4* mm *wavelength 2* m *and time period 1*s *and in phase at* $x = 0$, $t = 0$ *are travelling along the x-axis in opposite directions. Obtain the equation of the resultant wave and comment on its nature. Calculate the maximum displacement at* $x = 2{\cdot}3$ m. *Also locate the nodes and antinodes.* [IAS, 1990]

$$y_1 = a \sin \frac{2\pi}{\lambda} (vt - x) \quad \text{incident wave}$$

$$y_2 = a \sin \frac{2\pi}{\lambda}(vt + x) \qquad \text{reflected wave}$$

$$y = y_1 + y_2$$

$$y = a \sin \frac{2\pi}{\lambda}(vt - x) + a \sin \frac{2\pi}{\lambda}(vt + x)$$

$$y = \left(2a \cos \frac{2\pi x}{\lambda}\right) \sin \left(\frac{2\pi\, vt}{\lambda}\right)$$

Maximum Displacement, $A = 2a \cos \left(\frac{2\pi x}{\lambda}\right)$

$$a = 4 \text{ mm} = 4 \times 10^{-3} \text{ m}$$

$$x = 2{\cdot}3 \text{ m}$$

$$\lambda = 2 \text{ m}$$

$\therefore$ $$A = 2 \times 4 \times 10^{-3} \cos \left(\frac{2\pi \times 2{\cdot}3}{2}\right)$$

$$= 8 \times 10^{-3} \cos (2{\cdot}3\,\pi)$$

$$= 8 \times 10^{-3} \cos (0{\cdot}3\,\pi)$$

$$= 8 \times 10^{-3}\ 0{\cdot}5878$$

$$= \mathbf{4{\cdot}7 \times 10^{-3}\ mm}$$

$$\frac{dy}{dt} = \frac{4\pi\, av}{\lambda} \cos \left(\frac{2\pi\, vx}{\lambda}\right) \cos \left(\frac{2\pi\, vt}{\lambda}\right)$$

$$\frac{d^2y}{dt^2} = -\frac{8\pi^2 av^2}{\lambda^2} \cos \left(\frac{2\pi x}{\lambda}\right) \sin \left(\frac{2\pi\, vt}{\lambda}\right)$$

$$\frac{dy}{dx} = -\frac{4\pi a}{\lambda} \sin \frac{2\pi x}{\lambda} \sin \frac{2\pi\, vt}{\lambda}$$

(*i*) When

$$\sin \frac{2\pi x}{\lambda} = 0 \qquad \cos \frac{2\pi x}{\lambda} = \pm 1$$

Displacement $\quad y = \pm\ 2a \sin \frac{2\pi\, vt}{\lambda}$

Amplitude $\quad A = \pm\, 2a$

Velocity $\quad \frac{dy}{dt} = \pm \frac{4\pi\, av}{\lambda} \cos \left(\frac{2\pi\, vt}{\lambda}\right)$

Acceleration $\frac{d^2y}{dt^2} = \pm \frac{8\pi^2 av^2}{\lambda^2} \sin\left(\frac{2\pi\, vt}{\lambda}\right)$

Strain $\frac{dy}{dx} = 0$

These positions correspond to antinodes.

$$\therefore \qquad \sin\frac{2\pi x}{\lambda} = 0$$

$\sin m\pi = 0$, Here $m = 0, 1, 2, 3, \ldots$

$$m\pi = \frac{2\pi x}{\lambda}$$

$$x = \frac{m\lambda}{2}$$

$$x = 0, \frac{\lambda}{2}, \lambda, \frac{3\lambda}{2} \ldots \text{etc.}$$

(*ii*) When $\sin\left(\frac{2\pi x}{\lambda}\right) = \pm 1$ and $\cos\frac{2\pi x}{\lambda} = 0$

$$\cos\frac{2\pi x}{\lambda} = 0, \quad \cos\ (2m+1)\frac{\pi}{2} = 0$$

$$\frac{2\pi x}{\lambda} = (2m+1)\frac{\pi}{2}$$

$$x = (2m+1)\frac{\lambda}{4}$$

$$x = \frac{\lambda}{4}, \frac{3\lambda}{4}, \frac{5\lambda}{4} \ldots \text{etc.}$$

These correspond to nodes and

$$y = 0,\ A = 0,\ \frac{dy}{dt} = 0,\ \frac{d^2y}{dt^2} = 0$$

$$\frac{dy}{dx} = \pm\frac{4\pi a}{\lambda} \sin\frac{2\pi\, v\lambda}{\lambda}.$$

Example 6.2. *The equation of a transverse wave on a stretched string is $y = 2 \sin \pi\,(8t - 0{\cdot}005x - 0{\cdot}40)$. Write down the equation of a wave that would produce a stationary wave in the string on superposition with the given wave.*

[Bhagalpur, 1990]

Here $\qquad y = 2 \sin \pi\,(8t - 0{\cdot}005\,x - 0{\cdot}40)$

or $\qquad y_1 = 2 \sin \pi\,[8t - (0{\cdot}005\,x + 0{\cdot}40)]$ $\qquad \ldots (i)$

This is the incident wave

The reflected wave is

$$y_2 = 2 \sin \pi\, [8t + (0{\cdot}005x + 0{\cdot}40)] \qquad \ldots (ii)$$

The resultant of y_1 and y_2 will be a stationary wave.

6.6. Interference of Sound Waves

If two or more sound waves travel through the same medium, superposition of the waves takes place. The particles of the medium vibrate according to the vector sum (resultant) of the displacements due to interfering waves at that instant. As a result of the superposition, the distribution of energy in the medium is different from the distribution of energy due to the individual waves influencing the medium separately. The resultant vibration of the particles will depend upon the amplitude, time period, phase difference and direction of the interfering waves.

Consider two simple harmonic waves of the same frequency. Let a and b be the amplitudes of the two waves and phase difference ϕ. It is assumed that the two waves are plane and that they do not spread laterally. The displacement of a particle at any instant due to each wave is given by

$$y_1 = a \sin \frac{2\pi}{\lambda}(vt - x) \qquad \ldots (1)$$

$$y_2 = b \sin \left[\frac{2\pi}{\lambda}(vt - x) + \phi\right] \qquad \ldots (2)$$

Here v is the velocity of propagation of each wave and λ is the wave length.

The resultant displacement of the particle due to the superposition of the two waves,

$$y = y_1 + y_2$$

$$y = a \sin \frac{2\pi}{\lambda}(vt - x) + b \sin \left[\frac{2\pi}{\lambda}(vt - x) + \phi\right]$$

$$y = a \sin \frac{2\pi}{\lambda}(vt - x) + b\left[\sin \frac{2\pi}{\lambda}(vt - x) \cos \phi + \cos \frac{2\pi}{\lambda}(vt - x) \sin \phi\right]$$

$$y = \left[\sin \frac{2\pi}{\lambda}(vt - x)\right][a + b \cos \phi\,] + \left[\cos \frac{2\pi}{\lambda}(vt - x)\right][b \sin \phi]$$

Let $a + b\cos\phi = A\cos\theta$

$b\sin\phi = A\sin\theta$

$\therefore \quad A = \sqrt{a^2 + b^2 + 2\,ab\cos\phi} \qquad \ldots(3)$

and $\tan\theta = \dfrac{b\sin\phi}{a + b\cos\phi} \qquad \ldots(4)$

$$\therefore \quad y = \left[\sin\frac{2\pi}{\lambda}(vt - x)\right][A\cos\theta] + \left[\cos\frac{2\pi}{\lambda}(vt - x)\right][A\sin\theta]$$

$$y = A\sin\left[\frac{2\pi}{\lambda}(vt - x) + \theta\right] \qquad \ldots(5)$$

The resultant vibration has the same time period, (*i.e.* frequency and wave length). The amplitude and phase difference are different.

(1) When $\phi = 0, 2\pi, n\,(2\pi) \ldots$ etc.

and $a = b,$

$A = \sqrt{a^2 + a^2 + 2a^2}$

or $A = 2a$

Intensity of sound $I = A^2 = 4a^2$

Therefore, the intensity of sound is maximum at those points, where the two waves differ in phase by $n\,(2\pi)$ or the path difference is $n\,\lambda$

Here $n = 0, 1, 2, 3. \ldots$ etc.

(2) When $\phi = \pi, 3\pi, (2n + 1)\,\pi \ldots$ etc.

and $a = b$

$A = 0$

Intensity of sound $I = A^2 = 0$

Therefore, the intensity of sound is minimum at those points, where the two waves differ in phase by $(2n+1)\,\pi$ or the path difference is $(2n+1)\,\lambda/2$.

Here $n = 0, 1, 2, 3 \ldots$ etc.

6.7. Special Cases

(1) *When two waves of the same frequency and amplitude travel in the same direction having phase difference zero.*

Here $a = b$

$\phi = 0$

From equation (3)

$A = 2a$

From equation (4)

$$\tan \theta = 0 \text{ or } \theta = 0$$

From equation (5)

$$y = 2a \sin \frac{2\pi}{\lambda} (vt - x) \qquad \ldots (6)$$

The resultant vibration has an amplitude $2a$ as shown in Fig. 6.8.

In this condensations and rarefactions are coincident.

(2) *When two waves of the same frequency and amplitude travel in the same direction having phase difference π.*

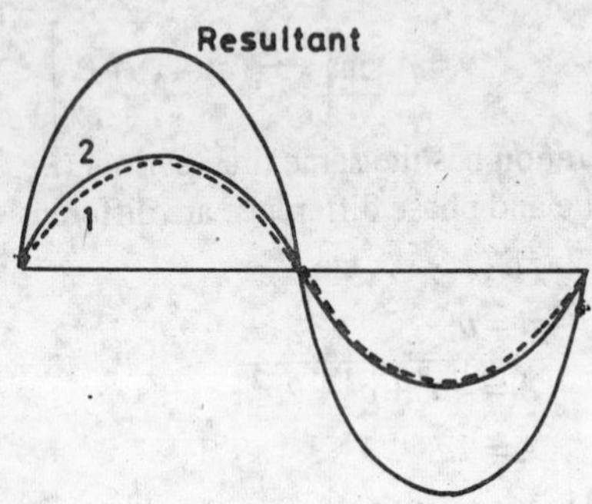

Fig. 6.8.

Here $$a = b$$

$$\phi = \pi$$

From equation (3)

$$A = 0$$

From equation (4)

$$\tan \theta = 0 \text{ or } \theta = 0$$

From equation (5)

$$y = 0$$

The resultant vibration has zero amplitude and medium remains undisturbed as shown in Fig. 6.9.

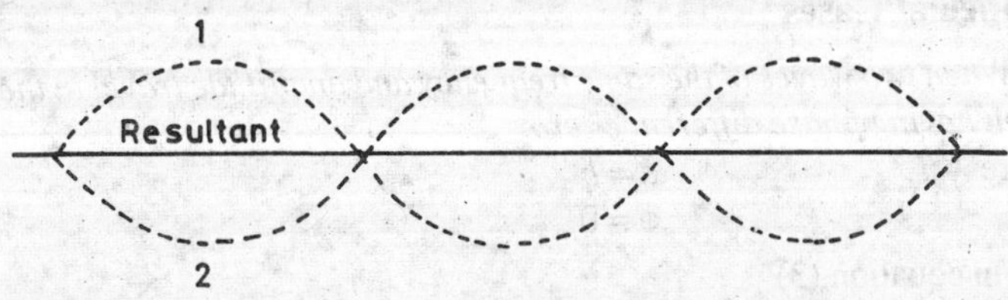

Fig. 6.9.

(3) When two waves of the same frequency and amplitude travel in opposite directions, the interference phenomenon gives rise to standing or stationary waves as discussed earlier.

6.8. Conditions for Interference of Sound Waves

(1) The two sources of sound must be coherent. They must emit waves of the same frequency and amplitude.

(2) The phase difference between the two sources must remain constant such that the phase difference between the two waves at any point does not change with time.

(3) The two wave trains must travel in the same direction.

Condition (1) is necessary so that the positions of maximum and minimum intensity of sound are distinct. If the difference between a and b is large, the minimum intensity positions are not distinct.

Condition (2) is necessary to have sustained maxima and minima. If the phase difference between the two waves continuously changes, the maximum and minimum intensity positions are not fixed.

Condition (3) is necessary because with increase in obliquity between the waves, the resultant intensity decreases.

6.9. Energy Distribution due to Interference of Sound Waves

From equation (3) of article 6.6,

$$A = \sqrt{a^2 + b^2 + 2ab \cos \phi}$$

When $a = b$

$$A^2 = 2a^2(1 + \cos \phi)$$

$$A^2 = 2a^2\left(2 \cos^2 \frac{\phi}{2}\right)$$

$$A^2 = 4a^2 \cos^2 \frac{\phi}{2} \qquad \ldots (1)$$

From equation (1), the intensity at the maxima, is $4a^2$ and the intensity at the minima is zero. According to the law of conservation of energy, the energy

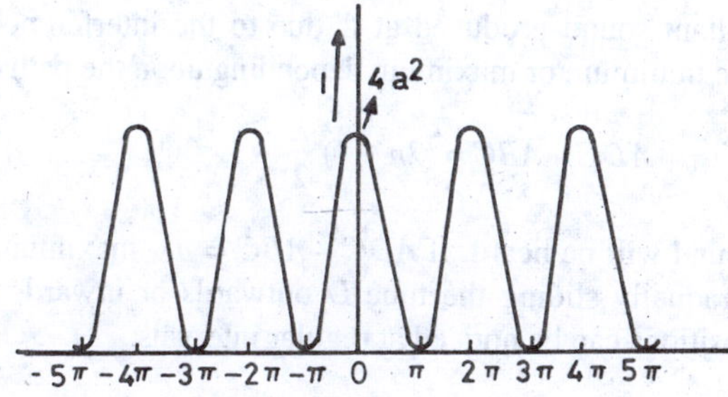

Fig. 6. 10.

cannot be destroyed. Here, also the energy is not destroyed but is only redistributed from the positions of minimum intensity to those of maximum intensity. For, at the maximum intensity positions the intensity due to the two waves should be $2a^2$ but it is actually $4a^2$. As shown in Fig. 6.10, the intensity varies from 0 to $4a^2$ and the average is still $2a^2$. It is equal to a uniform intensity of $2a^2$ which will be present in the absence of interference phenomenon between the two waves.

Hence, the formation of maxima and minima due to the interference of two sound waves is in accordance with the law of conservation of energy.

6.10. Quincke's Tube

The phenomenon of interference of sound waves can be demonstrated with a Quincke's tube. The tube D can slide inside the tube ABC as shown in Fig. 6.11. When a vibrating tuning fork is held near the opening at A, the sound waves travel through the tube along two paths ABC and ADC.

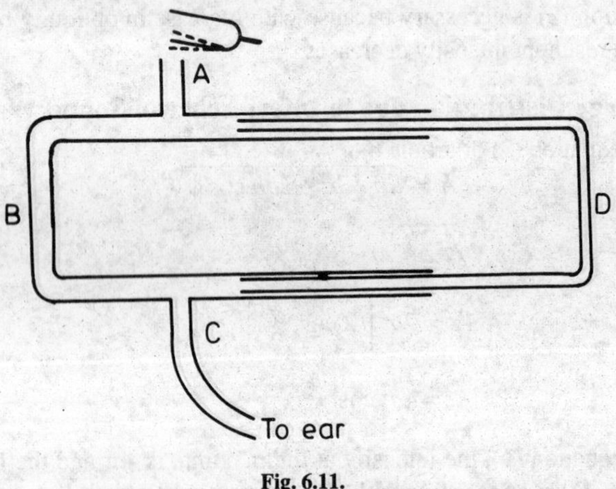

Fig. 6.11.

The resultant sound produced at C due to the interference of these two waves will be minimum or maximum depending upon the path difference.

If
$$ADC - ABC = (2n+1)\frac{\lambda}{2}$$

minimum sound will be heard. If $ADC - ABC = n\lambda$, maximum sound will be heard. By gradually sliding the tube D outwards or inwards maximum and minimum positions can be noticed at regular intervals.

6.11. Seebeck's Tube

The phenomenon of interference can also be demonstrated with the help of Seebeck's tube. It consists of a long glass tube open at both ends. A piston can slide in the tube. A small side tube is attached at C which is connected to a rubber tubing. The other end of the rubber tube is kept near the ear (Fig. 6.12).

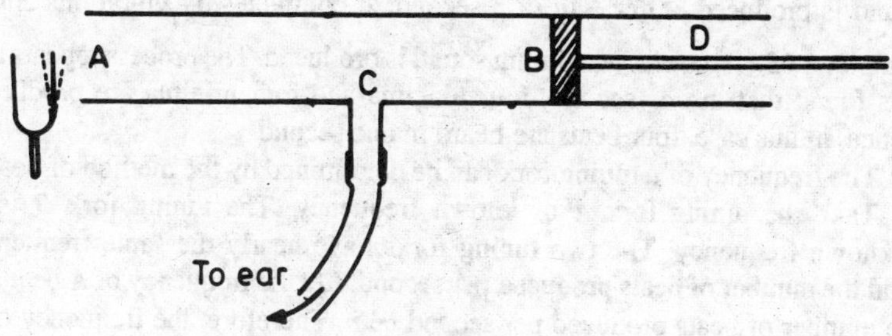

Fig. 6.12.

When a tuning fork is allowed to vibrate near the open end A, the ear perceives the resultant effect of two waves : the direct wave and the reflected wave from B. If the piston is gradually moved outwards, alternately maximum and minimum intensity of sound is heard. When the path difference between the direct and the reflected waves, is $(2n+1)\lambda/2$ due to interference of sound waves, the intensity of sound is minimum. When the path difference is $n\lambda$, maximum intensity of sound is heard.

It means that, when $BC = (2n+1)\lambda/4$, minimum intensity of sound is heard. When $BC = n\lambda/2$, maximum intensity of sound is heard.

This arrangement is suitable to determine the wavelength of sound waves in air or a gas medium. Knowing the frequency of the tuning fork, velocity of sound in a medium can also be calculated.

Suppose the frequency of the tuning fork is N and the distance between two successive maximum intensity positions is l.

Then, $$l = \frac{\lambda}{2}$$

or $$\lambda = 2l$$

$\therefore$ $$V = N\lambda$$

$$V = 2Nl.$$

6.12. Beats

When two sounding bodies of nearly the same frequency and amplitude are sounded together, the resultant sound consists of alternate maxima and minima. This phenomenon in which waxing and vaning of sound at regular intervals is heard is called *beats*. The number of beats heard per second is equal to the difference in frequency between the two sounding bodies. However, beats are heard only when the difference in frequency is not more than ten.

Suppose the frequency of one tuning fork $A = 256$ and a second tuning fork $B = 260$. When the two tuning forks are sounded together, after $\frac{1}{8}$ th of a second, A completes 32 vibrations and B completes 32·5 vibrations. Minimum sound is produced. After $\frac{1}{4}$ th of a second, A completes 64 vibrations and B completes 65 vibrations. Maximum sound is produced. The process repeats and it is found that in one second, four maxima and four minima are produced. Hence, in this case, four beats are heard in one second.

The frequency of a tuning fork can be determined by the method of beats.

Take the tuning fork A of known frequency. The tuning fork B is of unknown frequency. The two tuning forks have nearly the same frequency. Find the number of beats produced per second. Let the frequency of A be n and the number of beats produced per second be n. Therefore, the frequency of B will be $N \pm n$.

To find the correct value of B *viz.* $(N + n)$ or $(N - n)$, the tuning fork B is loaded with a little wax. When a tuning fork is loaded with wax, its frequency decreases. If the number of beats per second increases, the original frequency of B is $N - n$. If the number of beats per second decreases, the original frequency of B is $N + n$.

Note. (*i*) When a tuning fork is loaded, its frequency decreases.

(*ii*) When a tuning fork is filed, its frequency increases.

6.13. Analytical Treatment of Beats

Consider two wave trains of frequencies n_1 and n_2 where $(n_1 - n_2)$ is small. Let a and b be the amplitudes of the waves respectively. For the sake of simplicity, it is assumed that the two waves are in phase at any point in the medium at $t = 0$. The displacements y_1 and y_2 due to each wave are given by

$$y_1 = a \sin \omega_1 t$$

$$y_2 = b \sin \omega_2 t$$

Here $\omega_1 = 2\pi n_1$

and $\omega_2 = 2\pi n_2$

$\therefore$ $$y_1 = a \sin 2\pi n_1 t \qquad \dots (1)$$

and $$y_2 = b \sin 2\pi n_2' t \qquad \dots (2)$$

The resultant displacement is given by

$$y = y_1 + y_2$$

$$y = a \sin 2\pi n_1 t + b \sin 2\pi n_2 t$$

$$y = a \sin 2\pi n_1 t + b \sin 2\pi [n_1 - (n_1 - n_2)]t$$

$$y = a \sin 2\pi n_1 t + b [\sin 2\pi n_1 t \cos 2\pi (n_1 - n_2) t - \cos 2\pi n_1 t \sin 2\pi (n_1 - n_2) t]$$

$$y = a \sin 2\pi n_1 t + b \sin 2\pi n_1 t \cos 2\pi (n_1 - n_2) t - b \cos 2\pi n_1 t \sin 2\pi (n_1 - n_2) t$$

$$y = \sin 2\pi n_1 t [a + b \cos 2\pi (n_1 - n_2) t] - \cos (2\pi n_1 t) [b \sin 2\pi (n_1 - n_2) t]$$

Take $a + b \cos 2\pi (n_1 - n_2) t = A \cos \theta$

and $b \sin 2\pi (n_1 - n_2) t = A \sin \theta$

$\therefore$ $$y = A \sin (2\pi n_1 t - \theta) \quad \ldots (3)$$

Here $$\tan \theta = \frac{b \sin 2\pi (n_1 - n_2) t}{a + b \cos 2\pi (n_1 - n_2) t} \quad \ldots (4)$$

and $$A = \sqrt{a^2 + b^2 + 2ab \cos 2\pi (n_1 - n_2) t} \quad \ldots (5)$$

From equation (4), it is evident that the phase angle θ changes with respect to time. Similarly from equation (5), the amplitude of the resultant vibration also changes with time.

(1) Where $2\pi (n_1 - n_2) t = 2K\pi$

where $K = 0, 1, 2, 3 \ldots$ etc.,

the resultant amplitude,

$$A = \sqrt{a^2 + b^2 + 2ab}$$

$$A = (a + b) \quad \ldots (6)$$

The resultant amplitude is maximum, when

$$t = \frac{K}{(n_1 - n_2)}$$

i.e., at time instants, $0, \frac{1}{(n_1 - n_2)}, \frac{2}{(n_1 - n_2)}$ etc., the amplitude of the resultant is maximum. As the intensity of sound is directly proportional to the square of the amplitude, the maximum intensity of sound will be heard at the instants.

$$0, \frac{1}{(n_1 - n_2)}, \frac{2}{(n_1 - n_2)}, \frac{3}{(n_1 - n_2)} \ldots \text{etc.}$$

(2) When $2\pi (n_1 - n_2) t = (2K + 1) \pi$

where $K = 0, 1, 2, 3 \ldots$ etc.

the resultant amplitude

$$A = \sqrt{a^2 + b^2 - 2ab}$$

$$A = (a - b)$$

The resultant amplitude is minimum when,

$$t = \frac{(2K+1)}{2\,(n_1 - n_2)}$$

i.e., at time instants,

$$\frac{1}{2(n_1 - n_2)}, \frac{3}{2\,(n_1 - n_2)}, \frac{5}{2(n_1 - n_2)} \ldots \text{etc.},$$

the amplitude of the resultant is minimum.

Hence minimum intensity of sound will be heard at the instants,

$$\frac{1}{2\,(n_1 - n_2)}, \frac{3}{2\,(n_1 - n_2)}, \frac{5}{2\,(n_1 - n_2)} \ldots \text{etc.}$$

Thus the maxima and minima occur alternately after equal intervals of time $\frac{1}{2\,(n_1 - n_2)}$ · The time interval between two successive maxima or between two successive minima is $\frac{1}{(n_1 - n_2)}$ ·

The number of beats produced per second

$$= (n_1 - n_2).$$

If the amplitudes of the two waves trains are equal, the maximum amplitude of the resultant = $2a$ and the minimum amplitude = 0.

In this case, the positions of minima will be of zero intensity.

Example 6.3. *A tuning fork A produces 4 beats/second with a tuning fork B of frequency 256. A is filed and the beats occur at shorter intervals. What was its original frequency?*

Frequency of $B = 256$

Beats/s = 4

Frequency of $A = 256 + 4 = 260$ or $256 - 4 = 252$.

When A is filed its frequency increases. Suppose the original frequency of A is 260. When it is filed its frequency will be more than 260 and the beats/second increase *i.e.*, the beats occur at shorter intervals, which is given. Hence the original frequency of A = 260.

Example 6.4. *A note produces 4 beats/second with a tuning fork of frequency 512 and 6 beats/second with a tuning fork of frequency 514. Find the frequency of the note.*

In the first case

Frequency of the tuning fork = 512

Beats/second = 4

∴ Possible frequencies of the note are $512 + 4 = 516$

or $512 - 4 = \mathbf{508}$

In the second case

Frequency of the tuning fork = 514

Beats/second = 6

Possible frequencies of the note are $514 + 6 = 520$

or $514 - 6 = \mathbf{508}$.

Hence the frequency of the note is 508.

Example 6.5. *Two tuning forks A and B give 5 beats/second. The frequency of A = 512. When B is filed, 5 beats/second are again produced. Find the frequency of B before and after filing.*

Frequency of $A = 512$

Beats per second $= 5$

Frequency of B before filing is either,

$$512 + 5 = 517$$

or

$$512 - 5 = 507$$

Since, after filing 5 beats / second are again produced, frequency of B after filing is either

$$512 + 5 = 517$$

or

$$512 - 5 = 507$$

Let us consider the frequency of B before filing as 517. After filing the frequency increases. Therefore, after filing the frequency of B can not be equal to either 517 or 507. Therefore frequency of B cannot be equal to 517.

Consider the frequency of B before filing 507

After filing its frequency can be equal to 517

Therefore, this is the possible value.

Hence, frequency of B before filing **= 507**

Hence, frequency of B after filing **= 517**

Example 6.6. *Calculate the velocity of sound in a gas in which two waves of lengths* 1 metre *and* 1·01 metres *produce 10 beats in 3* seconds.

Number of beats in 3 s = 10

$$\therefore \quad \text{Beat/s} = \frac{10}{3}$$

Here $\lambda_1 = 100$ cm

$\lambda_2 = 1{\cdot}01$ = metres = 101 cm

Suppose the velocity $= V$

Frequency of the first wave $= n_1$

Frequency of the second wave $= n_2$

$$\therefore \qquad n_1 - n_2 = \text{beat/s} = \frac{10}{3}$$

$$n_1 - n_2 = \frac{10}{3}$$

But $$n_1 = \frac{V}{\lambda_1} = \frac{V}{100} \text{ and } n_2 = \frac{V}{\lambda_2} = \frac{V}{101}$$

Substituting the values of n_1 and n_2

$$\frac{V}{100} - \frac{V}{101} = \frac{10}{3} \text{ or } \frac{V}{10100} = \frac{10}{3}$$

$$V = \frac{101000}{3} = 33666{\cdot}67 \text{ cm/s}$$

$$= \mathbf{336{\cdot}67 \text{ m/s.}}$$

Example 6.7. *A tuning fork A of frequency 384* Hz *gives 6 beats per second when sounded with another tuning fork B. On loadi ng B with a little wax, the number of beats per second becomes 4. What is the frequency of B?* (Rajasthan, 1976)

Frequency of A = 384 Hz

Beats per second = 6

Frequency of B before loading is

either $384 + 6 = 390$ Hz

or $384 - 6 = 378$ Hz

After loading, beats per second = 4

Frequency of B after loading is

either $384 + 4 = 388$ Hz

or $384 - 4 = 380$ Hz

Since, after loading frequency decreases, the frequency of B before loading is **390 Hz.**

Example 6.8. *When two tuning forks were sounded together, 20 beats were produced in 8* seconds. *After loading one of the tuning forks with a little wax, they produce 32 beats in 8* seconds. *If the unloaded fork had a frequency of 512* Hz, *calculate the frequency of the other.* [Delhi (Supple.), 1976]

Frequency of $A = 512$ Hz

Frequency of $B = ?$

Before loading beats per second

$$= \frac{20}{8} = 2{\cdot}5$$

Frequency of B before loading is

either $$512 + 2{\cdot}5 = 514{\cdot}5$$

or $$512 - 2{\cdot}5 = 509{\cdot}5$$

After loading, beats per second

$$= \frac{32}{8} = 4$$

Frequency of B after loading is

either $$512 + 4 = 516$$

or $$512 - 4 = 508$$

Since, after loading the frequency of a tuning fork is lowered, the frequency of B, before loading is 509·5 Hz because after loading with a *little* wax, its frequency is slightly lowered to 508.

Hence frequency of B before loading

= 509·5 Hz.

6.14. Combination Tones

By mathematical analysis it can be shown that, when two strong notes of frequencies P and Q are sounded together, a series of tones of other frequencies are produced along with the original notes. These tones are called *combination tones.* Combination tones are different from the harmonics. The strongest of the combination tones is the **First Difference Tone** of frequency $(P - Q)$ or $(Q - P)$. The difference tones can be infinite in number, *viz.,* $P - 2Q, P - 3Q, P - 4Q$ . . . etc. It is also possible that two different tones of frequencies $(P - 2Q)$ and $(P - 4Q)$ can theoretically produce a difference tone of frequency $(P - 2Q) - (P - 4Q) = 2Q$.

In a similar manner one can expect summation tones of frequencies $(P + Q)$, $(P + 2Q)$, $(2P + Q)$ etc. In fact, an infinite number of combination tones, difference or summation type, are theoretically possible. Usually it is the first difference tones that is easily audible. The first difference tone is weaker than the two primaries. However, the second, third etc., difference tones and the summation tones are extremely weak in intensity.

6.15. Helmholtz Resonator

In various acoustical investigations, Helmholtz resonators are commonly used. Helmholtz particularly used them in the study of the quality of musical notes and hence these are called Helmholtz resonators.

The principle of a resonator is that the air contained in a hollow body with a narrow neck say a bottle, vibrates with a certain time period. If a tuning fork of the same frequency as that of natural frequency of a resonator is held near its mouth, the sound intensity increases and resonant vibrations are heard. Two common forms of Helmholtz resonators are shown in Fig. 6.13. Both the resonators are provided with a wide and a narrow mouth. The source of sound is kept at the wide end, and the ear is kept near the narrow end.

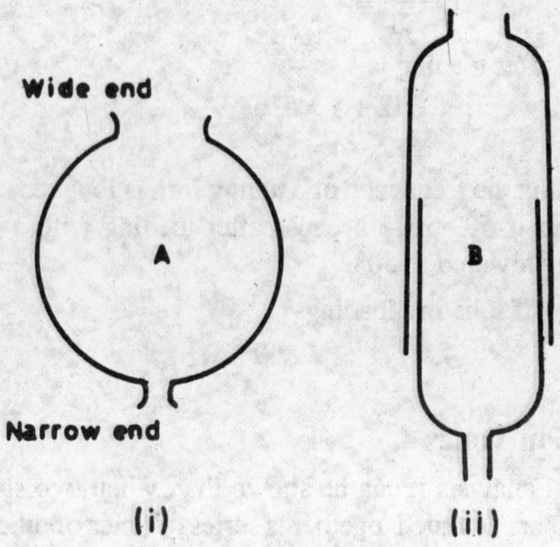

Fig. 6.13.

The natural frequency of resonator A is fixed because of its fixed volume of air. In the case of B, the volume of air enclosed can be altered by sliding the outer cylinder. The natural frequency can be varied as desired.

6.16. Theory of Resonator

Consider a flask containing air and of narrow cross section area a. Let the volume of the flask be V, and the mass of air contained in the neck be m. The pressure inside the flask is P and outside pressure is P_0 (Fig. 6.14).

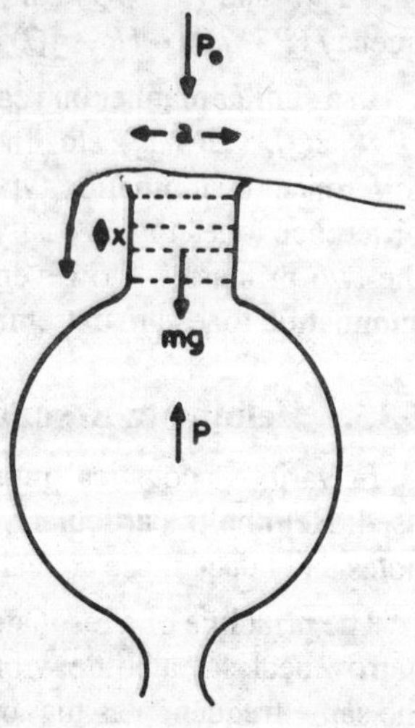

Fig. 6.14.

For equilibrium,

$$P = P_0 + \frac{mg}{a} \qquad \ldots (1)$$

When the air in the flask is in resonance with a particular frequency, the air contained in the neck moves up and down and acts like a piston on the large mass of air in the neck. Let, any instant, the mass of air in the neck move down through a distance x. If the compression is adiabatic, the new pressure P_1 in the vessel will be,

$$P_1 (V - ax)^{\gamma} = PV^{\gamma} \qquad \ldots (2)$$

$$\therefore \qquad P_1 = P\left[\frac{V}{V - ax}\right]^{\gamma}$$

or
$$P_1 = P\left[1 + \frac{ax}{V - ax}\right]^{\gamma}$$

$$P_1 = P\left[1 + \frac{\gamma\, ax}{V - ax}\right] \qquad \text{(approx.)}$$

$$\therefore \qquad P_1 - P = \frac{p\, \gamma ax}{V - ax} \qquad \ldots (3)$$

The net downward force F, on the air in the resonator,

$$F = \left[\left(P_0 + \frac{mg}{a}\right) - P_1\right] a$$

$$F = [P - P_1]\, a$$

Substituting the value of $(P - P_1)$ *from equation (3)*

$$F = -\left(\frac{P\gamma ax}{V - ax}\right) a \text{ or } F = -\frac{P\gamma\, a^2 x}{V - ax}$$

As ax is extremely small as compared to V,

$$\therefore \qquad F = -\frac{P\gamma\, a^2 x}{V}$$

But
$$F = \text{mass} \times \text{acceleration}$$

$$\therefore \qquad -\frac{P\gamma\, a^2 x}{V} = \text{mass} \times \text{acceleration}$$

$$\therefore \qquad \text{Acceleration} = \left(\frac{-P\gamma\, a^2}{mV}\right) x$$

$$\therefore \qquad \frac{\text{Acceleration}}{x} = -\left(\frac{P\gamma a^2}{mV}\right)$$

$$\therefore \qquad \frac{\text{Acceleration}}{\text{Displacement}} = -\left(\frac{-P\gamma\, a^2}{mV}\right) \qquad \ldots (4)$$

This represents simple harmonic motion.

$\therefore$ Time period,

$$t = 2\pi\sqrt{\frac{\text{displacement}}{\text{acceleration}}}$$

$$t = 2\pi\sqrt{\frac{mV}{P\gamma a^2}}$$

The frequency of oscillation,

$$f = \frac{1}{T} = \frac{1}{2\pi}\sqrt{\frac{P\gamma a^2}{mV}} \qquad \ldots (5)$$

The velocity of sound in air,

$$v = \sqrt{\frac{\gamma P}{\rho}} \quad \text{or} \quad P\gamma = v^2\rho$$

$$\therefore \qquad f = \frac{1}{2\pi}\sqrt{\frac{v^2\rho a^2}{mV}}$$

But mass of air, $m = al\rho$

Here l is the length of the neck.

$$\therefore \qquad f = \frac{1}{2\pi}\sqrt{\frac{v^2\rho a^2}{al\rho V}}$$

$$f = \frac{v}{2\pi}\sqrt{\frac{a}{lV}} \qquad \ldots (6)$$

$$\therefore \qquad f^2 V = \frac{v^2 a}{4\pi^2 l} \qquad \ldots (7)$$

As a, l and v remain constant,

$$f^2 V = \text{constant}$$

$$\therefore \qquad f \propto \frac{1}{\sqrt{V}}$$

Thus the frequency of the resonator is inversely proportional to the square root of its volume.

Example 6.9. *A Helmholtz resonator has a volume of 1* litre. *The radius of the neck is 1* cm *and the length of the neck is 0·2*cm. *Calculate the frequency at resonance. Take velocity of sound at room temperature = 350* m/s.

Here $V = 1000\ \text{cm}^3 = 10^{-3}\ \text{m}^3$

$l = 0{\cdot}2\ \text{cm} = 2 \times 10^{-3}\ \text{m}$

$r = 1\ \text{cm} = 10^{-2}\ \text{m}$

$$a = \pi r^2 = 3{\cdot}14 \times 10^{-4}\ \text{m}^2$$
$$v = 350\ \text{m/s}$$

The resonant frequency,

$$f = \frac{v}{2\pi}\sqrt{\frac{a}{lV}}$$

or $$f = \frac{350}{2 \times 3{\cdot}14}\sqrt{\frac{3{\cdot}14 \times 10^{-4}}{2 \times 10^{-3} \times 10^{-3}}}$$

$$f = \mathbf{698{\cdot}4\ Hz.}$$

6.17. Dependence of the Frequency of Resonator on the Size and the Shape of the Mouth

The frequency of a Helmholtz resonator also depends upon the size and shape of the mouth. The time taken by the excess of pressure inside to drive out air depends upon the size of the mouth of the resonator. If air can escape more easily (wider mouth), the time period is short *i.e.*, the frequency is high and *vice versa*. The property of the mouth to allow the air to escape less easily or more easily is denoted by the term acoustic conductivity of the mouth. The term acoustic conductivity depends upon the area of cross section *(a)* and the length *(l)* of the mouth.

The acoustic conductivity,

$$k = \frac{a}{l} \qquad \ldots (1)$$

The frequency of the resonator,

$$f = \frac{v}{2\pi}\sqrt{\frac{a}{lV}} = \frac{v}{2\pi}\sqrt{\frac{k}{V}} \qquad \ldots (2)$$

For a circular aperature in a thin wall l is very small. Releigh showed that in such cases, the acoustic conductivity k is approximately equal to the diameter.

$$\therefore \qquad k = \frac{a}{l} = 2r$$

or $$f = \frac{v}{2\pi}\sqrt{\frac{2r}{V}} \qquad \ldots (3)$$

Example 6.10. *A small hole is drilled in a spherical resonator of radius* 5×10^{-2} m. *The frequency at resonance is 300* hertz. *(a) Calculate the radius of the hole. Velocity of sound in air at room temperature = 350* m/s. *(b) Also calculate the resonant frequency if three additional holes of the same size are drilled in the resonator.*

Here $v = 350$ m/s

$$V = \frac{4}{3}\pi(5 \times 10^{-2})^3 \text{ m}^3$$

$$f = 300 \text{ hertz}$$

$$r = ?$$

(*a*) $$f = \frac{v}{2\pi}\sqrt{\frac{2r}{V}}$$

or $$r = \frac{2\pi^2 f^2 V}{v^2}$$

or $$r = \frac{2 \times (3{\cdot}14)^2 \times (300)^2 \times 1{\cdot}33 \times 3{\cdot}14 \times (5 \times 10^{-2})^3}{(350)^2}$$

$$r = 0{\cdot}007561 \text{ m.}$$

(*b*) For a total number of 4 holes of the same size, the total area of cross section is 4 times the original area. But the effective length and the volume of air in the resonator remain the same.

$\therefore$ $$f' = \frac{v}{2\pi}\sqrt{\frac{4a}{lV}}$$

$$f' = 2\left[\frac{v}{2\pi}\sqrt{\frac{a}{lV}}\right]$$

$$f' = 2\left[\frac{v}{2\pi}\sqrt{\frac{2r}{V}}\right]$$

$$f' = 2f$$

$$f' = 2 \times 300$$

$$f' = \mathbf{600 \text{ Hz.}}$$

Example 6.11. *A direct radiating loud speaker of radius 10* cm *is mounted in a rigid walled cabinet of inside dimensions* $40 \times 50 \times 60$ cm^3 *and wall thickness 2* cm. *Calculate the resonance frequency of the cabinet, considering that the cabinet works as a Helmholtz resonator. Velocity of sound in air at room temperature =350* m/s.

$$t = 2 \text{ cm} = 0{\cdot}02 \text{ m}$$

Here $r = 10$ cm $= 0{\cdot}1$ m

$$a = \pi r^2 = 3{\cdot}14 \times (0{\cdot}1)^2$$

or $$a = 0{\cdot}0314 \text{ m}^2$$

Effective length $l = \left(t + \frac{\pi r}{2}\right)$

$$= 0{\cdot}02 + 0{\cdot}157 = 0{\cdot}177 \text{ m}$$

$$V = 40 \times 50 \times 60 \text{ cm}^3$$

or $$V = 0{\cdot}12 \text{ m}^3$$

$$V = 350 \text{ m/s}$$

$$f = ?$$

$$f = \frac{V}{2\pi}\sqrt{\frac{a}{lV}} = \frac{350}{2 \times 3{\cdot}14}\sqrt{\frac{0{\cdot}0314}{0{\cdot}177 \times 0{\cdot}12}}$$

$$= \mathbf{67{\cdot}76 \text{ Hz.}}$$

EXERCISES

1. Explain the formation of stationary waves when two wave trains of exactly the same amplitude and wave length are travelling along the same string in opposite directions. Calculate the expression for the distance between two consecutive nodes.
2. Explain clearly the distinction between progressive waves and longitudinal waves.
3. Give briefly the theory of formation of stationary waves and discuss the characteristics of this wave motion.
4. Discuss analytically the formation of beats and show that the number of beats per second is equal to the difference in frequency of the two notes.
5. Discuss analytically the formation of stationary waves and explain how the characteristics change (*i*) with distance and (*ii*) with time.
6. Give a general account of distribution of energy in a stationary wave.
7. Explain fully the equation

$$y = 2a \cos \frac{2\pi x}{\lambda} \sin \frac{2\pi vt}{\lambda}.$$

8. Explain the principle of superposition. Use it to obtain the formation of stationary waves in an elastic medium. Discuss the property of the medium at the node and antinode positions. Show that in the case of a stationary wave, no energy is transferred across any section of the medium *(Delhi, 1973).*
9. Describe an experiment to demonstrate interference of sound waves. *(Delhi, 1973)*
10. Discuss analytically the formation of beats.

11. Two closed organ pipes 51 cm and 52 cm long produce three beats per second when they are simultaneously sounding their fundamental notes. Calculate the velocity of sound ignoring end correction. *(Delhi, 1973)*

[**Ans**. 318·24 m/s]

12. What are stationary waves and how are they formed? State their important properties. *(Delhi, 1973)*

13. What are beats? Discuss analytically the formation of beats. *(Delhi, 1972)*

14. Discuss the theory of the formation of stationary waves. State their important properties. *(Mysore, 1971; Delhi, 1971)*

15. Discuss the theory of Helmholtz resonator and give its uses. *(Culcutta, 1974)*

16. What are stationary waves and how are they formed ? *(Delhi, 1971)*

17. What is interference of sound ? How are beats produced? *(Delhi, 1972; Calcutta, 1973)*

18. What are beats ? How are they produced ? Derive an expression for the frequency of the beats. *(Delhi, 1974)*

19. Describe an experiment to demonstrate the formation of beats. *(Delhi, 1974)*

20. Describe Helmholtz resonator and obtain an expression for the natural frequency of such a resonator. Indicate its use in the analysis of sound. *[Madras, 1974; Behrampur (Hons.), 1972]*

21. Discuss how sound waves are reflected at the open and closed ends of a tube producing stationary waves in air. How are their properties studied analytically? *(Delhi, 1976)*

22. What are stationary waves? Investigate theoretically the formation of stationary waves and discuss their properties with regard to changes of displacement and pressure at different points. *[Delhi (Supple.), 1976]*

23. What are beats? Differentiate between the phenomenon of beats and interference. Show that the number of beats per second is eqaul to the difference between the frequencies of the two sources sounded together. *[Delhi (Supple.), 1976]*

24. What are beats? Explain graphically their formation. *(Rajasthan, 1975)*

25. Discuss the formation of beats. *(Kanpur, 1975)*

26. Deduce the equation for a stationary wave and from it derive its charactersitic properties. Compare these properties with those of a progressive wave. *(Kanpur, 1975)*

27. What are beats and how are they formed? *[Delhi, (Add/ Physics), 1976]*

28 Two organ pipes, one closed and the other open are found to have their first over tones of the same frequency. Find the ratio of their lengths. *(Calcutta, 1973)* [**Ans.** (3 : 4)]

29. Calculate the velocity of sound in a gas in which the waves of lengths 50 cm and 50·5 cm produce 6 beats/s. *(Calcutta, 1974)* [**Ans.** 303 m/s]

30. A tuning fork of unknown frequency gives 5 beats per second when sounded together with another tuning fork of frequency 384 hertz. The fork is now loaded with wax and three beats per second are heard. Calculate the unknown frequency. *(Delhi, 1972)* [**Ans.** 389]

31 *(a)* What are stationary waves and how are they formed ?

(b) Discuss important properties of stationary waves.

(c) Show that in the case of a stationary wave, no energy is transferred across any section of the medium. *(Delhi, 1990)*

32. Explain the formation of stationary waves and discuss their characteristics. How do they differ from progressive waves? *(Calcutta, 1992)*

33. Describe a Helmholtz's resonator. How can the frequency of sound be measured with it? *(Calcutta, 1992)*

34. What are stationary waves? Investigate theoretically the formation of such waves. Show that in such waves the net transmission of energy is zero. *(Bhagalpur, 1990)*

35. Explain analytically the formation of beats when two sources of nearly equal frequencies are sounded together. *(Gauhati,1992)*

36. Show that in the case of a stationary wave, no energy is transferred across any section of the medium. *(Delhi, 1991)*

37. Explain the principle of superposition. Use it to obtain the formation of stationary waves in an elastic medium. Discuss the property of the medium at the node and antinode positions. *(Delhi, 1992)*

CHAPTER 7

Vibrations in Strings and Air Columns

7.1. Velocity of Transverse Waves Along a Stretched String

First Method

Consider a string PQ stretched by a tension T. Let the string be plucked at the centre O and left free (Fig. 7.1). The string vibrates transversely. These vibrations are simple harmonic in nature. The force tending to bring any element of the string back to the equilibrium position is the component of tension acting at right angles to PQ.

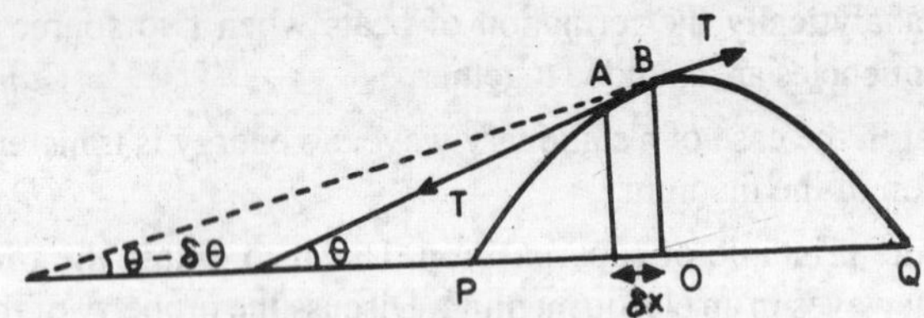

Fig. 7.1.

Let the undisplaced position of the string correspond to X-axis and let the displacement be along the Y-axis. Consider a small element AB of length δx. The tangents drawn at A and B make angles θ and $\theta - \delta\theta$ with the X-axis (Fig.7.1). The tension at A is resolved into two rectangular components. The downward component of tension at $A = T \sin \theta$.

As θ is small $\quad \sin \theta = \tan \theta$

$\therefore$ Downward component of tension

at $A = T \tan \theta$.

But $\quad \tan \theta = \dfrac{dy}{dx}$ at A (*i.e.* slope at A).

∴ Downward component of tension at A

$$= T\frac{dy}{dx} \quad \ldots (1)$$

The rate of change of slope with respect to the length of the element,

$$= \frac{d}{dx}\left(\frac{dy}{dx}\right) = \frac{d^2 y}{dx^2}$$

The change in slope for a distance δx

$$= \frac{d^2 y}{dx^2} \cdot \delta x$$

The slope at the point B

$$\tan(\theta - \delta\theta) = \frac{dy}{dx} - \frac{d^2 y}{dx^2} \cdot \delta x$$

The upward component of tension acting at B

$$= T \sin(\theta - \delta\theta)$$

As θ is small, $\sin(\theta - \delta\theta) = \tan(\theta - \delta\theta)$

∴ Upward component of tension at B

$$= T \tan(\theta - \delta\theta)$$

$$= T\left[\frac{dy}{dx} - \frac{d^2 y}{dx^2} \cdot \delta x\right] \quad \ldots (2)$$

From equations (1) and (2), resultant downward tension,

$$F = T\frac{dy}{dx} - T\left(\frac{dy}{dx} - \frac{d^2 y}{dx^2}\delta x\right)$$

$$F = T\frac{d^2y}{dx^2}\delta x \quad \ldots (3)$$

Let m be the mass of the string per unit length.

Mass of the element $= m\,\delta x$

Acceleration of the element in the direction of y-axis

$$\frac{d^2y}{dt^2}$$

Force acting on the element = mass × acceleration

$$F = m\,\delta x \frac{d^2 y}{dt^2} \quad \ldots (4)$$

From equations (3) and (4),

$$m\,\delta x \frac{d^2y}{dt^2} = T\frac{d^2y}{dx^2}\delta x$$

$$\frac{d^2y}{dt^2} = \frac{T}{m}\frac{d^2y}{dx^2} \qquad \ldots (5)$$

This is similar to the differential equation of wave motion,

$$\frac{d^2y}{dt^2} = v^2\frac{d^2y}{dx^2} \qquad \ldots (6)$$

Here v is the velocity of propagation of the wave.

Comparing equations (5) and (6),

$$v^2 = \frac{T}{m}$$

$$v = \sqrt{\frac{T}{m}} \qquad \ldots (7)$$

Also $\qquad v = n\lambda$

In case the string of length l vibrates in p segments, length of each segment $= l/p$ and each segment corresponds to half wave length.

$$\therefore \qquad \frac{l}{p} = \frac{\lambda}{2}$$

$$\lambda = \frac{2l}{p}$$

$$\therefore \qquad v = \frac{n(2l)}{p} \qquad \ldots (8)$$

Substituting this value of v in equation (7)

$$\frac{n(2l)}{p} = \sqrt{\frac{T}{m}}$$

$$n = \frac{p}{2l}\sqrt{\frac{T}{m}} \qquad \ldots (9)$$

In case the string vibrates in one segment

$$p = 1,$$

and

$$n = \frac{1}{2l}\sqrt{\frac{T}{m}} \qquad \ldots (10)$$

7.2. Alternative Method

Consider the portion $PABQ$ of a string in which a transverse wave is travelling from left to right with a velocity v. The string is also drawn towards the left with the same velocity so that the wave is stationary (Fig. 7.2).

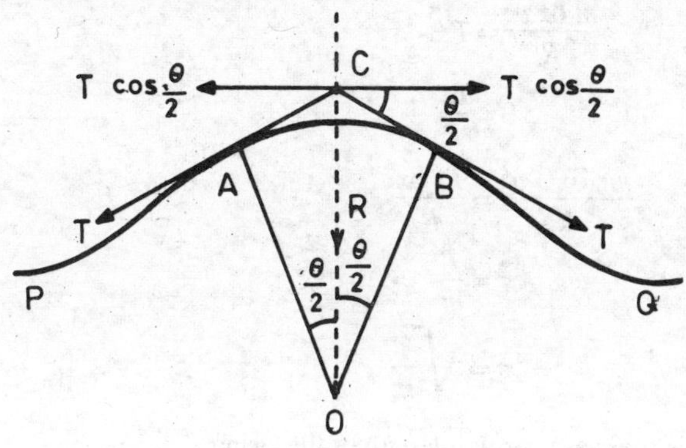

Fig. 7.2.

Let AB be a small element of the string. The centre of curvature of the element AB is O. The angle $AOB = \theta$. It is assumed that the curvature of AB is small and hence θ is small. Let T be the tension at A and B. The directions of these tensions are tangential to the element at A and B.

Resolving the tension at A into two rectangular components, $T \sin \frac{\theta}{2}$ and $T \cos \frac{\theta}{2}$ perpendicular and parallel to the string in the undisplaced portion.

Similarly the components at B will be $T \sin \frac{\theta}{2}$ and $T \cos \frac{\theta}{2}$ perpendicular and parallel to the string in the undisplaced portion.

The parallel components are equal and opposite. The perpendicular components are along CO.

The resultant tension along CO

$$= 2T \sin \frac{\theta}{2}$$

As θ is small, $\sin \frac{\theta}{2} = \frac{\theta}{2}$

$$\therefore \quad \text{Resultant tension} = 2T \frac{\theta}{2} = T\theta \qquad \ldots (1)$$

For equilibrium position, this resultant tension provides the necessary centripetal force,

$$= \frac{(m\,\delta x)\,v^2}{R} \qquad \ldots (2)$$

$$\therefore \qquad \frac{m\,\delta x\,v^2}{R} = T\,\theta$$

$$\text{But} \qquad \theta = \frac{\delta x}{R}$$

$$\therefore \qquad \frac{m\,.\,\delta x\,.\,v^2}{R} = \frac{T\,.\,\delta x}{R}$$

$$\text{or} \qquad v^2 = \frac{T}{m}$$

$$v = \sqrt{\frac{T}{m}} \qquad \ldots (3)$$

Here m is the mass per unit length of the string.

$$\text{Also} \qquad v = n\lambda$$

In case the string of length l vibrates in p segments, length of each segment $= \frac{l}{p}$ and each segment corresponds to half wave length.

$$\therefore \qquad \frac{l}{p} = \frac{\lambda}{2}$$

$$\text{or} \qquad \lambda = \frac{2l}{p}$$

$$\therefore \qquad v = \frac{n(2l)}{p} \qquad \ldots (4)$$

Substituting this value of v in equation (3)

$$\frac{n(2l)}{p} = \sqrt{\frac{T}{m}}$$

$$n = \frac{p}{2l}\sqrt{\frac{T}{m}} \qquad \ldots (5)$$

In case the string vibrates in one segment

$$p = 1$$

$$\text{and} \qquad n = \frac{1}{2l}\sqrt{\frac{T}{m}} \qquad \ldots (6)$$

7.3. Laws of Transverse Vibration of Strings

There are three laws of transverse vibration of strings :

(1) The fundamental frequency is inversely proportional to the length of the string

$$n \propto \frac{1}{l}$$

(2) The fundamental frequency is directly proportional to the square root of the stretching force or tension

$$n \propto \sqrt{T}$$

(3) The fundamental frequency is inversely proportional to the square root of the mass per unit length

$$n \propto \frac{1}{\sqrt{m}}$$

Combining the above three laws,

$$n \propto \frac{1}{l}\sqrt{\frac{T}{m}}$$

or

$$n = \frac{k}{l}\sqrt{\frac{T}{m}}$$

The value of the constant

$$k = \frac{1}{2}$$

$$\therefore \qquad n = \frac{1}{2l}\sqrt{\frac{T}{m}} \qquad \ldots (1)$$

If D is the diameter of the wire and d is the density of the material of the wire, then

$$m = \frac{\pi D^2}{4} \times 1 \times d$$

$$n = \frac{1}{2l}\sqrt{\frac{4T}{\pi D^2 d}}$$

$$n = \frac{1}{lD}\sqrt{\frac{T}{\pi d}} \qquad \ldots (2)$$

7.4. Verification of the Laws of Transverse Vibration of Strings

(1) $$n \propto \frac{1}{l} \qquad \text{or } nl = \text{constant}$$

Take a sonometer and a tuning fork of frequency 256 (Fig. 7.3). Keeping a load of 1 kg, find the position of the two bridges when the wire is in unison with the tuning fork. This can be found by using a small paper rider on the sonometer wire. Suppose the length= l_1. Now, using a tuning fork of frequency 512 find the resonating length, keeping tension the same. Let this length be l_2. It is found that l_2 is half l_1. Also $n_1 l_1 = n_2 l_2$.

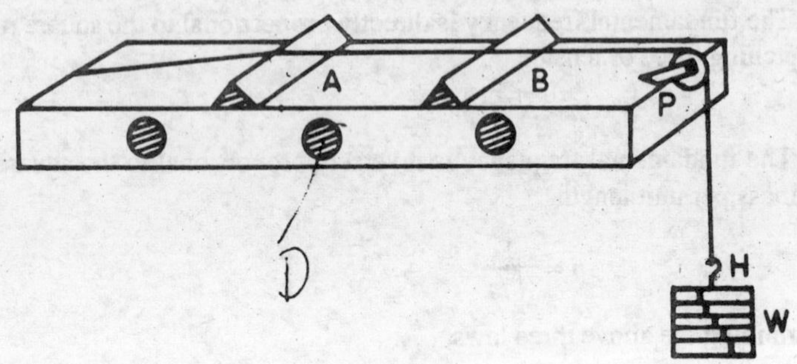

Fig. 7.3.

$$n \propto \frac{1}{l} \qquad \text{or} \qquad nl = \text{constant}$$

(2) $$n \propto \sqrt{T}$$

For a tuning fork of frequency 256, find the resonating length l_1 for a load of 1 kg. It will be found that when the load is increased to 4 kg, the tuning fork of frequency 512 is in resonance with the same length of the wire l_1. Here

$$\frac{n_2}{n_1} = \sqrt{\frac{T_2}{T_1}}, \qquad \text{or } n \propto \sqrt{T}$$

Also $$n^2 \propto T \qquad \text{or } T \propto n^2$$

(3) $$n \propto \frac{1}{\sqrt{m}}$$

Take two wires of the same material and of diameters in the ratio of 1 : 2.

Stretch the first wire with a tension of 1 kg wt. Find the resonating length with a tuning fork of frequency 512. Now stretch the second wire with the same tension of 1 kg and find the resonating length with a tuning fork of frequency 256. It is found that the two lengths are equal.

Here $\frac{n_1}{n_2} = \frac{D_2}{D_1}$.

$$m_1 = \frac{\pi D_1^2 \times d}{4} \quad \text{and } m_2 = \frac{\pi D_2^2 \times d}{4}$$

But $\frac{m_2}{m_1} = \frac{D_2^2}{D_1^2}$ or $\frac{D_2}{D_1} = \sqrt{\frac{m_2}{m_1}}$

$\therefore$ $\frac{n_1}{n_2} = \sqrt{\frac{m_2}{m_1}}$ or $n \propto \frac{1}{\sqrt{m}}$.

Example 7.1. *Two strings each of length* 60 cm *are stretched, one by a force of* 4 kg *wt and the other by a force of* 9 kg wt. *What is the interval between the two notes that are produced?*

$$\frac{n_2}{n_1} = \sqrt{\frac{T_2}{T_1}} = \sqrt{\frac{9}{4}} = \frac{3}{2}$$

The interval between the two notes is the fifth.

Example 7.2. *A wire of length* 108 cm *produces a fundamental note of frequency 256 when stretched by a weight of* 1 kg. *By how much its length should be increased so that its pitch is raised by a major tone if it is now stretched by a weight of* 4 kg.

$$l_1 = 108 \text{ cm}, \quad T_1 = 1 \text{ kg}, n_1 = 256$$

$$l_2 = ? \qquad T_2 = 4 \text{ kg}$$

$$\frac{n_2}{n_1} = \frac{9}{8}$$

or $n_2 = \frac{9}{8} \times 256 = 288$

$$\frac{n_2}{n_1} = \frac{l_1}{l_2}\sqrt{\frac{T_2}{T_1}}$$

$$\frac{9}{8} = \frac{108}{l_2}\sqrt{\frac{4}{1}}$$

$$l_2 = 192 \text{ cm}$$

The increase in length

$$= 192 - 108 = \mathbf{84\ cm.}$$

Example 7.3. *A wire of specific gravity 7, one metre long and 1* mm *in diameter is stretched by a weight of 11* kg. *Calculate the pitch of the fundamental note.*

Here, $d = 7, \qquad l = 100$ cm

$$D = 1 \text{ mm} = \frac{1}{10} \text{ cm}$$
$$T = 11 \text{ kg} = 11 \times 1000 \times 98 \text{ dynes}$$
$$n = ?$$
$$n = \frac{1}{lD}\sqrt{\frac{T}{\pi d}}$$
$$n = \frac{1}{100 \times \frac{1}{10}}\sqrt{\frac{11 \times 1000 \times 980 \times 7}{22 \times 7}} = \mathbf{70.}$$

Example 7.4. *A string when stretched by a weight of* 4 kg *gives a note of frequency 250. What weight will produce an octave of this note?*

Here, $n_1 = 256,\ n_2 = 2 \times 256 = 512$

$T_1 = 4$ kg wt ; $T_2 = ?$

$$\frac{n_2}{n_1} = \sqrt{\frac{T_2}{T_1}} \quad \text{or} \quad \frac{512}{256}\sqrt{\frac{T_2}{T_1}}$$

or $$\sqrt{\frac{T_2}{4}} = 2, \quad T_2 = \mathbf{16\ kg\ wt.}$$

Example 7.5. *Calculate the frequency of the fundamental note of a string, 1* metre *long and weighing* 2 grams *when stretched by a weight of* 400 kg.

Here, $l = 1$ m

Total mass = 2 g = 0·002 kg

$$m = \frac{0{\cdot}002}{1} = 0{\cdot}002 \text{ kg/m}$$
$$T = 400 \text{ kg} = 400 \times 9{\cdot}8 \text{ newtons}$$
$$n = \frac{1}{2l}\sqrt{\frac{T}{m}}$$
$$= \frac{1}{2 \times 1}\sqrt{\frac{400 \times 9{\cdot}8}{0{\cdot}002}}$$
$$n = \frac{1}{2}\sqrt{400 \times 4900} = \mathbf{700.}$$

Example 7.6. *A tuning fork of frequency 160 is sounded along with a sonometer wire of length* 25 cm *stretched to a tension of* 1·25 kg wt. *Calculate the number of beats/s. (Mass per unit length* = 0·025 g.*)*

Here, $l = 25$ cm, $m = 0{\cdot}025$ g/cm

$T = 1{\cdot}25$ kg wt $= 1{\cdot}25 \times 1000 \times 980$ dynes

$$n = \frac{1}{2l}\sqrt{\frac{T}{m}} = \frac{1}{2 \times 25}\sqrt{\frac{1{\cdot}25 \times 1000 \times 980}{0{\cdot}025}}$$

$$n = \frac{7000}{50} = 140$$

Frequency of the tuning fork $N = 160$

$\therefore$ Beats/s $= N - n = 160 - 140 = \mathbf{20.}$

Example 7.7. *A wire gives out a fundamental note of 256 cycles per second when it is under a tension of* 10 kg wt.

(*i*) *Under what tension will the string emit a frequency of 512 cycles per second ?*

(*ii*) *How would you make the wire emit a note of 768 cycles per second keeping the tension at* 10 kg wt.

(*i*) Here, $n_1 = 256$, $T_1 = 10$ kg wt.

$n_2 = 512$, $T_2 = ?$

$$\frac{n_2}{n_1} = \sqrt{\frac{T_2}{T_1}}$$

$$\frac{512}{256} = \sqrt{\frac{T_2}{10}}$$

or $T_2 = \mathbf{40\ kg\ wt.}$

(*ii*) Here, in the first case

$n_1 = 256$

Length of the wire $= l_1$

In the second case $n_2 = 768$

Length of the wire $= l_2$

$$\frac{n_2}{n_1} = \frac{l_1}{l_2} \text{ or } \frac{768}{256} = \frac{l_1}{l_2}$$

or
$$l_2 = \frac{l_1}{3}.$$

Therefore, by changing the length to $\frac{1}{3}$ of the length in the first case, the wire will emit a note of frequency 768.

Example 7.8. *A flexible string of length* 0·99 metre *and mass one* gram *is stretched by a tension of* T newtons. *The string vibrates in three segments with a frequency of* 500 hertz. *Calculate the tension.*

$$n = \frac{p}{2l}\sqrt{\frac{T}{m}}$$

or
$$n^2 = \frac{p^2 T}{4l^2 m}$$

or
$$T = \frac{4n^2 l^2 m}{p^2}$$

Here
$$n = 500 \text{ hertz}$$
$$l = 0{\cdot}99 \text{ m}$$
$$m = \frac{10^{-3}}{0{\cdot}99} \text{ kg/m}$$
$$p = 3$$

∴
$$T = \frac{4 \times (500)^2 \times (0{\cdot}99)^2 \times 10^{-3}}{9 \times 0{\cdot}99}$$
$$T = \mathbf{110 \text{ newtons.}}$$

Example 7.9. *A flexible string of length 0·88 m is stretched by a force of* 55 newtons. *The mass of the string is* 1 gram. *Calculate the frequency of vibration of the string if it vibrates in* 5 segments.

$$n = \frac{p}{2l}\sqrt{\frac{T}{m}}$$

Here
$$p = 5$$
$$T = 55 \text{ newtons}$$
$$l = 0{\cdot}88 \text{ m}$$
$$m = \frac{10^{-3}}{0{\cdot}66} \text{ kg/m}$$
$$n = \frac{5}{5 \times 0{\cdot}88}\sqrt{\frac{55 \times 0{\cdot}88}{10^{-3}}}$$
$$n = \mathbf{625 \text{ Hz.}}$$

Example 7.10. *Two similar sonometer wires of the same material, under the same tension, produce* 2 beats/s. *The length of one wire is* 50 cm *and that of the other is* 50·1 cm. *Calculate the frequencies of the two wires.*

(Delhi, 1974)

Here
$$n_1 - n_2 = 2 \qquad \ldots (1)$$
$$l_1 = 50 \text{ cm}$$
$$l_2 = 50{\cdot}1 \text{ cm}$$

As the tensions and material of the wires are the same
$$n_1 l_1 = n_2 l_2$$

or
$$\frac{n_1}{n_2} = \frac{l_2}{l_1}$$
$$\frac{n_1}{n_2} = \frac{50{\cdot}1}{50}$$

or
$$n_1 = \left(\frac{50{\cdot}1}{50}\right) n_2 \qquad \ldots (2)$$

Substituting this value of n_1 in equation (1),

$$\left(\frac{50{\cdot}1}{50}\right) n_2 - n_2 = 2$$

$$n_2 = \mathbf{1000\ Hz.}$$

$$\therefore \quad n_1 = n_2 + 2$$

$$n_1 = 1000 + 2$$

$$n_1 = \mathbf{1002\ Hz.}$$

7.5. Melde's Experiment

In the Melde's Experiment, one end of the string is connected to the prong of an electrically maintained tuning fork. The other end is connected to the scale pan. The string passes over a smooth frictionless pulley. The distance between the tuning fork and the pulley can be adjusted. There are two modes of vibration (*i*) transverse mode of vibration and (*ii*) longitudinal mode of vibration (Fig. 7.4).

Transverse mode of vibration. The tuning fork vibrates at right angles to the length of the string [Fig. 7.4(*i*)]. In this case the frequency of vibration of the string, is equal to the frequency of the tuning fork. Suppose N is the frequency of the tuning fork and the string of length l vibrates in p_1 segments.

$$\therefore \quad N = \frac{p_1}{2l}\sqrt{\frac{T}{m}} \qquad \ldots (1)$$

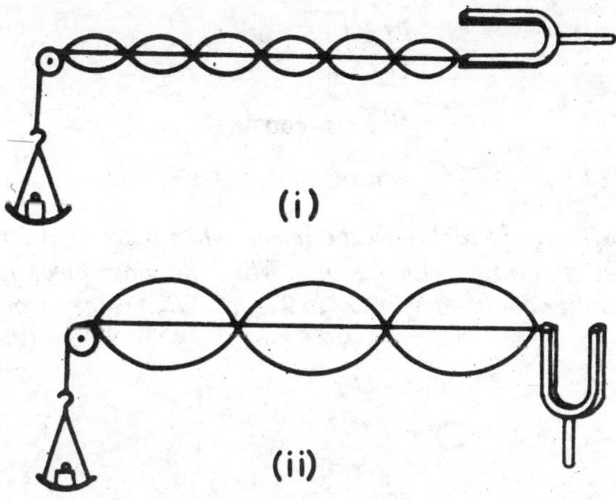

Fig. 7.4.

Longitudinal mode of vibration. The tuning fork vibrates along the direction of the length of the string [Fig. 7.4 (*ii*)]. In this case for one complete vibration of the tuning fork the string completes half vibration. Suppose the frequency of the tuning fork is N. Therefore, the frequency of vibration of the string is $N/2$. If the string of length l vibrates in p_2 segments, the frequency of vibration of the string,

$$\frac{N}{2}=\frac{p_2}{2l}\cdot\sqrt{\frac{T}{m}} \quad \ldots (2)$$

Special Case

(*i*) Suppose the string vibrates in p_1 segments in the transverse mode, then for the same tuning fork and the same tension, the string will vibrate in half the number of segments in longitudinal mode of vibration.

From equations (1) and (2),

$$p_2=\frac{p_1}{2}$$

(*ii*) From equation (1), for the transverse mode of vibration

$$N=\frac{p_1}{2l}\sqrt{\frac{T}{m}}$$

$$N^2=\frac{p_1^2 T}{4l^2 m}$$

or $$Tp_1^2=4N^2 l^2 m=\text{constant.}$$

Similarly from equation (2), for longitudinal mode of vibration

$$\frac{N}{2}=\frac{P_2}{2l}\sqrt{\frac{T}{m}}$$

$$Tp_2^2=N^2 l^2 m=\textbf{constant}$$

or in general $$Tp^2=\textbf{constant.}$$

Example 7.11. *In Melde's experiment, when a string is stretched by a piece of glass it vibrates with 7 loops. When the glass piece is completely immersed in water the string vibrates in 9 loops. What is the specific gravity of glass?* (Punjab, 1973)

Here $$T_1p_1^2=T_2 p_2^2$$

$$T_1\times(7)^2=T_2(9)^2$$

$$\frac{T_1}{T_2}=\frac{81}{49}$$

$$\frac{T_2}{T_1} = \frac{49}{81}$$

$$1 - \frac{T_2}{T_1} = 1 - \frac{49}{81} = \frac{32}{81}$$

$$\frac{T_1 - T_2}{T_1} = \frac{32}{81}$$

or
$$\frac{T_1}{T_1 - T_2} = \frac{81}{32} = 2{\cdot}531.$$

Therefore the specific gravity of glass = **2·531.**

Example 7.12. *Find the tension needed to produce stationary waves with four loops in a string one metre long and 0·5 gram in weight, fixed to a tuning fork of frequency* 200 hertz, *when the prongs of the fork are vibrating perpendicular to the string.* (Delhi, 1972)

Here
$$N = \frac{p_1}{2l}\sqrt{\frac{T}{m}} \quad \text{or } T = \frac{4l^2 N^2 m}{p_1^2}$$

Here
$$l = 100 \text{ cm} = 1 \text{ metre}$$

$$p_1 = 4$$

$$N = 200 \text{ hertz}$$

$$m = \frac{0{\cdot}5}{100} = 0{\cdot}005 \text{ g/cm} = 5 \times 10^{-4} \text{ kg/m}$$

$$T = \frac{4 \times (1)^2 \times (200)^2 \times 5 \times 10^{-4}}{(4)^2}$$

$$T = \mathbf{5 \text{ newtons.}}$$

Example 7.13. *A flexible string of length* 0·99 m *and mass 1* g *is stretched by a tension of T* newtons. *The string vibrates in three segments with a frequency of* 500 Hz. *Calculate the tension.* (Delhi, 1992)

Here,
$$l = 0{\cdot}99 \text{ m}$$

$$M = 10^{-3} \text{ kg}$$

$$m = \frac{M}{l} = \frac{10^{-3}}{0{\cdot}99} \text{ kg/m}$$

$$n = 500 \text{ Hz}$$

$$p = 3$$

$$n = \frac{p}{2l}\sqrt{\frac{T}{m}}$$

$$T=\frac{4n^2 l^2 m}{p^2}$$

$$T=\frac{4\times(500)^2\times(0{\cdot}99)^2\times 10^{-3}}{9\times 0{\cdot}99}$$

$$T=\mathbf{110\ N.}$$

Example 7.14. *Two wires of the same material and diameter are subjected to the same tension. They produce 3 beats per second when sounded together. Calculate their frequencies if their lengths are* 42·5 cm *and* 43·0 cm *respectively.* (Gauhati, 1992)

Here, $l_1 = 42{\cdot}5$ cm

$l_2 = 43$ cm

$$n_1=\frac{1}{2l_1}\sqrt{\frac{T}{m}}$$

$$n_2=\frac{1}{2l_2}\sqrt{\frac{T}{m}}$$

Here T and m are the same.

$$\therefore \quad \frac{n_1}{n_2}=\frac{l_2}{l_1}=\frac{43}{42{\cdot}5}$$

$$n_1=\left(\frac{43}{42{\cdot}5}\right)n_2 \qquad \ldots(1)$$

Also $n_1 - n_2 = 3$

$$\left(\frac{43}{42{\cdot}5}\right)n_2 - n_2 = 3$$

$$n_2 = \mathbf{255}$$

$$\therefore \quad n_1 = 255 + 3 = \mathbf{258.}$$

Example 7.15. *Two strings of the same material and of the same cross-section are suspended on a sonometer. One is loaded with* 12 kg *and the other with* 3 kg. *The first string is tuned to the second harmonic of the second string. If the second string is* 100 cm *in length, what is the length of the first string?* (Bhagalpur 1990)

Here $$n_1=\frac{1}{2l_1}\sqrt{\frac{T_1}{m}}$$

$$n_2=\frac{1}{2l_2}\sqrt{\frac{T_2}{m}}$$

$$\frac{n_1}{n_2} = \frac{l_2}{l_1}\sqrt{\frac{T_1}{T_2}}$$

Here $n_1 = 3\,n_2$

and $\dfrac{T_1}{T_2} = \dfrac{12}{3} = 4$

$\therefore$ $3 = \dfrac{l_2}{l_1}\sqrt{4}$

$$l_2 = 1{\cdot}5\,l_1$$

Here $l_1 = 1$ m

$\therefore$ $l_2 = \mathbf{1{\cdot}5\ m.}$

Example 7.16. *An addition of* 24 kg *to the tension of a stretched string changed its frequency to three times the original frequency. What was the original tension?* (IAS, 1984)

Frequency,

$$f_1 = 2\pi\sqrt{\frac{T}{m}} \qquad \ldots (i)$$

$$f_1 = 2\pi\sqrt{\frac{T+24}{m}} \qquad \ldots (ii)$$

$$\frac{f_2}{f_1} = \sqrt{\frac{T+24}{T}}$$

$$(3)^2 = \frac{T+24}{T}$$

$$T = \mathbf{3\ kg\ wt}$$

$$= 3 \times 9{\cdot}8 = \mathbf{29{\cdot}4\ N.}$$

Example 7.17. *A movable bridge divides a sonometer wire into two parts, which differ in length by* 1 cm *and produce 4 beats per second when sounded together. If the whole length is* 100 cm, *find the frequencies of the parts.* (IAS, 1989)

Here $l_1 = 49{\cdot}5$ cm

$l_2 = 50{\cdot}5$ cm

As $n = \dfrac{1}{2l}\sqrt{\dfrac{T}{m}}$

$$nl = \text{constant}$$

$\therefore$ $n_1 l_1 = n_2 l_2$

$$49{\cdot}5\, n_1 = 50{\cdot}5\, n_2 \qquad \ldots (i)$$

$$n_2 = \left(\frac{49{\cdot}5}{50{\cdot}5}\right) n_1 \qquad \ldots (ii)$$

Also

$$n_1 - n_2 = 4 \qquad \text{(given)}$$

$$n_1 - \left(\frac{49{\cdot}5}{50{\cdot}5}\right) n_1 = 4$$

$$n_1 = \mathbf{202} \qquad \ldots (iii)$$

$$202 - n_2 = 4$$

$$n_2 = \mathbf{198.} \qquad \ldots (iv)$$

7.6. Vibrations of Air Columns

Stationary longitudinal waves are produced in wind instruments which are provided with a column of air called the resonator. The vibrations are caused by the mouth piece and the mouth piece is different for different instruments. The mouth piece acts as a source and supplies the necessary energy to maintain the vibrations in the column of air. Depending upon the nature of excitation, the organ pipes can be classified into two categories :

(*i*) **Flute or Flue Pipes and** (*ii*) **Reed Pipes.**

Both these categories may have many varieties which differ in shape and material. The tubes which are normally used have one end open and the other end either open end or closed. The nature of the reflected wave will be different in the open end and closed end organ pipes. In the case of the closed end organ pipe the reflection takes place at a rigid wall while in the case of open end pipe the reflection takes place at a yeilding wall. In the case of a closed end organ pipe the ends have always antinodes. In the case of a closed end organ pipe, the open end has an antinode and the closed end has always a node.

These sounding bodies *viz.* the organ pipes can vibrate in a number of ways, giving notes of different pitch. The resultant sound is a complex note. The lowest vibrating frequency of the instrument is called the *fundamental frequency*.

The rest of the notes are the integral multiples of the fundamental frequency and are called *harmonics* or *overtones.*

Suppose the fundamental frequency is *n*, then the overtones are 2*n*, 3*n*, 4*n*, etc.

7.7. Resonance

Free vibration. When a vibrating body is set to vibrate it begins to vibrate with its natural frequency. In the case of a simple pendulum

$$n = \frac{1}{2\pi}\sqrt{\frac{g}{l}}$$

and in the case of a stretched string

$$n = \frac{1}{2l}\sqrt{\frac{T}{m}}$$

Such vibrations are called free vibrations.

Forced vibrations. A vibrating body can set or force another body into vibration even if their natural frequencies are different. In such a case, a body which is forced to vibrate, does not vibrate for a long time. Its vibration dies off quickly.

Resonance. It is a phenomenon of vibratory motion in which a body is set into vibration with its natural frequency by another body vibrating with the same frequency.

7.8. Velocity of Sound in Air by Resonance Method

A resonance tube apparatus is taken. It is properly levelled and the reservoir R is filled with water. A tuning fork of known frequency is struck gently on a rubber pad and brought near the top end of the tube (Fig. 7.5).

With the help of a reservoir R and a pinch cock, the position of the water level is changed. The position of maximum intensity of sound is found. This corresponds to a position of resonance. The particles of the air column in the tube vibrate with the same frequency as that of the tuning fork.

When the prong of the tuning fork moves from a to b, it sends a wave of compression. When the prong moves from b to a, a wave of compression comes back after reflection from the water surface. Similarly when the prong moves from a to c, it sends a wave of rarefaction through the air column in the tube. When the prong moves from c to a, the wave of rarefaction is reflected back through the air column. It means, that when the

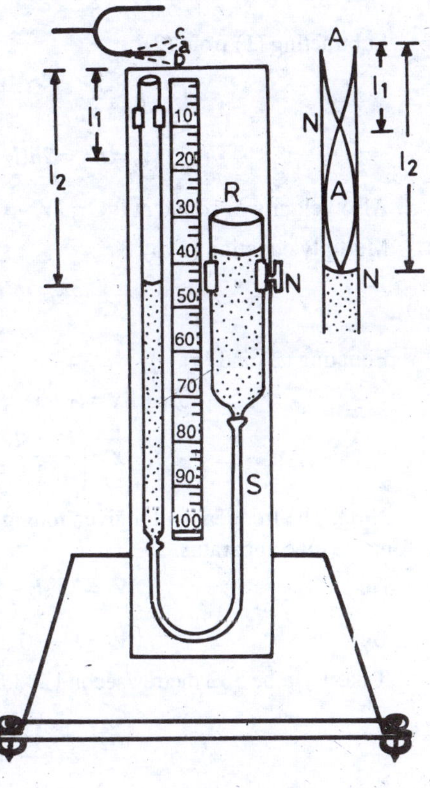

Fig. 7.5.

tuning fork completes one-fourth of the vibration, the sound wave moves from the top end to the level of water in the tube. Therefore, the length of the air column.

$$l_1 = \frac{\lambda}{4}.$$

End correction. In practice l_1 is less than $\lambda/4$ because the reflection of sound does not take place just at the end of the tube but a small distance above the open end. The end correction $x = 0{\cdot}3$ where D is the diameter of the tube.

Actually, the open end of the tube behaves as an antinode and the closed end behaves as a node. The distance between an adjacent node and an antinode is $\lambda/4$.

Suppose, the frequency of the tuning fork is n. Let the first resonance occur at l_1 and the second resonance at a length l_2 of the air column.

Then,
$$l_1 + x = \frac{\lambda}{4} \qquad \ldots (1)$$

$$l_2 + x = \frac{3\lambda}{4} \qquad \ldots (2)$$

Subtracting (1) from (2)

$$l_2 - l_1 = \frac{\lambda}{2}$$

$$V_t = n\lambda = 2n(l_2 - l_1) \qquad \ldots (3)$$

Also velocity of sound at 0°C = $(V_t - 0{\cdot}61\ t)$ m/s ... (4)

Multiply equation (1) by 3

$$3l_1 + 3x = \frac{3\lambda}{4} \qquad \ldots (5)$$

Equating (2) and (5)

$$3l_1 + 3x = l_2 + x$$

$$x = \left(\frac{l_2 - 3l_1}{2}\right).$$

Note. The frequency of a given tuning fork can also be determined using resonance tube apparatus.

Here,
$$V_t = 2n\,(l_2 - l_1)$$

But
$$V_t = V_0 + 0{\cdot}61t \text{ m/s}$$

Take V_0 to be 332 metres/second and find V_t.

$$n = \frac{V_t}{2\,(l_2 - l_1)}.$$

As V_t, l_2 and l_1 are known, n can be calculated.

Example 7.18. *Resonance was observed with an air column when its length was* 16 cm *and again when it was* 50 cm. *If the frequency of the tuning fork is 512, calculate (i) the velocity of sound and (ii) end correction.*

Here $l_1 = 16$ cm, $l_2 = 50$ cm $n = 512$

$$V_t = 2n\,(l_2 - l_1) = 2 \times 512\,(50 - 16)$$

$$= \mathbf{34816\ cm/s}$$

$$n = \frac{l_2 - 3l_1}{2} = \frac{50 - 48}{2} = \mathbf{1\ cm.}$$

Example 7.19. *If the lengths of the first and second resonating air columns are* 16·5 cm *and* 51·5 cm *respectively with a tuning fork of frequency* 512 Hz, *calculate the velocity of sound in air.* (Osmania, 1992)

Here $l_1 = 16{\cdot}5$ cm

$$l_2 = 51{\cdot}5 \text{ cm}$$

$$l_2 - l_1 = 51{\cdot}5 - 16{\cdot}5 = 35 \text{ cm}$$

$$= 0{\cdot}35 \text{ m}$$

$$n = 512 \text{ Hz}$$

$$V = 2n\,(l_2 - l_1)$$

$$= 2 \times 512 \times 0{\cdot}35$$

$$= \mathbf{358{\cdot}4\ m/s.}$$

7.9. Organ Pipe

It is a wind instrument. It consists of the mouthpiece M, a slot A and a lip L (Fig. 7.6). If the other end of the pipe is closed it is called a closed end organ pipe. If the other end is open it is called an open end organ pipe. Air is blown

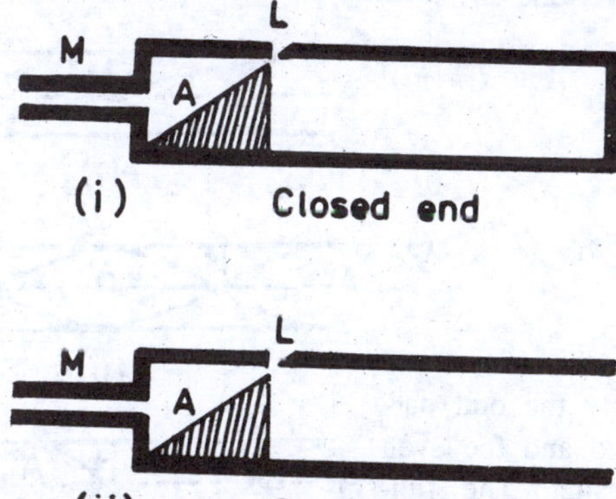

Fig. 7.6.

through the mouthpiece M and is ejected out through the slot A. The lip L vibrates. The air column acts as a resonator. The air column resonates with a particular fundamental frequency depending on its length. The open end behaves as an antinode and the closed end as a node.

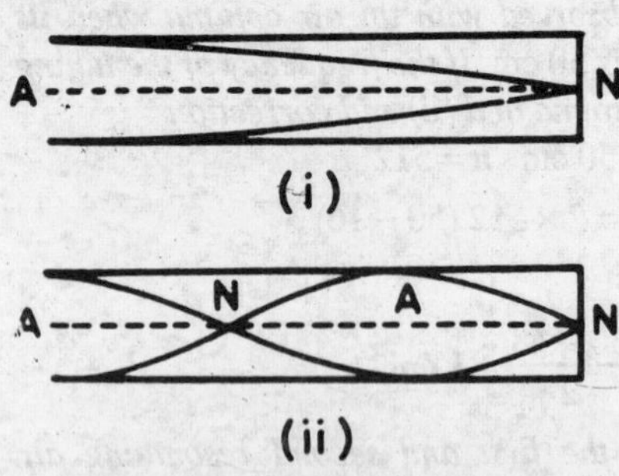

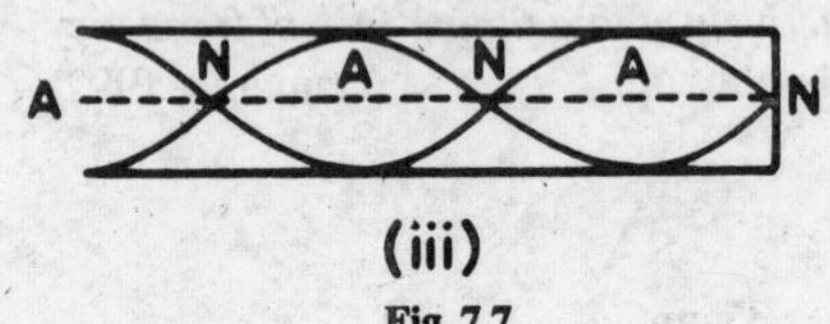

Fig. 7.7.

7.10. Closed End Organ Pipe

Suppose the length of the pipe is l. For the fundamental frequency of vibration of the air column

$$l = \frac{\lambda_1}{4} \qquad \text{[Fig. 7.7 (i)]}$$

Frequency

$$n_1 = \frac{V}{\lambda_1} = \frac{V}{4l} \qquad \ldots (1)$$

For the first overtone

$$l = \frac{3\lambda_2}{4} \qquad \text{[Fig. 7.7 (ii)]}$$

Frequency

$$n_2 = \frac{V}{\lambda_1} = \frac{3V}{4l}$$

$$= 3n_1 \qquad \ldots (2)$$

For the second overtone

$$l = \frac{5\lambda_3}{4} \qquad \text{[Fig. 7.7 (iii)]}$$

Frequency

$$n_3 = \frac{V}{\lambda_3} = \frac{5V}{4l} = 5n_1 \qquad \ldots (3)$$

Therefore, in the case of a closed end organ pipe, only the odd harmonics are produced and the even harmonics are missing. The fundamental frequency = $V/4l$.

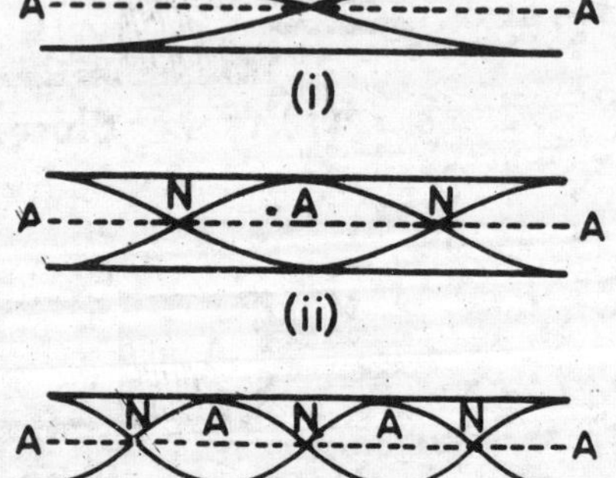

Fig. 7.8.

7.11. Open End Organ Pipe

In an open end pipe antinodes are present at both the ends.

Suppose the length of the pipe is l.

For the fundamental frequency of vibration of the air column

$$l = \frac{\lambda_1}{2} \qquad \text{[Fig. 7.8 (}i\text{)]}$$

$$n_1 = \frac{V}{\lambda_1} = \frac{V}{2l} \qquad \ldots (1)$$

For the first overtone

$$l = \lambda_2 \qquad \text{[Fig. 7.8. (}ii\text{)]}$$

Frequency

$$n_2 = \frac{V}{\lambda} = \frac{V}{l}$$

$$= 2n_1 \qquad \ldots (2)$$

For the second overtone

$$l = \frac{3\lambda_3}{2} \qquad \text{[Fig. 7.8 (}iii\text{)]}$$

Frequency $\quad n_3 = \frac{V}{\lambda_3} = \frac{3V}{2l} = 3n_1 \qquad \ldots (3)$

Therefore the harmonics having frequencies $n_1, 2n_1, 3n_1$ . . . etc. are produced. Hence all the harmonics are present.

Example 7.20. *Compare the lengths of an open end and closed end pipes which emit the same fundamental note.*

For a closed pipe

$$n_c = \frac{V}{4l_c} \qquad \ldots (1)$$

For an open pipe

$$n_0 = \frac{V}{2l_0} \qquad \ldots (2)$$

Here $\quad n_0 = n_c$

$$\therefore \quad \frac{V}{4l_c} = \frac{V}{2l_0} \text{ or } \frac{l_0}{l_c} = 2.$$

Example 7.21. *Compare the fundamental frequencies of an open end and closed end pipes of the same length.*

$$n_0 = \frac{V}{2l_0}, \quad n_c = \frac{V}{4l_c}$$

$$\frac{n_o}{n_c} = \frac{2l_c}{l_0} \qquad \text{But } l_0 = l_c$$

$$\frac{n_o}{n_c} = 2.$$

Example 7.22. *Two open organ pipes of lengths* 50 cm *and* 50·5 cm *produce 3* beats/second. *Calculate the velocity of sound in air.*

Here, $l_1 = 50$ cm, $l_2 = 50·5$ cm

Suppose the velocity of sound = V

$$n_1 = \frac{V}{2l_1} = \frac{V}{100}$$

$$n_2 = \frac{V}{2l_2} = \frac{V}{2 \times 50·5} = \frac{V}{101}$$

Beats/second = 3

$$n_1 - n_2 = 3$$

$$\therefore \quad \frac{V}{100} - \frac{V}{101} = 3 \text{ or } \frac{V}{10100} = 3$$

or $\qquad$ $V =$ **30,300 cm/second.**

Example 7.23. *Two tuning forks A and B give 6* beats/second. *A resounds with a closed column of air* 15 cm *long and B with an open column 30·5* cm *long. Calculate their frequencies.*

Suppose, the frequency of $A = n_1$

$$n_1 = \frac{V}{4l_1} = \frac{V}{4 \times 15} = \frac{V}{60}$$

Frequency of B $= n_2$

$$n_2 = \frac{V}{2l_2} = \frac{V}{2 \times 30·5} = \frac{V}{61}$$

$\therefore$ Beats/second $= 6$ or $n_1 - n_2 = 6$

$$\frac{V}{60} - \frac{V}{61} = 6 \text{ or } \frac{V}{3660} = 6$$

or $\qquad$ $V =$ **21960 cm/second.**

Example 7.24. *An organ pipe filled with air has a fundamental frequency of 500* vibrations/s. *The first harmonic of another organ pipe closed at one end and filled with carbon dioxide has the same frequency as that of the first harmonic of the open organ pipe. Calculate the length of each pipe. (Velocity of sound in air = 330* m/s, *velocity of sound in CO_2 = 264* m/s.)

(1) **Open end pipe** containing air

$$n = \frac{V}{2l}$$

Here, $n = 500, \quad V = 330$ m/s

$$500 = \frac{330}{2l} \text{ or } l = \mathbf{0{\cdot}33 \text{ metre.}}$$

(2) **Closed end pipe** containing carbon dioxide

$$n = \frac{V}{4l}$$

Here $n = 500, \quad V = 264$ m/s

$$500 = \frac{264}{4l} \text{ or } l = \mathbf{0{\cdot}132 \text{ metre.}}$$

7.12. Vibrations in Rods

Sound waves travel through solids, liquids and gases. The velocity of a sound wave in an isotropic medium is given as

$$u = \sqrt{\frac{E}{\rho}}$$

The elasticity of a solid is higher as compared to liquids and gases.

It has been found that velocity of sound through iron is higher than the velocity of sound through water or air.

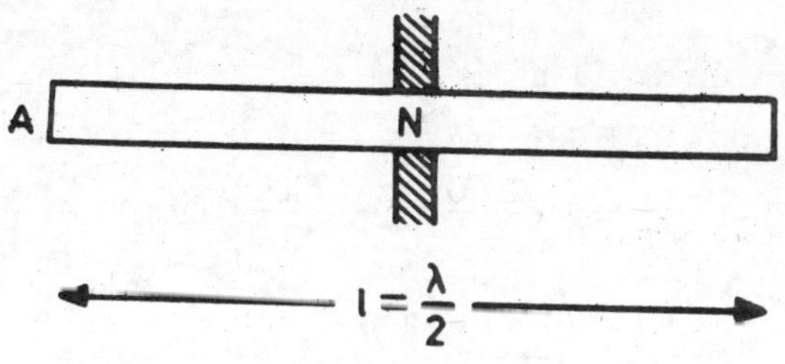

Fig. 7.9.

Consider a metal rod rigidly clamped at the mid point (Fig. 7.9). If one end of the rod is gently rubbed with a resined cloth, longitudinal stationary vibrations are set up in the rod. At the clamped position, there is a node (compression). At the free ends, there are antinodes (rarefactions). The distance between the two ends of the rod is equal to half the wave length of sound through the rod.

Here $l = \frac{\lambda}{2}$ or $\lambda = 2l$

The longitudinal vibrations in rods has been used in experiments for determining the velocity of sound through iron, glass, wood, brass, copper etc.

Example 7.25. *The frequency of the note next higher to the fundamental, as given by a rod of an alloy,* 1 m *long and clamped at its mid-point is 1,000. If the density of the alloy be* $7{\cdot}5 \times 10^3$ kg-m^{-3}. *Calculate the value of the Young's modulus for it.*

(Calicut, 1992)

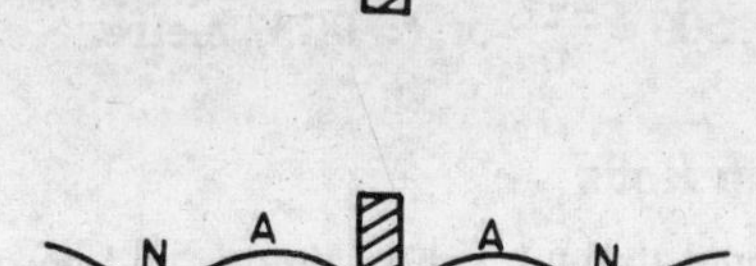

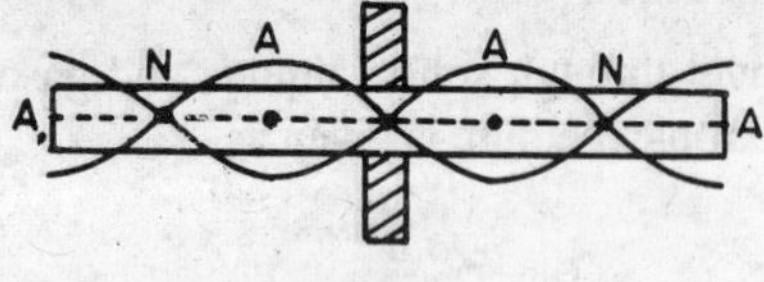

Fig. 7.9. A

Here $n_2 = 3n_1$

$$n_2 = 1000$$

$$n_1 = \left(\frac{1000}{3}\right)$$

$$l = 1 \text{ m}$$

$$v = n_1\lambda_1 = 2n_1 l$$

$$= 2 \times \frac{1000}{3} \times 1 = \left(\frac{2000}{3}\right) \text{ m/s}$$

and $\rho = 7{\cdot}5 \times 10^3$ kg/m^3

Also $v = \sqrt{\frac{E}{\rho}}$

$$E = \rho\, v^2$$

$$E = 7{\cdot}5 \times 10^3 \times \left[\frac{2000}{3}\right]^2$$

$$\mathbf{E = 3{\cdot}3 \times 10^9 \ N/m^2.}$$

7.13. Kundt's Tube

It consists of a glass tube. A cork B is attached to a wooden handle and can be moved inside the tube. The rod of the required material for which the

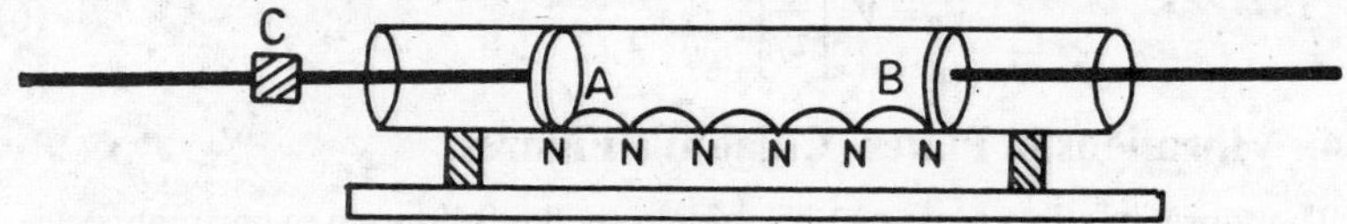

Fig. 7.10.

velocity of sound to be determined is clamped in the middle at C. The front end of the rod has an aluminium disc A attached to it. The tube contains air at room temperature and a little lycopodium powder is put along the length of the tube (Fig.7.10).

Velocity of sound in solids. When the rod is rubbed with a resined cloth gently, longitudinal vibrations are set up in the rod. The free ends of the rod behave as antinodes and the clamped point C acts as a node. These vibrations are transferred to the air column and the air column vibrates. The cork B is adjusted so that the air column resonates with the vibrations of the rod. At this stage, the lycopodium powder sets itself in the form of heaps at regular intervals of distance.

Let the length of the rod be l_r and the mean distance between two consecutive heaps be l_a.

$$\therefore \qquad l_r = \frac{\lambda_r}{2} \text{ and } l_a = \frac{\lambda_a}{2}$$

For the rod,
$$n = \frac{V_r}{\lambda_r} = \frac{V_r}{2l_r} \qquad \ldots (1)$$

For the air column
$$n = \frac{V_a}{\lambda_a} = \frac{V_a}{2l_a} \qquad \ldots (2)$$

From (1) and (2)

$$V_r = V_a \left[\frac{l_r}{l_a}\right] \qquad \ldots (3)$$

Velocity of sound in gases. Here the experiment is performed first with air and then with the required gas, using the same rod.

Suppose, the distance between two consecutive heaps in the air column = l_a and the mean distance between two consecutive heaps in the gas column = l_g.

From equation (3)

$$V_r = V_a \left(\frac{l_r}{l_a}\right) \qquad \ldots (4)$$

and $$V_r = V_g \left(\frac{l_r}{l_g} \right) \quad \ldots (5)$$

From (4) and (5)

$$V_g = V_{\bar{a}} \left[\frac{l_g}{l_a} \right] \quad \ldots (6)$$

7.14. Vibrations in Plates (Chladni's Figures)

The vibrations in a plate can be studied by the following experiment.

Fix a glass plate on a stand (Fig. 7.11). Sprinkle sand on the plate to form a uniform thin layer. Hold the plate at one corner and bow the middle of one side. It is noticed that the particles of sand arrange themselves into a symmetrical pattern as shown in Fig. 7.11.

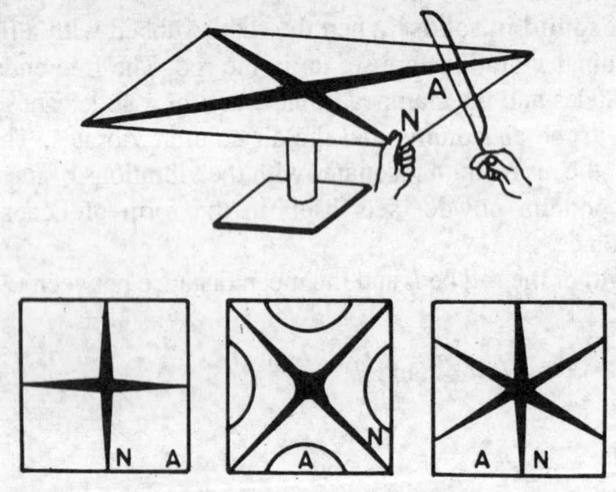

Fig. 7.11.

By holding the plate firmly at other points and bowing at different positions, different symmetrical patterns are obtained. In Fig. 7.11, the point N refers to the positions where the plate is firmly held and the point A refers to the positions where the plate is bowed. The position N corresponds to the node and the position A corresponds to the antinode. These patterns are called Chladni's Figures.

Each pattern of the Chladni's Figures, corresponds to a particular mode of vibration. However, these modes are not harmonically related in frequency.

Example 7.26. *A brass rod of length* 3 metres *is clamped in the middle. Calculate the frequency of the fundamental note emitted by it, when it vibrates longitudinally. Density of brass* = $8{\cdot}3$ g/ cm^3 *and Young's modulus of brass* = $10{\cdot}76 \times 10^{11}$ dynes/cm^2. (Delhi, 1974)

Here $v = \sqrt{\frac{Y}{\rho}}$ and $n = \frac{v}{2l}$

$\therefore$ $n = \frac{1}{2l}\sqrt{\frac{Y}{\rho}}$

Here $l = 3$ metres

$\rho = 8{\cdot}8$ g/cm^2 = 8300 kg/m^3

$Y = 10{\cdot}76 \times 10^{11}$ dyns/cm^2

$= 10{\cdot}76 \times 10^{10}$ newtons/m^2

$\therefore$ $n = \frac{1}{2 \times 3}\sqrt{\frac{10{\cdot}76 \times 10^{10}}{8300}}$

$n =$ **600 hertz.**

7.15. Experimental Demonstration of Nodes and Antinodes in Organ Pipes

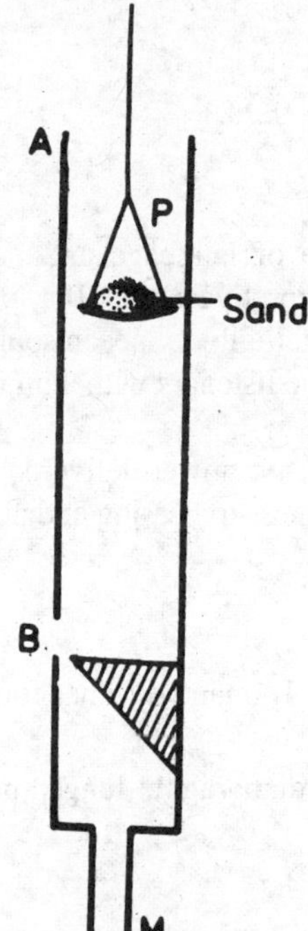

Fig. 7.12.

Take an organ pipe made of glass and held vertically. Air is blown through the opening at *M*. The air column vibrates and nodes and antinodes are formed.

Take a scale pan *P* having a small quantity of sand of fine grain (Fig. 7.12). Lower the scale pan, gradually inside the pipe. It is observed that at some positions, the sand particles remain completely at rest while at some other positions the sand particles vibrate. The positions where the sand particles are at rest, correspond to nodes. The positions where the sand particles vibrate correspond to positions of antinodes.

Initially air is blown with low pressure so that the air column vibrates with its fundamental frequency. In this case the open end at *A* will be an antinode and the closed end at *B* will be a node.

7.16. Musical Sound and Noise

Musical sound. The sound that produces a pleasing effect or sensation on the ear is called a musical sound. Moreover, the musical sound succeeds at regular intervals in quick succession and without any sudden change in loudness. The sounds produced by plucking a string of an instrument and by a harmonium are musical sounds. [Fig. 7.13 (*a*)].

Noise. The sound that produces jarring effect or displeasing effect on the ear is called a noise. The noise succeeds at irregular intervals and there is sudden change in loudness. The sounds produced by the gun and by a plate falling on the ground are examples of noise [Fig. 7.13 (*b*)].

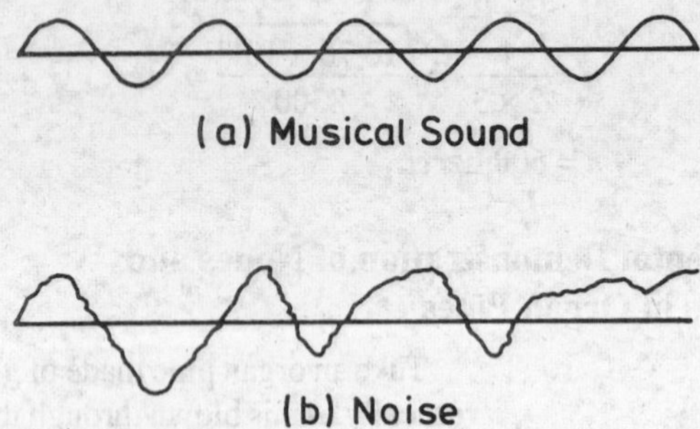

Fig. 7.13.

7.17. Speech

The speech basically refers to the structure of language and its main characteristics are loudness, pitch, timbre and interpretive aspect. If a speech is well recognized and understood it is said to be intelligible. Speech sounds are complex audible acoustic waves which provide the listeners with a number of clues for understanding.

An intelligible speech depends upon the acoustic power delivered during the speech, the characteristics of speech, sensitiveness to hearing and noises in the surroundings.

7.18. Human Voice

The human voice is a natural source of sound. Human voice has four main parts :

(1) **Power generator.** This comprises of diaphragm, lungs, bronchi, trachea and muscles.

(2) **Vibrator.** It is called larynx.

(3) **Resonators**. The acoustic resonators in the human system are nose, mouth, throat, other empty spaces in the mouth and sounding board *e.g.*, head, chest etc.

(4) **Articulators.** These are lips, tongue, teeth etc.

The loudness of sound in a human voice primarily depends upon the stream of air forced through the vocal cords from the lungs. The frequency of the human voice is dependent upon the elasticity and vibrations of the vocal cords. The quality of sound depends on the resonators.

7.19. Human Ear

Human ear is a natural sound receiver. The human voice and the human ear together form a fundamental and natural sound system.

The hearing mechanism is a highly sensitive electro-acoustic transducer. The human ear responds to sound waves of a wide range of frequencies, wave forms and intensity. It communicates acoustic pressure variation of the ear drum into pulses in the auditory nerve system. These pulses are communicated to the brain which in turn identifies and interprets these pulses. The brain converts these pulses into aural sensations *viz.* perception of sound.

The human ear responds to frequencies in the range 20 to 20,000 hertz. The range of sound intensity over which the ear is sensitive is 1 watt/m^2 to 10^{-12} watt/m^2. The human ear is more sensitive to variations in frequencies as compared to variations in sound intensities. The human ear is comparatively more sensitive to sounds of low intensity.

7.20. Characteristics of Musical Sound

There are three characteristics of musical sounds : (1) loudness or intensity, (2) pitch and (3) quality or timbre.

1. Loudness or Intensity. The amount of sound energy crossing per unit area around a point in one second is known as intensity of sound. Loudness depends upon intensity and also upon the sensitiveness of the ear. Loudness and intensity are related to each other by the relation

$$L \propto \log I$$

where L represents the sensation of loudness and I, the intensity of sound.

Loudness or intensity depends upon the following factors :

(*i*) **Amplitude**. Loudness is directly proportional to the square of the amplitude of the sounding body. The amplitude of sound produced by men is large and hence loud sound is produced. The amplitude of sound produced by ladies or children is small, therefore, the sound produced is feeble. Mosquito also produces a wave of small amplitude, therefore, the sound produced by a mosquito is also feeble.

(*ii*) **Surface area.** Loudness is directly proportional to the surface area of the sounding body. A tuning fork of large size produces a loud sound as compared to a tuning fork of small size. Beating drums with large surfaces produce a loud sound as compared to the beating drums with small surface area. A tuning fork ordinarily produces a feeble sound. When its stem is pressed against a table, a loud sound is produced. The particles of the table are forced to vibrate with the frequency of the tuning fork and the apparent surface area increases, hence a loud sound is produced.

(*iii*) **Distance between the source and the listener.** The intensity of sound is inversely proportional to the square of the distance between the source and the listener, provided the source produces sound waves in all directions. Therefore, the sound becomes feeble and feeble with increase in distance between the listener and the source.

(*iv*) **Density of the medium.** The greater the density of the medium, the louder is the sound. When the density of the medium is decreased, the sound becomes feeble.

(*v*) **Motion of air.** If air is blowing in the direction of propagation of the sound waves, loudness increases. If air is blowing in a direction opposite to the direction of propagation of the sound waves, loudness decreases.

2. Pitch. It is a sensation that depends upon the frequency. A shrill sound is produced by a source of high frequency whereas the pitch is lower if the frequency is lower. Pitch does not depend upon loudness or quality. The voice produced by ladies and children has high pitch because the frequency is high. The voice of an old man has low pitch and is hoarse because the frequency of sound is low. The frequency of the sound produced by a mosquito is of high pitch due to high frequency. The pitch of sound changes due to Droppler's principle when either the source or the observer or both are in motion.

3. Quality or Timbre. It depends on the presence of overtones. The quality of sound enables us to distinguish between two sounds having the same loudness and pitch. A sounding body produces waves of frequency $2n$, $3n$, $4n$, etc., where n is the fundamental frequency. Nature has provided different overtones in the voice of different persons. Due to the quality of sound one can recognise his friend from his voice without seeing him.

Suppose a person is calling you. The loudness and pitch will tell us whether it is a voice from a man, a lady or a child. The quality will further help us, to find the particular person producing the sound, man, woman or child. That is why quality plays a very important part. Otherwise, voices produced by all men would have been similar.

The roaring of a lion has high amplitude but low frequency. Therefore, the roaring of a lion can be heard even at far away places, but it is not a shrill sound.

The humming of a mosquito or a bee has low amplitude but *high frequency.* The sound is not heard when the mosquito is even a few metres away from the ear but it is a shrill sound.

7.21. Intensity of Sound

The intensity of sound is defined as the average rate of transfer of energy per unit area, the area being perpendicular to the direction of propagation of sound. Determination of the intensity of sound is important in practical acoustics.

Amount of energy transfer per unit area per second

$$I = 2\pi^2 \rho n^2 a^2 v \qquad \ldots (1)$$

Here ρ is the density of the medium, n is the frequency, a is the amplitude and v is velocity of sound.

Velocity of sound,

$$v = \sqrt{\frac{E}{\rho}}$$

and

$$E = -\frac{p}{dV/V}$$

dV is the change in volume, V the original volume and p is the excess of pressure

$$v = \sqrt{\frac{-p}{\left(\frac{dV}{V}\right)\rho}}$$

Taking $\frac{dV}{V} = \frac{dy}{dx}$ and simplifying

$$p = -v^2 \rho \frac{dy}{dx} \qquad \ldots (2)$$

A simple harmonic wave is represented by the equation

$$y = a \sin \frac{2\pi}{\lambda}(vt - x)$$

$$\frac{dy}{dx} = -\frac{2\pi a}{\lambda} \cos \frac{2\pi}{\lambda}(vt - x)$$

Substituting this value of $\frac{dy}{dx}$ in equation (2)

$$p = \frac{2\pi a v^2 \rho}{\lambda} \cos \frac{2\pi}{\lambda}(vt - x) \qquad \ldots (3)$$

The maximum excess of pressure

$$p_{max} = \frac{2\pi a v^2 \rho}{\lambda} \qquad \ldots (4)$$

and

$$p = p_{max} \cos \frac{2\pi}{\lambda}(vt - x) \qquad \ldots (5)$$

$$p_{max} = 2\pi a \rho v \left(\frac{v}{\lambda}\right)$$

$$p_{max} = 2\pi a \rho v n \qquad \ldots (6)$$

From equations (1) and (6)

$$I = 2\pi^2 \rho n^2 a^2 v$$

$$I = \frac{(2\pi \rho n av)^2}{2\rho v}$$

$$I = \frac{p^2{}_{max}}{2\rho v} \qquad \ldots (7)$$

Equation (7) shows that the intensity of sound varies directly as the square of the excess of pressure. Therefore, in acoustics it is important to measure the excess of pressure to obtain the value of intensity of sound. For ordinary conversation $p_{max} = 0{\cdot}1$ newton/m^2.

In the case of ordinary conversation, the sound output per square metre is $1{\cdot}13 \times 10^{-5}$ watt. Human ear is an extremely sensitive organ and can detect sound intensity lower than this value. It has been found that ear can detect intensities as low as 10^{-12} watt/m^2.

7.22. Measurement of Intensity of Sound—Decibel and Phon

The intensity of sound is defined as the quantity of energy propagating through a unit area per unit time, the direction of propagation being perpendicular to the area. The unity of intensity in the CGS system is ergs/cm^2-s and in SI system it is joules/m^2-s. The amount of power transmitted per unit area is measured in watts/m^2. A convenient unit is micro-watts/m^2.

The loudness of sound is just an aural sensation and it is physiological phenomenon rather than a physical phenomenon. The intensity of sound refers to the external or the objective measurement and the loudness refers to the internal or subjective aspect. Intensity of sound is a definite physical quantity and loudness is merely a degree of sensation. Loudness of sound increases with the intensity of sound according to Weber-Fecher law in physiology. According to this law, the loudness produced is proportional to the logarithm of intensity. This law does not hold good near the upper and the lower limits of audibility. According to this law,

$$S \propto \log I$$

or

$$S = K \log I$$

where K is a constant

or

$$\frac{dS}{dt} = \frac{K}{I}$$

The quantity dS/dI is called the sensitiveness of the ear. The sensitiveness of the ear decreases with increase in the intensity of sound.

In the case of all practical measurements, it is the relative intensity that is important and not the absolute value. Hence, the intensity of sound is often measured as its ratio to a standard intensity I_0. The intensity level is equal to I/I_0.

The standard intensity taken is 0·01 watt/m^2.

Suppose a person speakes in a normal conversational tone, he emits energy at the rate 10^{-5} joule/s. The mouth aperture is about 10^{-3} m^2 while speaking. If he speaks into a short tube the whole of the sound energy spreads along the length of the tube and the intensity of sound is 10^{-2} watt/m^2. This value of sound intensity is the standard intensity I_0. The person hearing at the other end of the tube gets the feeling of standard intensity. If a person shouts into the tube as loud as he can, the intensity will be $100I_0$. When the intensity is $100\ I_0$, up to $1000\ I_0$, the listener feels pain.

The faintest sound that can be heard depends also upon the frequency of the note. The average person's threshold of audibility is about $10^{-10}\ I_0$ for a frequency of 440 hertz. Thus, the range of hearing for a human ear is from $10^{-10}\ I_0$ upto $100\ I_0$. Hence the human ear has a dynamic range of 10^{12} in intensity.

7.23. Bel

Whenever the intensity of sound increases by a factor of 10, the increase in intensity is said to be 1 **bel.** Therefore, the dynamic range of audibility of the human ear is 12 **bels** or 120 decibles. When intensity increases by a factor $10^{0·1}$, the increase in intensity is 0·1 bel or 1 decibel.

Suppose the loudness is S for intensity I and S_0 for intensity I_0:

$$\therefore \quad S = K \log_{10} I \quad \dots (1)$$

$$S_0 = K \log_{10} I_0 \quad \dots (2)$$

The intensity level L is the difference in loudness.

$$L = S - S_0$$

$$L = K \log_{10} I - K \log_{10} I_0$$

$$L = K \log_{10} \left(\frac{I}{I_0} \right)$$

When the value of $K = 1$,

$$L = \log_{10} \left(\frac{I}{I_0} \right)$$

Here the unit of L is bel.

If L is measured in decibels,

$$L = 10 \log_{10}\left(\frac{I}{I_0}\right) \text{decibels.}$$

Suppose the intensity level changes by 1 decibel, then

$$1 = 10 \log_{10}\left(\frac{I}{I_0}\right)$$

or
$$\frac{I}{I_0} = 1{\cdot}26.$$

It means that the intensity level alters by 1 decibel when the intensity of sound changes by 26%.

The lowest change in intensity level that can be detected by the human ear is 1 decibel.

Example 7.27. *Calculate the change in intensity level when the intensity of sound increases* 100 times *its original intensity.*

Here, Initial intensity = I_0

Final intensity = I

$$\frac{I}{I_0} = 100$$

Increase in intensity level $= L$

$$L = 10 \log_{10}\left(\frac{I}{I_0}\right)$$

$$L = 10 \log_{10} 100 = \mathbf{20\ decibels.}$$

Example 7.28. *Calculate the change in intensity level when the intensity of sound increases by 10^6 times its original intensity.*

Here, Initial intensity = I_0

Final intensity = I

$$\frac{I}{I_0} = 10^6$$

Increase in intensity level, $L = 10 \log_{10}\left(\frac{I}{I_0}\right)$

$$L = 10 \log_{10} (10^6)$$

$$L = \mathbf{60\ decibels.}$$

Table 7.2

Intensity Levels of Different Sounds

Source of Sound	*Intensity Level in decibels*
Threshold of hearing	0
Rustle of leaves	10
Whisper	15-20
Normal conversation	60-65
Heavy traffic	70-80
Roaring of a lion (at a distance of 6 m)	90
Thunder	100-110
Painful Sounds	130 and above

7.24. Phon

The intensity levels given in the above table refer to the loudness in decibels with the assumption that the threshold of audibility is the same, irrespective of the pitch of the sound. However, the sensitivity of the ear and the threshold of audibility vary over wide ranges of frequency and intensity. Hence, the intensity level will be different at different frequencies even for the same value of I/I_0. For measuring the intensity level, a different unit called the *phon* is used. The measure of loudness in phons of any sound is equal to the intensity level in decibels of an equally loud pure tone of frequency 1000 hertz. Thus, the phon scale and the decibel scale agree for a frequency of 1000 hertz but the two values differ at other frequencies.

Suppose the intensity level of a note of frequency 480 hertz is to be determined. A standard source of frequency 1000 hertz is sounded and the intensity of the standard source is adjusted so that it is equal to the loudness of the given note of frequency 480. The intensity level of the standard source in decibels is numerically equal to the loudness of the given source in phons.

Suppose a note of frequency 3000 hertz and intensity level 70 decibels gives the same loudness as a standard source of frequency 1000 hertz at intensity level 67 decibels. The intensity level of the note of frequency 3000 hertz is 67 *phons.*

7.25. Musical Scale

Chord. When two or more notes are sounded together, the combined note produced is called *chord.* If the combined note produces a pleasing effect on the ear, it is called *concord.* If the combined note produces a displeasing effect on the ear it is called *discord.*

Harmony. When two or more notes are sounded simultaneously, the combined note, producing a pleasing effect on the ear, is called *harmony.*

In English music, where a number of instruments play simultaneously, harmony is produced.

Melody. When two or more notes are sounded one after the other, the combined note producing pleasing effect on the ear, is called *melody.* In Indian folk music, melody is produced.

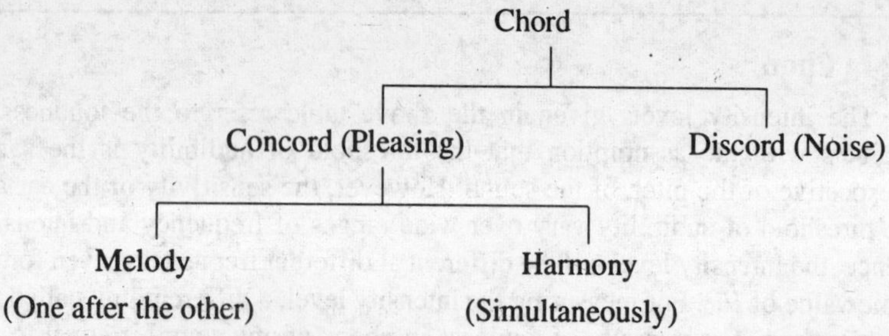

Interval. The ratio between the frequencies of two notes is called a musical interval. It is equal to 1 or more than 1. It is never less than 1.

Unison. Here the musical interval =1

$$\frac{N_2}{N_1} = 1 \text{ or } N_2 = N_1.$$

The two notes are said to be in unison, when their frequencies are equal.

Octave. Here the musical interval = 2

$$\frac{N_2}{N_1} = 2 \text{ or } N_2 = 2N_1.$$

Here N_2 is the octave of N_1.

Major tone. Here the musical interval $= \frac{9}{8}$

Minor tone. Here the musical interval $= \frac{10}{9}$

Semi tone. Here the musical interval $= \frac{16}{15}$

Fifth tone. Here the musical interval $= \frac{3}{2}\cdot$

Diatonic musical scale. This scale has eight keys and seven intervals (Fig. 7.14).

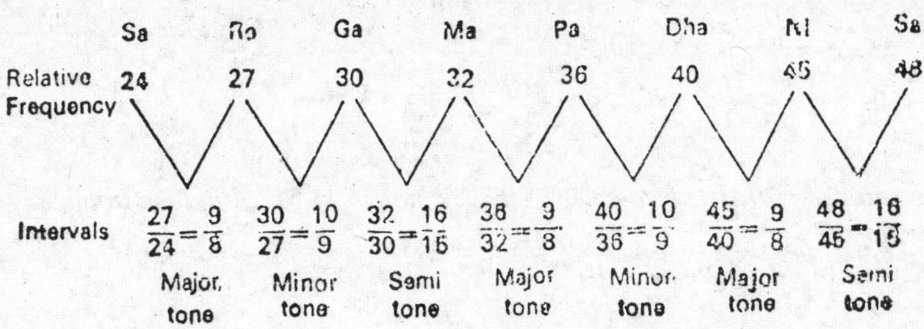

Fig. 7.14.

In an *equally tempered scale,* there are thirteen keys and 12 intervals (Fig.7.15). The intervals are equal and each interval = $2^{1/12}$

C	C_s	D	D_s	E	F	F_s	G	G_s	A	A_s	B	C′
$2^{\frac{0}{12}}$	$2^{\frac{1}{12}}$	$2^{\frac{2}{12}}$	$2^{\frac{3}{12}}$	$2^{\frac{4}{12}}$	$2^{\frac{5}{12}}$	$2^{\frac{6}{12}}$	$2^{\frac{7}{12}}$	$2^{\frac{8}{12}}$	$2^{\frac{9}{12}}$	$2^{\frac{10}{12}}$	$2^{\frac{11}{12}}$	2

Fig. 7.15.

In this scale five notes have been introduced. They are C_s, D_s, F_s, G_s, and A_s. The main advantage of this scale is that the interval is the same between the consecutive notes and a singer can conveniently use any key as his foundamental.

Suppose $C = 256$

$$D = 256 \times \frac{9}{8} = 288$$

$$E = 288 \times \frac{10}{9} = 320$$

$$F = 320 \times \frac{16}{15} = \frac{1024}{3} = 341 \cdot 33$$

$$G = \frac{1024}{3} \times \frac{9}{8} = 384$$

$$A = 384 \times \frac{10}{9} = \frac{1280}{3} = 426{\cdot}66$$

$$B = \frac{1280}{3} \times \frac{9}{8} = 480$$

$$C' = 480 \times \frac{16}{15} = 512.$$

Example 7.29. *The frequency of the key note C = 512. Find the frequency of the note G.*

$$C = 512$$

$$G = 512 \times \frac{36}{24} = 768 \text{ (Because, if } C = 24,\ G = 36).$$

7.26. Limits of Audibility

The limits of audibility of sound depend upon the *intensity* and *frequency* of sound. In order that a sound is audible, it must have a certain minimum intensity and a certain minimum frequency. The minimum intensity of sound necessary for the sound to be audible, is called *threshold intensity* of audibility. The minimum audible frequency is called *lower pitch limit* of audibility.

There is also a maximum intensity limit, beyond which the sound produces a sensation of pain on the ear. Similarly, there is a maximum frequency limit, beyond which the sound is not audible. The maximum audible intensity limit is called the threshold intensity of feeling. The maximum audible frequency limit is called upper pitch limit of audibility. The minimum audible intensity and the maximum audible intensity vary with the frequency of sound.

The limits of audibility can be understood with the help of an audiogram drawn by Wegel (Fig. 7.16). In an audiogram, the frequency is plotted along the *X*-axis on a logarithmic scale and the intensity (pressure in newton/m^2) is plotted along the *Y*-axis on a logarithmic scale. In the figure, curve 1 represents the threshold of audibility and curve 2 represents the threshold of feeling. These two curves when extrapolated, enclose an area on the audiogram. This enclosed area is called the auditory sensation area. This audiogram represents the auditory sensation area for a normal human ear. If the pitch and the intensity of any sound are beyond the limits set by the auditory sensation area, the sound is not heard. The sound wave may be felt by the ear as a pressure but not as an audible sound.

The human ear is sensitive for frequencies in the range 500 hertz to 7000 hertz. This range of frequency represents the range of ordinary speech. The peak sensitiveness of the ear ranges from 2000 to 2500 hertz. Beyond a frequency of 7000 hertz the human ear is insensitive. The frequency range used in music extends from 40 to 4000 hertz.

The threshold intensity increases both at high and low frequency levels. The increase is more at the low frequency level. The intensity that causes painful sensation in the ear is maximum at a frequency of about 800 hertz. However, at this frequency, the threshold of audibility is low. At a frequency of 1000 hertz, the ratio between the two pressure amplitudes is $10^7 : 1$. The corresponding intensity ratio is $10^{14} : 1$. This ratio is enormously high.

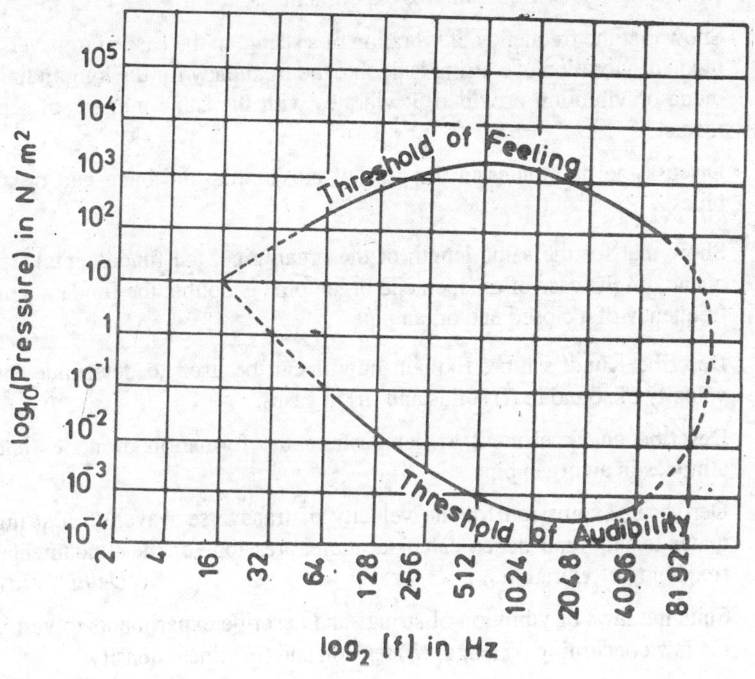

Fig. 7.16.

The lowest audible frequency is 30 hertz whereas the highest is 20000 hertz. The audible frequency range for the ear between 30 hertz and 20000 hertz consitutes a range of eleven octaves. However, in the case of the human eye, the range of frequency in the visible region ($3{\cdot}75 \times 10^{14}$ hertz to $7{\cdot}5 \times 10^{14}$ hertz) constitutes only one octave. Out of the eleven audible octaves, seven are available in music.

EXERCISES

1. Derive an expression for the velocity of a simple harmonic wave in a stretched string.
2. Show that the fundamental frequency of vibration of a stretched string is given by
$$n = \frac{1}{2l}\sqrt{\frac{T}{m}}.$$
3. Discuss the laws of transverse vibration of a stretched string and explain how the diameters of two wires can be compared using a sonometer.
4. Describe Melde's experiment and explain how the laws of vibration of strings can be verified with this experiment.
5. Show that the frequency of vibration of a string, in the case of transverse mode of vibration of a string is double the frequency for the longitudinal mode of vibration provided, it vibrates with the same number of segments.
6. Discuss the harmonics in the case of closed end and open end organ pipes.
7. Show that for the same length of the organ pipe, the fundamental frequency in the case of an open end organ pipe is double the fundamental frequency of a closed end organ pipe.
8. Describe Kundt's tube. Explain how it can be used to determine the velocity of sound in (*i*) solids and (*ii*) in gases.
9. Describe an experiment to demonstrate the formation of nodes and atinodes in an organ pipe.
10. Derive an expression for the velocity of transverse waves in a string under tension and hence calculate the expression for the fundamental frequency of vibration. *(Delhi, 1973)*
11. State the laws of vibration of strings and describe experiments to verify the law concerning (*i*) length (*ii*) tension and (*iii*) linear density. *(Delhi, 1973)*
12. Describe Melde's experiment. Show that the frequency of vibration of a string in the longitudinal mode is half that in the transverse mode of vibration. *(Delhi, 1971)*
13. Derive an expression for the frequency of transverse waves in a stretched string. *(Delhi, 1972)*
14. Discuss the production of harmonic in an open pipe. *(Delhi, 1972)*

15. Give the construction and working of a Kundt's tube. Explain, how it can be used to determine the velocity of sound in a metallic rod. *(Delhi, 1974)*

16. Discuss the characteristics of musical sound. *(Calcutta, 1973)*

17. Explain the use of Kundt's apparatus for the different acoustical determinations. *(Delhi, 1976)*

18. Derive an expression for the velocity of transverse waves in a stretched string and deduce the laws of vibrating strings. Describe how you would verify these laws with a sonometer. *[Delhi (Supple.), 1976]*

19. Discuss the vibrations of an air column in an open organ pipe. *[Delhi (Supple.), 1976]*

20. Describe and explain Melde's experiment. *[Delhi (Supple.), 1976]*

21. In Melde's experiment in the longitudinal mode of vibration a string vibrates in 4 loops with a load of 8 grams. How much load will be required in order that the string may vibrate in 8 loops in the transverse mode. *(Delhi, 1971)* [**Ans.** 8 grams]

22. What is resonance ? Explain through an experiment how this phenomena is used to determine the velocity of sound in air. *(Osmania, 1992)*

23. Deduce an expression for the velocity of transverse waves in a string. Hence find the frequency of the first octave for a vibrating string fixed at both ends. Point out the position of nodes and antinodes for this particular mode of vibration. *(Calcutta, 1992)*

24. Derive an expression for the velocity of transverse waves along a stretched string. *(Delhi, 1990)*

25. Describe the method of determining the velocity of sound by using a Kundt's tube. *(Gauhati, 1992)*

26. Discuss the theory of vibrations of an air column in an open organ pipe. Compare (*i*) the length of an open end pipe with that of a closed end pipe which emit the same fundamental note, and (*ii*) the fundamental frequencies of an open end and a closed end pipe of the same length. *(Delhi, 1991)*

27. Derive an expression for the velocity of transverse waves along a stretched string. *(Delhi, 1992)*

CHAPTER 8

Reflection, Refraction and Diffraction

8.1. Reflection of Sound

In every day life we come across many acoustic phenomena, *e.g.,* echoes, roaring of thunder, reverberation in buildings etc. All these phenomena are based on the reflection of sound waves. Reflection of sound plays an important role in many acoustic applications *e.g.,* speaking tubes, horns and trumpets, sounding boards etc. The stationary waves produced in stretched strings and the resonant vibrations produced in open and closed end organ pipes are also based on the principle of reflection of sound.

8.2. Reflection of a Plane Wave at a Plane Surface

Let XY be the plane reflecting surface and AMB the incident plane wavefront. All the particles on AB will be vibrating in phase. Let i be the angle of incidence (Fig. 8.1).

In the time the disturbance at A reaches C, the secondary waves from the point B must have travelled a distance BD equal to AC. With the point B as centre and radius equal to AC construct a sphere. From the point C draw tangents CD and CD'. Then $BD = BD'$.

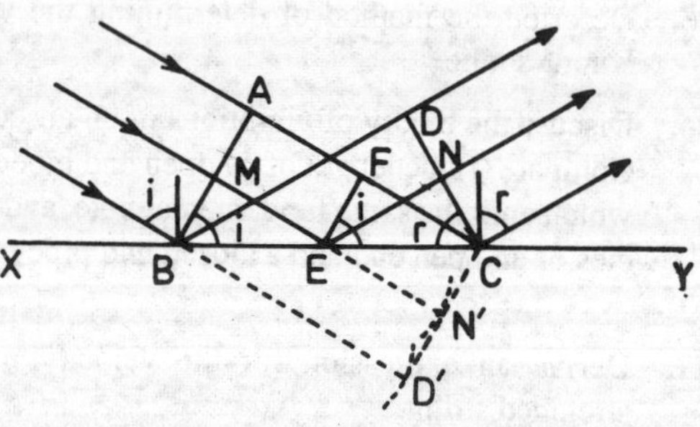

Fig. 8.1.

In the Δs BAC and BDC

BC is common

$BD = AC$

and $$\angle BAC = \angle BDC = 90°$$

The two triangles are congruent.

$$\therefore \quad \angle ABC = i = \angle BCD = r$$

$$i = r.$$

Thus the angle of incidence is equal to the angle of reflection. Hence, CD forms the reflected plane wavefront. It can be shown that all the points on CD form the reflected plane wavefront. In the time the disturbance from F reaches the point C, the secondary waves from E must have travelled a distance $EN = FC$. With E as centre and radius FC draw a sphere and draw tangents CN and CN' to the sphere. It can be shown that the triangles EFC and ENC are congruent.

$$AC = AF + FC$$

But $$AF = ME$$

and $$FC = EN$$

$$\therefore \quad AC = ME + EN.$$

Thus, all the secondary waves from the different points on AB reach the corresponding points on CD at the same time. Therefore, CD forms the reflected plane wavefront and also the angle of incidence is equal to the angle of reflection.

8.3. Experimental Demonstration of Reflection of Sound

In Fig. 8.2, AB is a plane reflecting surface *i.e.*, a polished metal plate.

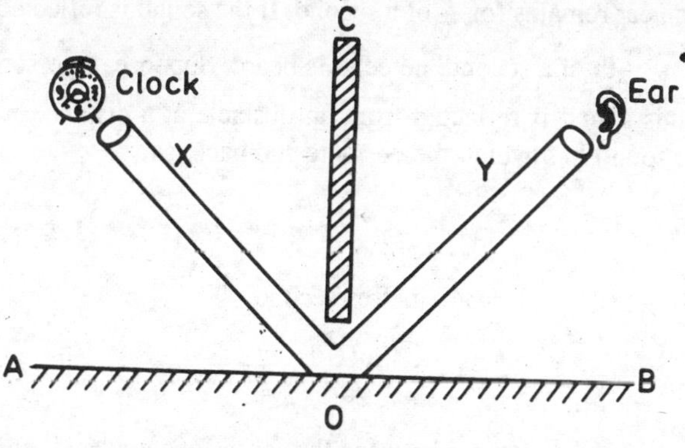

Fig. 8.2.

X and Y are two long tubes hinged at O. The two tubes can be moved in a vertical plane. C is an obstacle which prevents the propagation of sound through it.

An alarm clock is placed at the top end of the tube X and the ear is placed near the top end of the tube Y. Keeping X fixed, the position of the tube Y is gradually changed. At one position, maximum sound is heard. It is noticed that the tubes X and Y are equally inclined to the normal at O *i.e.*, the angle of incidence is equal to the angle of reflection.

8.4. Whispering Galleries

S is the position of a speaker and L is the position of an observer (Fig. 8.3). Even a very low intensity sound made by S (*i.e.*, a whisper) can be heard by the observer at L when he keeps his ear close to the wall. This is due to the multiple reflection of the sound waves from the curved wall of the gallery.

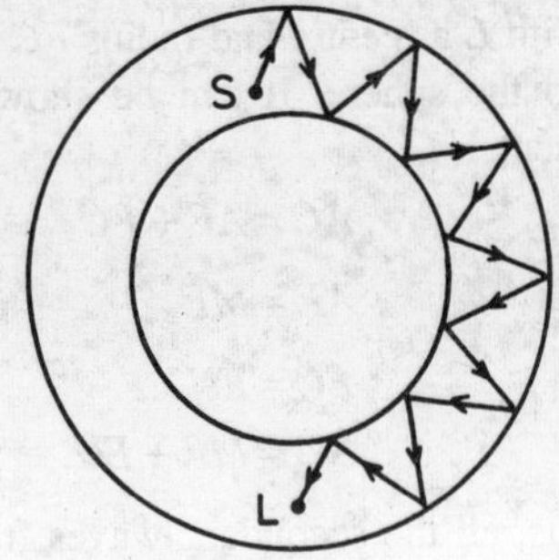

Fig. 8.3.

In **Salt Lake City**, the walls of a hall are elliptical in shape. Even if a small pin is dropped at one of its foci, the sound can be heard by the listener at the second focus.

8.5. Echo

The repetition of the sound produced due to reflection by a distant extended surface like a cliff, hill, well, building etc. is called an *echo*. The effect of sound on the human ear remains for $\frac{1}{10}$ of a second. If the sound is reflected back in a time less than $\frac{1}{10}$th of a second, no echo is heard. Suppose, a person produces sound and this sound is reflected from an obstacle at a distance x. The time taken by the sound to travel to the obstacle and back is t.

$$t = \frac{2x}{V}$$

Here $\qquad V = 340$ metres/second

$$x = \frac{Vt}{2} = \frac{340 \times 0{\cdot}1}{2} = 17 \text{ metres.}$$

It means that if the distance between the source of sound and the obstacle is less than 17 metres (56 ft) no echo (repetition) is heard. But echo is heard if the distance is more than 17 metres.

If a person is standing in between two hills and produces a sound, the sound is reflected again and again from the two hills and successive echoes are heard.

Example 8.1. *A ship is moving towards a cliff with a constant speed and the siren on the ship is sounded every minute. The echo of the first whistle is heard after* 20 seconds, *and that of the second after* 16·5 second. *Compute the original distance of the ship from the cliff and also its speed. Velocity of sound is* 350 m/s.

Let the distance between the ship and the cliff be x metres and the speed of the ship y metres/s.

The total distance travelled by sound in 20 s

$$= (2x - 20y) = 350 \times 20$$

$$\therefore \quad 2x - 20y = 7000 \qquad \ldots (i)$$

The second echo is heard after 16·5 s.

The distance travelled by sound in 16·5 s

$$= 2(x - 60y) - 16{\cdot}5\,y = 350 \times 16{\cdot}5$$

$$2x - 120y - 16{\cdot}5\,y = 5775$$

$$= 2x - 136{\cdot}5\,y = 5775 \qquad \ldots (ii)$$

Subtracting (*ii*) from (*i*)

$$116{\cdot}5\,y = 1225$$

$$y = 10{\cdot}52 \text{ m/s} \qquad \ldots (iii)$$

From equation (*i*)

$$2x - 20 \times 10{\cdot}52 = 7000$$

$$2x = 7210{\cdot}4$$

$$x = \mathbf{3605{\cdot}2} \text{ m.}$$

Example 8.2. *A man standing between two parallel cliffs fires a rifle. He hears one echo after* $1\frac{1}{2}$ s, *another* $2\frac{1}{2}$ s *and a third one after 4* s. *If the distance between the two cliffs is 700* m *what is the velocity of sound and what is the distance of the man from one of the cliffs.*

The third echo is heard after the sound has been reflected from the first and the second cliffs.

$\therefore$ Total distance between the two cliffs

$$= 700 \text{ m}$$

Total distance travelled by sound

$$= 2 \times 700 = 1400 \text{ m}$$

$$\text{Time } = 4 \text{ s}$$

$$\therefore \text{ Velocity of sound} = V = \frac{S}{t} = \frac{1400}{4} = 350 \text{ m/s}$$

The first echo is heard after $1\frac{1}{2}$ s $= \frac{3}{2}$ s

∴ Time taken by the sound to travel from the man to the first cliff

$$= \frac{3}{4} \text{ s}$$

Distance between the man and the first cliff

$$S_1 = 350 \times \frac{3}{4} = \mathbf{262{\cdot}5 \text{ m}}.$$

Example 8.3. *A man stationed between two parallel cliffs fires a gun. He hears the first echo after* 2 seconds *and the next after* 5 seconds. *What is the distance between the two cliffs? Velocity of sound in air is 350* m/s.

The first echo is heard after 2 seconds

∴ Time taken by sound to travel from the man to the first cliff

$$= \frac{2}{2} = 1 \text{ s}$$

$$V = 350 \text{ m/s}$$

Distance between the man and the first cliff $S_1 = 350 \times 1$

$$= 350 \text{ m}$$

The second echo is heard after 5 seconds

∴ Time taken by sound to travel from the man to the second cliff

$$= \frac{5}{2} \text{ seconds}$$

Distance between the man and the second cliff

$$= S_2 = 350 \times \frac{5}{2} = 875 \text{ m}$$

∴ Total distance between the two cliffs

$$= S_1 + S_2 = 350 + 875 = \mathbf{1225 \text{ m}}.$$

8.6. Application of Reflection of Sound

1. The principle of reflection of sound is used in an auditorium. A large plane sounding board (cylindrical or parabolic) is placed behind the speaker (Fig. 8.4). The speaker stands at the focus of the board. When the speaker speaks, the sound waves get reflected from the sounding board and the reflected sound travels towards the audience. This enables the increase in the intensity of sound and also uniform distribution of sound.

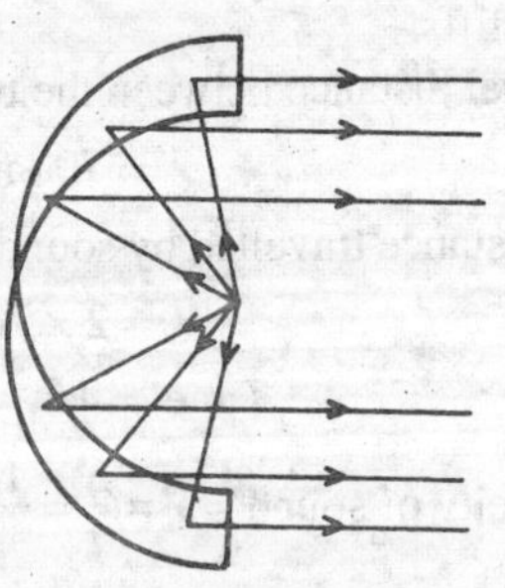

Fig. 8.4.

2. **Speaking tube or megaphone.** The speaking tube or megaphone consists of a conical metal horn (Fig. 8.5). At the narrow end of the tube, the speaker speaks. The sound waves get reflected from the inside wall of the tube and are propagated practically along the axis of the tube. This provides higher intensity of sound in a particular direction. Here, the sound energy is not spread in all directions but it is concentrated in a desired direction.

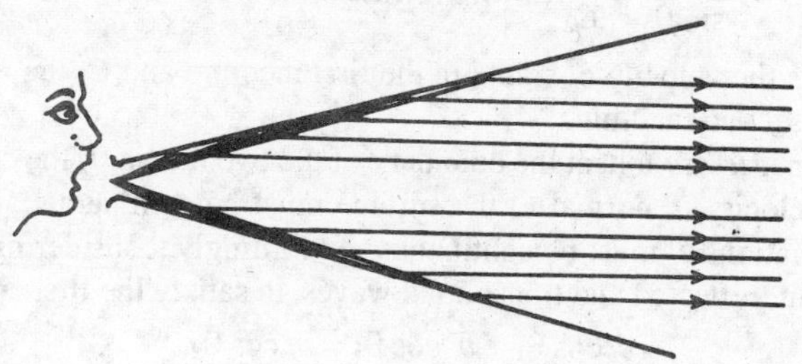

Fig. 8.5.

3. **Ear trumpet**. An ear trumpet is a useful device for persons who are hard of hearing. The sound waves that enter the wide end of the trumpet suffer multiple reflections. The sound entering the ear has large amplitude and thus higher intensity. This enables the person to hear even low intensity sounds.

8.7. Phase Change on Reflection

According to the laws of reflection and refraction, the velocity in each medium is independent of the direction. Let AB be a surface separating two media.

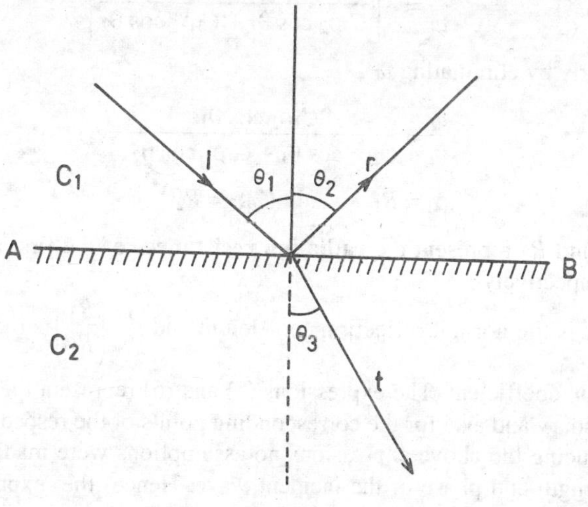

Fig. 8.6.

Here θ_1, θ_2 and θ_3 are the angles of incidence, reflection and refraction respectively (Fig. 8.6).

Also, $$\frac{\sin \theta_1}{C_1} = \frac{\sin \theta_2}{C_1} = \frac{\sin \theta_3}{C_2} \quad \ldots (1)$$

$\therefore$ $$\theta_1 = \theta_2 \text{ (law of reflection)}$$

and $$\frac{\sin \theta_1}{\sin \theta_3} = \frac{C_1}{C_2} \text{ (law of refraction).}$$

Here C_1 is the velocity of sound in the first medium and C_2 the velocity of sound in the second medium.

It is also necessary that at the boundary of the two media *(i)* the component of particle velocity v, normal to the surface must be continuous and *(ii)* the pressure, variation ∂P must be continuous. Accordingly, considering i, r and t as the incident, reflected and transmitted waves, to satisfy the first condition,

$$v_i \cos \theta_1 + v_r \cos \theta_1 = v_t \cos \theta_2 \quad \ldots (2)$$

To satisfy the second condition

$$\partial P_i + \partial P_r = \partial P_t \quad \ldots (3)$$

But $$\partial P = \pm C \rho v$$

$\therefore$ $$C_1 \rho_1 v_i - C_1 \rho_1 v_r = C_2 \rho_2 v_t \quad \ldots (4)$$

The negative sign for the second term indicates the reversed direction of the reflected wave.

Eliminating v_t from equations (2) and (4)

$$\frac{v_r}{v_i} = -\left[\frac{C_2\rho_2 \cos \theta_1 - C_1\rho_1 \cos \theta_2}{C_2\rho_2 \cos \theta_1 + C_1\rho_1 \cos \theta_2}\right] = r \quad \ldots (5)$$

Similarly by eliminating v_r ,

$$\frac{v_t}{v_i} = \frac{2C_1\rho_1 \cos \theta_1}{C_2\rho_2 \cos \theta_1 + C_1\rho_1 \cos \theta_2} = t_{12} \quad \ldots (6)$$

Take $$C_1\rho_1 = R_1 \quad \text{and} \quad C_2\rho_2 = R_2$$

where R_1 and R_2 represent the **radiation resistances** of the first and second medium respectively.

Here r^2 is the normal reflection coefficient and $t_{12}^2 \left(\frac{R_3}{R_1}\right)$ is the normal transmission coefficient. The expressions (5) and (6) represent the conditions at the boundary and also for the corresponding points of the respective waves. While deducing the above expressions no assumptions were made regarding the wavelength and phase of the incident wave. Hence the expressions are applicable to waves of any type.

Normal incidence. In the case of normal incidence

$$\theta_1 = \theta_2 = 0$$

$$\therefore \qquad \cos\theta_1 = \cos\theta_2 = 1$$

Substituting these values in equations (5) and (6),

$$\frac{v_r}{v_i} = \frac{R_1 - R_2}{R_1 + R_2} = r \qquad \ldots (7)$$

and

$$\frac{v_t}{v_i} = \frac{2R_1}{R_1 + R_2} = t_{12} \qquad \ldots (8)$$

In accordance with the principle of reversibility, if all the velocities in a dynamical system are reversed, the whole previous motion has to be obtained, provided there is no dissipation. Consequently, if the sound wave is transmitted from the second to the first medium,

$$t_{21} = \frac{2R_2}{R_1 + R_2}$$

and

$$t_{12}\, t_{21} + r^2 = 1 \qquad \ldots (9)$$

Equation (9) is independent of any assumption as regards change of phase at reflection.

Also incident energy must be equal to tø sum of the reflected and the transmitted energy.

$$\therefore \qquad E_i = E_r + E_t$$

But

$$E \propto C\rho\, v^2$$

$$\therefore \qquad C_1\rho_1 v_i^2 = C_1\rho_1 v_r^2 + C_2\rho_2 v_t^2$$

$$\therefore \qquad R_1 v_i^2 = R_1 v_r^2 + R_2 v_t^2$$

$$\therefore \qquad v_i^2 = v_r^2 + \left(\frac{R_2}{R_1}\right) v_t^2$$

Dividing by v_i^2

$$1 = \frac{v_r^2}{v_i^2} + \left(\frac{R_2}{R_1}\right)\frac{v_t^2}{v_i^2}$$

$$1 = r^2 + \left(\frac{R_2}{R_1}\right) t_{12}^2$$

or

$$t_{12}^2\left(\frac{R_2}{R_1}\right) + r^2 = 1 \qquad \ldots (10)$$

Special cases

(1) From equation (7), if $R_1 = R_2$

$$\frac{v_r}{v_i} = 0.$$

It means that there is no reflected wave and the whole of the energy is transmitted to the second medium.

(2) If $R_2 > R_1$ (*i.e.* wave travelling from air to water)

i.e. $C_2\rho_2 > C_1\rho_1$

$$\frac{v_r}{v_i} \text{ is } -\text{ve}$$

$\therefore$ v_r is –ve

It means that the particle velocity of the reflected wave is in reversed phase.

However, the quantity $\frac{v_r}{C} = S_r$, the phase of the condensation in the reflected wave remains unaltered because both v_r and C are reversed in sign.

(3) If $R_2 < R_1$ (*i.e.* wave travelling from water to air)

i.e. $C_2\rho_2 < C_1\rho_1$

$$\frac{v_r}{v_i} \text{ is } +\text{ ve}$$

It means that the phase of the particle velocity remains unchanged. However, the phase of the condensation is *reversed.*

(4) From equations (7) and (8),

$$\frac{S_r}{S_i} = -\left(\frac{v_r/C}{v_i/C}\right) = -\left(\frac{v_r}{v_i}\right) = \left(\frac{R_2 - R_1}{R_1 + R_2}\right)$$

$$\frac{S_r}{S_i} = \frac{R_2 - R_1}{R_1 + R_2} \qquad \ldots (11)$$

Also

$$\frac{S_t}{S_i} = \frac{2R_1}{R_1 + R_2} \qquad \ldots (12)$$

This means that the phase of the transmitted wave remains unchanged in all cases.

(5) If the velocity of sound C_1 in the first medium is less than the velocity of sound C_2 in the second medium, there will be a critical angle of incidence θ_1, when θ_2 is equal to 90°. In this case

$$\sin\theta_1 = \frac{C_1}{C_2}$$

For an angle of incidence greater than θ_1, there will be total internal reflection. Since the velocity of sound in water is greater than the velocity of sound in air, total internal reflection will occur only when the waves are incident from air on the water surface. Here, it must be remembered that total internal reflection in sound occurs only when the incident wave is travelling from rarer to denser medium and not from denser to rarer medium.

8.8. Transmission of Sound from Air to Water

From equations (7) and (8) of Article 8.7

$$\frac{v_r}{v_i} = \frac{R_1 - R_2}{R_1 + R_2} = r \qquad \ldots (1)$$

and

$$\frac{v_r}{v_i} = \frac{2R_1}{R_1 + R_2} = t_{12} \qquad \ldots (2)$$

From equations (11) and (12) of Article 8.7

$$\frac{S_r}{S_i} = \frac{R_2 - R_1}{R_1 + R_2} \qquad \ldots . (3)$$

$$\frac{S_t}{S_i} = \frac{2R_1}{R_1 + R_2} \qquad \ldots (4)$$

The above equations are applicable for normal incidence.

For air $\quad R_1 = \rho_1 C_1 = 0{\cdot}001293 \times 33200$

$= 40$ CGS units (app.)

and water $\quad R_2 = \rho_2 C_2 = 1 \times 15 \times 10^4$ CGS units

Substituting these values in equation (3)

$$\frac{S_r}{S_i} = 0{\cdot}9994.$$

It indicates that the incident wave is totally internally reflected almost completely. The phase of the condensation is not changed but the particle velocity is reversed (*i.e.* nodal condition).

To conclude, in general, in all cases of sound transmission, from gaseous to a solid or a liquid medium or *vice versa*, *i.e.* when the radiation resistance R_1 and R_2 are widely different, there is almost complete total internal reflection.

8.9. Refraction of a Plane Wavefront at a Plane Surface

Let XY represent the surface separating the media 1 and 2. v_1 and v_2 are the velocities of sound in the two media (Fig. 8.7). APB is the incident plane wavefront. By the time the distrubance at B reaches C, the secondary waves from A must have travelled a distance $AD = v_2 t$, where t is the time taken by the waves to travel the distance BC.

$\therefore \quad BC = v_1 \,.\, t$

and $\quad AD = v_2 \,.\, t$

With A as centre and radius AD (= $v_2 . t$) draw a sphere. Draw the tangent CD to the sphere from the point C. Then, CD represents the refracted plane wavefront. To prove that CD is the common wave front, it is enough to show

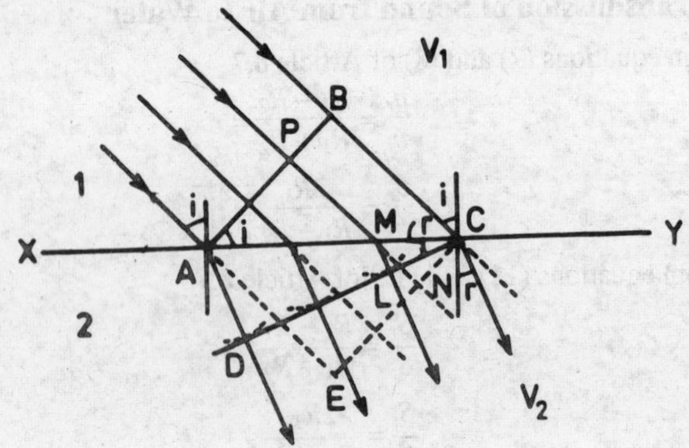

Fig. 8.7.

that in the time the disturbance travels from B to C or A to D, the disturbance at P reaches L. With the point M as centre draw a sphere such that CD happens to be the tangent to the sphere. From the Δs ACD and MCL.

$$\frac{AD}{ML} = \frac{AC}{MC} \qquad \ldots (1)$$

Similarly from the Δs ACE and MCN

$$\frac{AE}{MN} = \frac{AC}{MC} \qquad \ldots (2)$$

From equations (1) and (2)

$$\frac{AD}{ML} = \frac{AE}{MN}$$

or

$$\frac{AE}{AD} = \frac{BC}{AD} = \frac{v_1 t}{v_2 t} = \frac{MN}{ML}$$

or

$$\frac{BC}{AD} = \frac{MN}{ML} = \frac{v_1}{v_2} \qquad \ldots (3)$$

Hence, if AD is the radius of the secondary wavefront for the point A then ML is the radius of the secondary wavefront for the point M.

Let i and r be the angles of incidence and refraction respectively. From the Δs ABC and ACD

$$\frac{\sin i}{\sin r} = \frac{BC}{AC} \Big/ \frac{AD}{AC}$$

or $$\frac{\sin i}{\sin r} = \frac{BC}{AD} = \frac{v_1 t}{v_2 t} = \frac{v_1}{v_2}$$

Thus the bending or refraction of sound waves will depend upon the velocities of sound in the two media.

8.10. Experimental Demonstration of Refraction of Sound

The principle of refraction of sound waves can be demonstrated with a simple experiment. A rubber balloon *L* in the shape of a convex lens is taken and it is filled with carbon dioxide (Fig. 8.8). The density of carbon dioxide is more than air. Therefore, velocity of sound in carbon dioxide is less than in air. A source of sound is kept at *A*. The incident spherical wavefront *CPD*, after refraction through the lens *L*, gives rise to the refracted spherical wavefront *EQF*. The sound converges to the point *B* just similar to light waves. *B* is the sound image corresponding to the source *A*. The ear placed at *B* receives sound of maximum intensity.

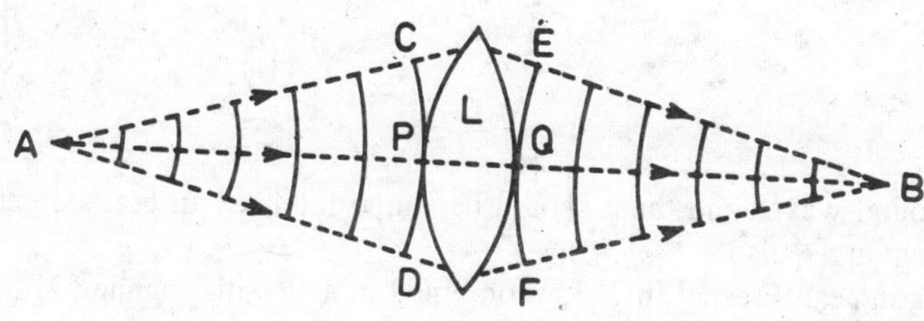

Fig. 8.8.

8.11. Diffraction of Sound

It is a matter of common experience that the sound is heard behind an obstacle placed in the path of sound. This bending of sound round an obstacle is called diffraction of sound. In the case of light, even for a very small obstacle, a shadow is observed and it is difficult to observe the illumination in the geometrical shadow. Both sound and light are propagated as waves in the medium and are diffracted.

8.12. Explanation of Diffraction of Sound

Each progressive wave, according to Huygens wave theory produces secondary waves, the envelope of which forms the secondary wavefront. In Fig. 8.9 (*i*), *S* is a source of sound and *MN* is an aperture. *XY* is a screen placed in the path of sound. *AB* is the portion of the screen where sound energy is received and above *A* and below *B*, it is the region of the geometrical shadow. Considering *MN* as the primary wavefront, then according to Huygens construction, if the secondary wavefronts are drawn one would expect encroachment of sound in the geometrical shadow. Thus, the shadows formed by obstacles are not sharp. **This bending of sound waves round the edges of an obstacle or the encroachment of sound within the geometrical shadow is called diffraction of sound**. Similarly if an opaque obstacle *MN* is placed in the path of sound [Fig. 8.9 (*ii*)], the back portion of the obstacle should also

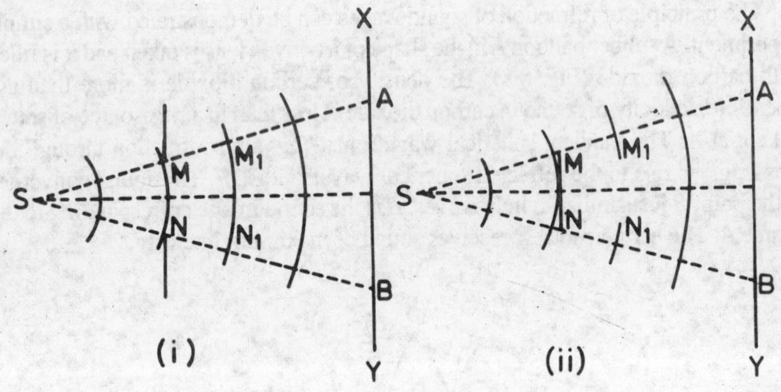

Fig. 8.9.

receive sound waves and there should be sound intensity in the geometrical shadow region *AB* also.

Augustin Jean Fresnel in 1815, combined in a striking manner, Huygens wavelets with the principle of interference and could satisfactorily explain the bending of light and sound waves round obstacles.

8.13. Fresnel's Assumptions

According to Fresnel, the resultant effect at an external point due to a wavefront will depend on the factors discussed below.

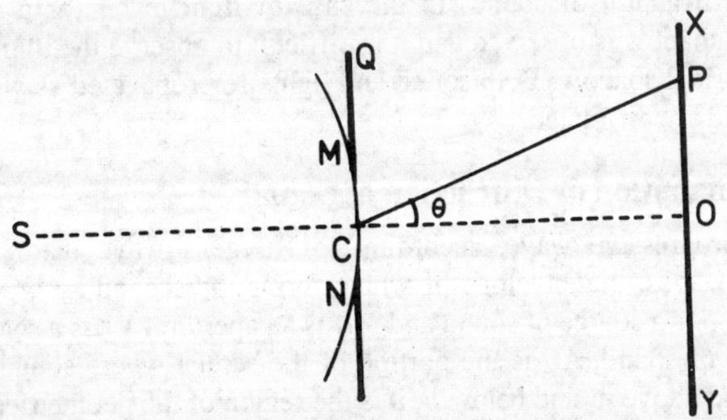

Fig. 8.10.

In Fig. 8.10, *S* is a source of sound and *MN* is an aperture. *XY* is the screen and *SO* is perpendicular to *XY*. *MCN* is the incident spherical wavefront due to the point source *S*. To obtain the resultant effect at a point *P* on the screen, Fresnel assumed that : (1) a wavefront can be divided into a large number of strips or zones called Fresnel's zones of small area and the resultant effect at any point will depend on the combined effect of all the secondary waves emanating from the various zones ; (2) the effect at a point due to any particular zone will depend on the distance of the point from the zone ; (3) the effect at *P* will also depend on the obliquity of the point with reference to the zone under consideration, *e.g.* due to the part of the wavefront at *C*, the effect will be maximum at *O* and decreases with increasing obliquity. It is maximum in a direction radially outwards from *C* and it decreases in the opposite direction. The effect at a point due to the obliquity factor is proportional to $(1 + \cos\theta)$ where $\angle PCO = \theta$. Considering an elementary wavefront at *C*, the effect is maximum at *O* because $\theta = 0$ and $\cos\theta = 1$. Similarly in a direction tangential to the primary wavefront at *C* (along *CQ*) the resultant effect is one-half of that along *CO* because $\theta = 90°$ and $\cos 90 = 0$. In the direction *CS* the resultant effect is zero since $\theta = 180°$ and $\cos 180 = -1$ and $1 + \cos 180 = 1 - 1 = 0$. This property of the secondary waves eliminates one of the difficulties experienced with the simpler form of Huygens principle *viz.*, that if the secondary waves spread out in all directions from each point on the primary wavefront, they should give a wave travelling forward as well as backward. As the amplitude at the rear of the wave is zero there will evidently be no back wave.

8.14. Intensity of Sound at a Point due to a Plane Wavefront

ABCD is a plane wavefront perpendicular to the plane of the paper [Fig. 8.11 (*i*)] and *P* is an external point at a distance *b* ($OP = b$) from the wavefront.

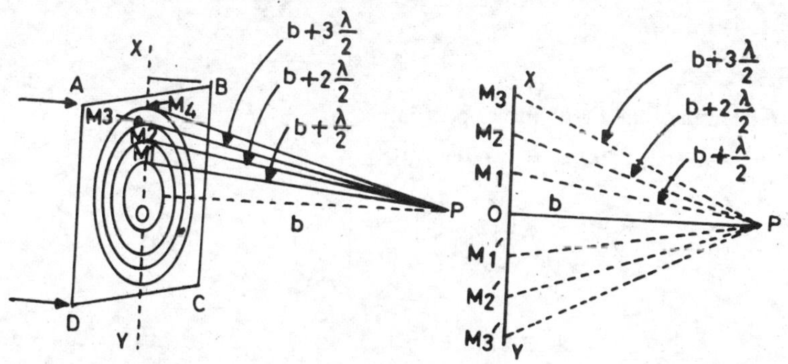

Fig. 8.11.

λ is the wavelength of sound. OP is perpendicular to $ABCD$. To find the resultant intensity at P due to the wavefront $ABCD$, Fresnel's method consists in dividing the wavefront into a number of half period elements or zones called Fresnel's zones and to find the effect of all the zones at the point P.

With P as centre and radii equal to $b+\lambda/2$, $b+2\lambda/2$, $b+3\lambda/2$ etc., construct spheres which will cut out circular areas of radii OM_1, OM_2, OM_3, etc., on the wavefront. These circular zones are called half period zones or half period elements. Each zone differs from its neighbour by a phase difference of π or a path difference of $\lambda/2$. Thus, the secondary waves starting from the points O and M_1 and reaching P will have a phase difference of π or a path difference of $\lambda/2$. A Fresnel half period zone with respect to an external point P is a thin annular zone (or a thin rectangular strip) of the primary wavefront in which the secondary waves from any two corresponding points of neighbouring zones differ in path by $\lambda/2$.

In Fig. 8.11 *(i)*, O is the pole of the wavefront XY with reference to the external point P. OP is perpendicular to XY. In Fig. 8.12, 1, 2, 3, etc. are the half period zones constructed on the primary wavefront XY. OM_1 is the radius of the first zone, OM_2 is the radius of the second zone and so on. P is the point at which the resultant intensity has to be calculated.

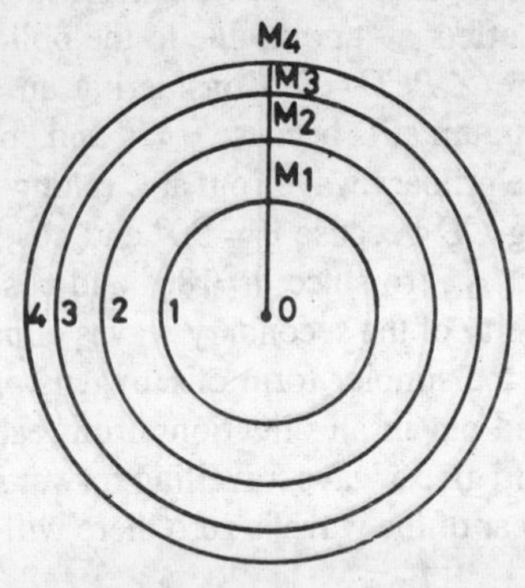

Fig. 8.12.

$$OP = b,\ OM_1 = r_1,\ OM_2 = r_2,\ OM_2 = r_3 \text{ etc.}$$

and
$$M_1P = b+\frac{\lambda}{2},\quad M_2P = b+\frac{2\lambda}{2},$$
$$M_3P = b+\frac{3\lambda}{2} \text{ etc.}$$

The area of the first half period zone is,

$$\pi \,.\, OM_1^2 = \pi\ [M_1P^2 - OP^2]$$
$$= \pi\left[\left(b+\frac{\lambda}{2}\right)^2 - b^2\right]$$
$$= \pi\left[b\lambda + \frac{\lambda^2}{4}\right]$$
$$= \pi b\lambda \text{ approximately.} \qquad \text{...(1)}$$

(As λ is small, in comparison to b, λ^2 term is negligible).

The radius of the first half period zone

$$r_1 = OM_1 = \sqrt{b\lambda} \quad \text{... (2)}$$

The radius of the second half period zone is

$$OM_2 = [M_2P^2 - OP^2]^{1/2}$$
$$= [(b+\lambda)^2 - b^2]^{1/2}$$
$$= \sqrt{2b\lambda} \text{ approximately.}$$

The area of the second half period zone

$$= \pi\,[OM_2^2 - OM_1^2]$$
$$= \pi\,[2b\lambda - b\lambda] = \pi b\lambda.$$

Thus, the area of each half period zone is equal to $\pi b\lambda$. Also, the radii of the 1st, 2nd, 3rd etc. half period zones are $\sqrt{1b\lambda}$, $\sqrt{2b\lambda}$, $\sqrt{3b\lambda}$ etc. Therefore, the radii are proportional to the square roots of the natural numbers. However, it should be remembered that the areas of the zones are not constant but are dependent on (*i*) λ, the wavelength of sound and (*ii*) b, the distance of the point from the wavefront. The area of the zone increases with increase in the wavelength of sound and with increase in the distance of the point P from the wavefront.

The effect at a point P will depend on (*i*) the distance of P from the wavefront, (*ii*) the area of the zone, and (*iii*) the obliquity factor. Here, the area of each zone is the same. The secondary waves reaching the point P are continuously out of phase and in phase with reference to the central or the first half period zone. Let m_1, m_2, m_3 etc. (Fig. 8.13) represent the amplitudes of vibration of the particles of the medium at P due to secondary waves from the 1st, 2nd, 3rd, etc. half period zones. As we consider the zones outwards from O, the obliquity increases and hence the quantities m_1, m_2, m_3 etc. are of continuously decreasing order. Thus, m_1 is slightly greater than m_2, m_2 is slightly greater than m_3 and so on. Due to the phase difference of π between any two consecutive zones, if the displacements of the particles due to odd numbered zones is in the positive direction, then due to the even numbered zones the displacement will be in the negative direction at the same instant. As the amplitudes are of gradually

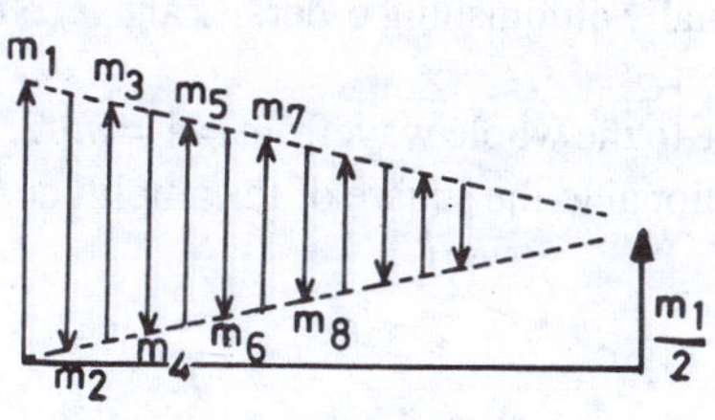

Fig. 8.13.

decreasing magnitude, the amplitude of vibration at P due to any zone can be approximately taken as the mean of the amplitudes due to the zones preceding and succeeding it.

e.g. $$m_2 = \frac{m_1 + m_3}{2}$$

The resultant amplitude at P at any instant is given by

$$A = m_1 - m_2 + m_3 - m_4 \ \ldots \ + m_n \text{ if } n \text{ is odd.}$$

(If n is even, the last quantity is $- m_n$).

$$\therefore \quad A = \frac{m_1}{2} + \left[\frac{m_1}{2} - m_2 + \frac{m_3}{2}\right] + \left[\frac{m_3}{2} - m_4 + \frac{m_5}{2}\right] + \ldots \frac{m_n}{2}$$

But $$m_2 = \frac{m_1}{2} + \frac{m_3}{2} \text{ and } m_4 = \frac{m_3}{2} + \frac{m_5}{2}$$

$$\therefore \quad A = \frac{m_1}{2} + \frac{m_n}{2} \quad \text{.... if } n \text{ is odd}$$

and $$A = \frac{m_1}{2} + \frac{m_{n-1}}{2} - m_n \quad \text{.... if } n \text{ is even.}$$

If the whole wavefront $ABCD$ is unobstructed, the number of half period zones that can be constructed with reference to the point P is infinite *i.e.* $n \to \infty$. As the amplitudes are of gradually diminishing order, m_n and m_{n-1} tend to zero.

$\therefore$ The resultant amplitude at P due to the whole wavefront $= A = m_1/2$. The intensity of sound at a point is proportional to the square of the amplitude.

$$\therefore \quad I \propto \frac{m_1^2}{4}.$$

Thus the intensity at P is only one fourth of that due to the first half period zone alone. Here, only half the area of the first half period zone is effective in providing the intensity of sound at the point P. An obstacle of the size of half the area of the first half period zone placed at O will screen the effect of the whole wavefront and the intensity at P due to the rest of the wavefront will be zero. In the case of sound waves, the wavelengths are far greater than the wavelength of light and hence the area of the first half period zone for a plane wavefront of sound is very large. *If the effect of sound at a point beyond an obstacle is to be shadowed, an obstacle of very large size has to be used to get no sound effect.*

EXERCISES

1. Discuss the reflection of a plane wavefront of sound at a plane surface.
2. Describe an experiment to demonstrate the reflection of sound.
3. What is an echo? How is it produced? Explain, why an echo cannot be heard if the distance between the source of sound and the obstacle is less than 17 metres. Also explain the formation of successive echoes.
4. Describe briefly some practical applications of reflection of sound.
5. Discuss the phase change on reflection when (*i*) sound waves travel from air to water and (*ii*) from water to air.
6. Discuss the refraction of sound waves at the interface of two media and explain, how most of the sound energy is totally internally reflected when sound waves are travelling from rarer to denser medium.
7. Discuss the refraction of a plane wavefront of sound at a plane surface.
8. Describe an experiment to demonstrate refraction of sound.
9. What do you understand by diffraction of sound?
10. Discuss the intensity of sound at a point due to a progressive plane wavefront.
11. Distinguish between the diffraction of sound with diffraction of light.
12. Explain, why the diffraction effects are easily noticed in the case of sound whereas in the case of light even an obstacle of small size produces a geometrical shadow.
13. Deduce the laws of reflection with the help Huygen's theory of secondary wavelets. *(Rajasthan, 1975)*
14. Write short notes on :

    (*i*) Reflection and Refraction of sound waves. *(Delhi, 1973)*

    (*ii*) Diffraction of sound.

    (*iii*) Echo.

    (*iv*) Whispering Galleries.

CHAPTER 9

Doppler Effect

9.1. Doppler Effect

It is commonly observed that the pitch of a note apparently changes when either the source or the observer are in motion relative to each other. When the source approaches the observer or when the observer approaches the source or when both approach each other the apparent pitch is higher than the actual pitch of the sound produced by the source. Similarly, when the source moves away from the observer or when the observer moves away from the source or when both move away from each other, the apparent pitch is lower than the actual pitch of the sound produced by the source.

Suppose a person is standing on a platform. The apparent pitch of the whistle of the engine increases, when the engine is approaching the person. When the engine moves away from the person, the apparent pitch of the whistle of the engine decreases. This apparent change in the pitch due to the relative motion between the source and the observer is called **Doppler Effect.**

Doppler effect in sound is asymmetric. When the source moves towards the observer with a certain velocity, the apparent pitch is different to the case when the observer is moving towards the source with the same velocity. But it is not so in the case of light. Doppler effect in light is symmetric. The apparent pitch in different cases is calculated in the subsequent articles.

9.2. Observer at Rest and Source in Motion

(*a*) When the source moves towards the stationary observer

Suppose a source S is producing sound of pitch n and wavelength λ. The velocity of sound is v. (Fig. 9.1).

Let the source move with a velocity a towards the observer. In one second, n waves will be contained in a length $(v - a)$ and the apparent wavelength,

$$\lambda' = \frac{(v-a)}{n}$$

The apparent pitch,

$$n' = \frac{v}{\lambda'}$$

$$\therefore \quad n' = \left(\frac{v}{v-a}\right)n \qquad \ldots (1)$$

Thus the apparent pitch of the note increases when the source moves towards a stationary observer.

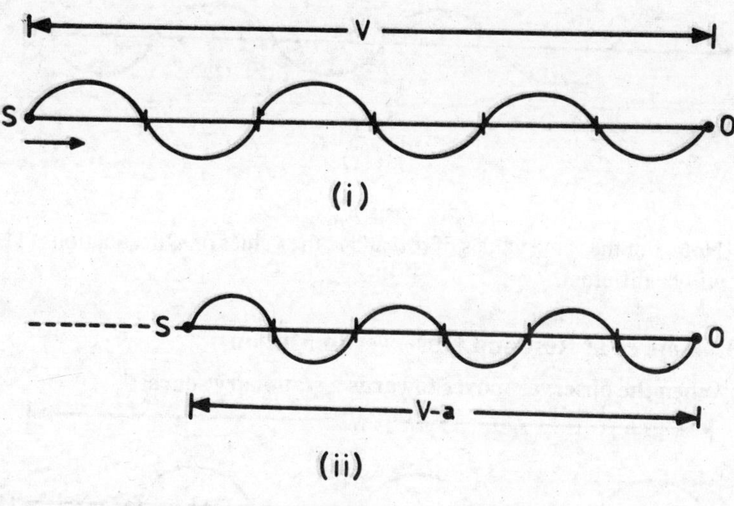

Fig. 9.1.

(*b*) When the source moves away from the stationary observer

Suppose a source S is producing sound of pitch n and wavelength λ. The velocity of sound is v (Fig. 9.2). Let the source move with a velocity a away from the observer. In one second, n waves will be contained in a length $(v + a)$ and the apparent wavelength,

$$\lambda' = \frac{v+a}{n}$$

The apparent pitch,

$$n' = \frac{v}{\lambda'}$$

$$\therefore \quad n' = \left(\frac{v}{v+a}\right)n \qquad \ldots (2)$$

Thus the apparent pitch of the note decreases when the source moves away from a stationary observer.

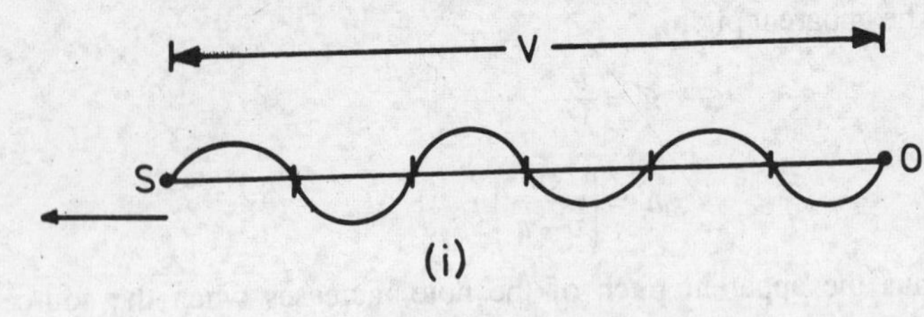

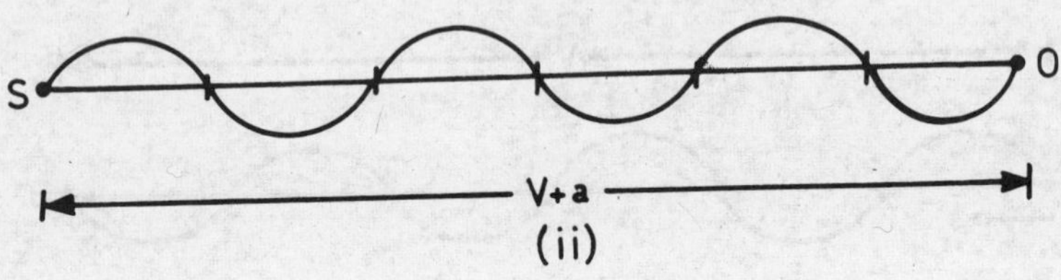

Fig. 9.2.

Note. For the same values of v, a and n, the values of n' in equations (1) and (2) will be different.

9.3. Source at Rest and Observer in Motion

(a) **When the observer moves towards a stationary source**

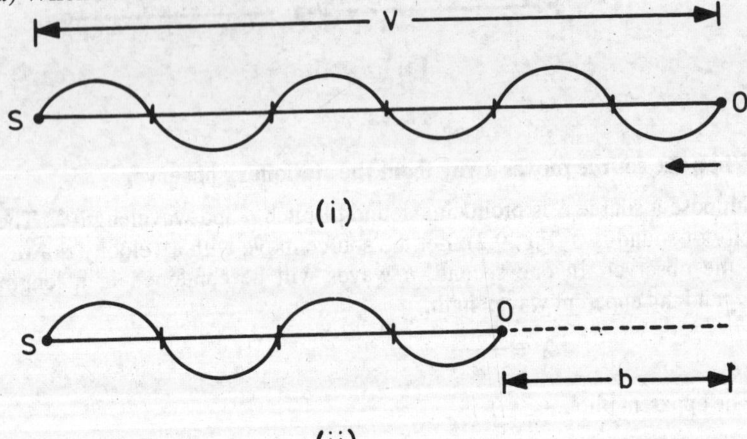

Fig. 9.3.

Suppose a source S is producing sound of pitch n and wavelength λ. The velocity of sound is v (Fig. 9.3). Let the observer move with a velocity b towards a stationary source. In this case the observer receives more number of waves in one second. The apparent wavelength remains the same. The apparent frequency

$$n' = n + \frac{b}{\lambda}$$

$$n' = \frac{v}{\lambda} + \frac{b}{\lambda}$$

$$n' = \left(\frac{v+b}{\lambda}\right)$$

But $$\lambda = \frac{v}{n}$$

$$\therefore \quad n' = \left(\frac{v+b}{v}\right)n \quad \ldots(1)$$

Thus the apparent pitch of the note increases when the observer moves towards the stationary source.

(b) **When the observer moves away from a stationary source**

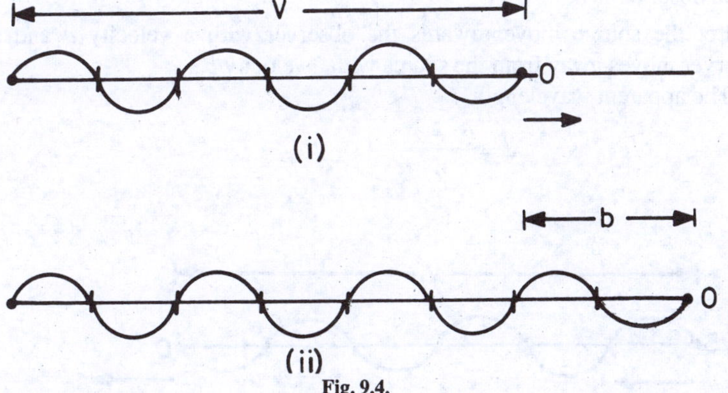

Fig. 9.4.

Suppose a source S is producing sound of pitch n and wavelength λ (Fig. 9.4). The velocity of sound is v. Let the observer moves with a velocity b away from a stationary source. In this case, the observer receives less number of waves in one second. The apparent wavelength remains the same. The apparent frequency

$$n' = n - \frac{b}{\lambda}$$

$$n' = \frac{v}{\lambda} - \frac{b}{\lambda}$$

$$n' = \left(\frac{v-b}{\lambda}\right)$$

But $$\lambda = \frac{v}{n}$$

$\therefore$ $$n' = \left(\frac{v-b}{v}\right) n \qquad \ldots (2)$$

Thus the apparent pitch of the note decreases when the observer moves away from a stationary source.

[**Note.** For the same values of v, b and n, the values of n' in equations (1) and (2) will be different .]

9.4. When Both the Source and the Observer are in Motion

When the source moves towards the observer and the observer moves away from the source

Suppose a source S is producing a sound of pitch n and wavelength λ. The velocity of sound is v (Fig. 9.5). The velocity of the source is a and the velocity of the observer is b.

Let the source move towards the observer with a velocity a and the observer moves away from the source with a velocity b.

The apparent wavelength

$$\lambda' = \frac{(v-a)}{n}.$$

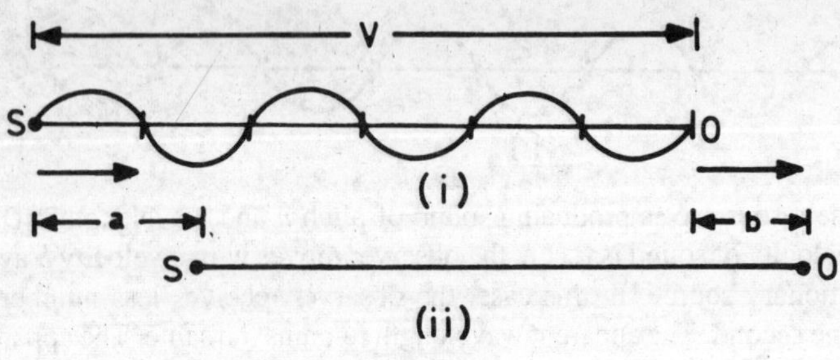

Fig. 9.5.

and $$n' = \left(\frac{v-b}{\lambda'}\right)$$

$\therefore$ $$n' = \left(\frac{v-b}{v-a}\right) n \qquad \ldots (1)$$

Special cases

(a) **When the source and observer move towards each other**

In equation (1), taking b to be negative,

$$n' = \left[\frac{v-(-b)}{v-a}\right] n$$

$$n' = \left(\frac{v+b}{v-a}\right) n \qquad \ldots (2)$$

(b) **When the source and observer move away from each other**

In equation (1), taking a to be negative

$$n' = \left[\frac{v-b}{v-(-a)}\right] n$$

$$n' = \left(\frac{v-b}{v+a}\right) n \qquad \ldots (3)$$

(c) **Source moving away from the observer and the observer moving towards the source**

In equation (1), taking both a and b negative,

$$n' = \left[\frac{v-(-b)}{v(-a)}\right] n$$

$$n' = \left(\frac{v+b}{v+a}\right) n \qquad \ldots (4)$$

[**Note.** While solving numerical problems, the general formula

$$n' = \left(\frac{v-b}{v-a}\right) n$$

should be applied.]

The general relation refers to the case when the source moves towards the observer and the observer moves away from the source. When any of these two directions change, the signs of a and b have to be changed. The modified formula should be used for calculating the apparent pitch.

9.5. Effect of Wind Velocity

Suppose the wind is moving with a velocity w in the direction of propagation of sound, the apparent velocity of sound will be $(v + w)$. In all relations, in place of v, $(v+w)$ should be used. If the wind is blowing in a direction opposite to the direction of propagation of sound, the apparent velocity of sound will be $(v-w)$. In all relations, in place of v, $(v-w)$ should be used.

The general relation will be

$$n' = \left[\frac{(v+w)-b}{(v+w)-a}\right]n \qquad \ldots (5)$$

Here the wind direction is the same as the direction of propagation of sound.

When the direction of wind is opposite to the direction of propagation of sound,

$$n' = \left[\frac{(v-w)-b}{(v-w)-a}\right]n \qquad \ldots (6)$$

9.6. Tracking of Artificial Satellites

The Doppler effect provides a convenient method for tracking an earth satellite. Let the earth satellite emit a radio signal of constant frequency n (Fig. 9.6). The apparent frequency n_1 of the signal as received on the earth decreases as the satellite moves from position 1 to position 2. This is due to the fact that the component velocity of the satellite towards the earth station A decreases from position 1 to position 2. Also this velocity component points away from the earth as the satellite moves from position 2 to position 3. If the signal viewed from the satellite is combined with a signal of constant frequency at the earth station, a beat frequency of audible note can be heard. This beat frequency decreases as the satellite passes over head.

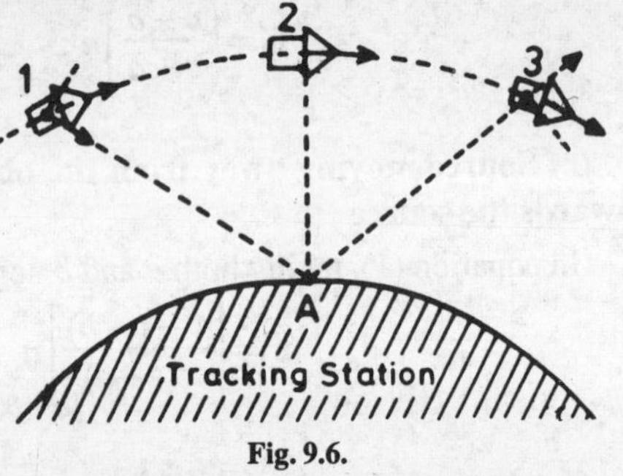

Fig. 9.6.

A similar principle is used in detecting the speed of the automobiles by traffic police. Here, an electromagnetic wave of constant frequency is emitted by a source attached to the police van. This wave is reflected by the moving car, which in turn acts as a moving source. There is a Doppler shift in frequency of the reflected wave. From the shift in frequency, the speed of the car can be measured.

9.7. Doppler Effect in Light

There is a change in the frequency of light radiation when the source or the observer move with respect to one another. This phenomenon is known as

Doppler effect. It is similar to the apparent change in the pitch of sound when either the source or the observer is in motion with respect to one another.

According to the theory of relativity, a material medium is not necessary for the propagation of light waves, whereas for sound waves material medium is a necessity. In sound, the Doppler effect is asymmetric *i.e.*, the apparent frequencies are different when the source is moving towards a stationary observer and when the observer is moving towards a stationary source. In the first case

$$n' = \left(\frac{v}{v-a}\right)n$$

and in the second case

$$n' = \left(\frac{v+b}{v}\right)n$$

Here n is the frequency and v is the velocity of sound. n' is the apparent frequency.

But it is not so in the case of light. In light, the Doppler effect is symmetric. The apparent frequency is the same when either the source moves towards a stationary observer or the observer moves towards a stationary source.

(i) Suppose, the observer is stationary and the source is moving towards the observer with a velocity v. Let the velocity of light be c.

$$c = \nu\lambda$$

Here, ν is the frequency and λ is the wavelength.

As there is no medium, the observer will receive more waves due to the motion of the source. The wavelength is not changed.

The apparent frequency

$$\nu' = \nu + \frac{v}{\lambda}$$

$$\nu' = \nu + \frac{\nu v}{c}$$

$$\nu' = \nu\left[1 + \frac{v}{c}\right] \quad \ldots(1)$$

(ii) When the source is stationary and the observer is moving towards the source with a velocity v.

$$\nu' = \nu + \frac{v}{\lambda}$$

$$\nu' = \nu + \frac{\nu v}{c}$$

$$\nu' = \nu\left[1 + \frac{v}{c}\right] \qquad \ldots (2)$$

Therefore, equations (1) and (2) are similar.

(iii) When the source moves away from the stationary observer or the observer moves away from the stationary source with a velocity v.

$$\nu' = \nu - \frac{v}{\lambda}$$

$$\nu' = \nu - \frac{\nu v}{c}$$

$$\nu' = \nu\left[1 - \frac{v}{c}\right] \qquad \ldots (3)$$

(iv) When the source and the observer move towards each other and each is moving with a velocity v.

$$\nu' = \nu\left[1 + \frac{v}{c}\right]^2 \qquad \ldots (4)$$

(v) When the source and the observer move away from each other and each is moving with a velocity v.

$$\nu' = \nu\left[1 - \frac{v}{c}\right]^2 \qquad \ldots (5)$$

The phenomenon of Doppler effect is of great importance in light. It shows that there is a shift in the frequency when the source of light moves with a velocity comparable to the velocity of light. In the field of astronomy, Doppler effect has been found to be very useful.

Applications

(1) **Velocity and rotation of the sun.** By the study of Doppler shift from the light received from the western and eastern edges of the sun, it has been found that the shift is due to a velocity of about 2 km/s. Moreover, no such shift has been observed from the light received from the north and south edges. This shows that the sun rotates about the north-south axis.

(2) **Discovery of double stars.** By the constant observation of the sky, it has been found that some of the stars that appear to be single are actually double stars and are known as spectroscopic binaries. These stars revolve about each other. When one is approaching the earth, and the other is going away from the earth, there is a shift in their spectral lines and a single spectral line is split up into two lines whose separation depends upon time and the time period is equal to the time-period of revolution of the stars. By this method a number of double-star systems have been found.

(3) **Red shift.** It has been observed that some distant nebulae are moving away with a velocity greater than 20×10^3 km/s and the important spectral lines appear to shift towards the red end of the spectrum by 200 Å. This gives the idea that the universe is expanding.

(4) **Saturn's rings.** The planet Saturn has been found to be surrounded by concentric rings. With the help of Doppler effect it has been found that these rings are not solids but consist of a number of 'satellites' moving around the Saturn in these orbits. If the rings were solids, the outer edge of the ring should have greater velocity than the inner edge. But, according to the principle of a satellite

$$\frac{mV^2}{R} = \frac{GMm}{R^2}$$

or

$$V^2 = \frac{GM}{R}$$

i.e., the velocity of the satellite in the inner orbit is more than that in the outer orbit. This fact has been established by the Doppler shift. Thus, the rings of the Saturn are not solids but there are a large number of satellites moving in these orbits called the rings of Saturn.

Example 9.1. *A person is standing on a railways platform. An engine while approaching the platform blows a whistle of pitch* 660 hertz. *The speed of the engine is* 72 km/hr, *velocity of sound* = 350 m/s. *Calculate the apparent pitch of the whistle as heard by the person.*

Here $v = 350$ m/s

$a = 72$ km/hr $= 20$ m/s

$n = 660$ hertz

The source is moving towards a stationary observer.

$$\therefore \quad n' = \left(\frac{V}{V-a}\right)n$$

$$n' = \left(\frac{350}{350-20}\right) \times 660$$

$n' =$ **700 hertz.**

Example 9.2. *A person is standing on a platform. A railway engine moving away from the person with a speed of* 72 km/hr *blows a whistle of pitch* 740 hertz. *Calculate the apparent pitch of the whistle as heard by the person. The velocity of sound* = 350 m/s.

Here $v = 350$ m/s

$a = 72$ km/hr $= 20$ m/s

$n = 740$ hertz.

The source is moving away from a stationary observer.

$$\therefore \qquad n' = \left(\frac{V}{V+a}\right)n$$

$$n' = \left(\frac{350}{350+20}\right)\times 740$$

$$n' = \mathbf{700\ hertz.}$$

Example 9.3. *A person is standing near a railway track and a train moving with a speed of* 36 km/hr *is approaching him. The apparent pitch of the whistle as heard by the person is* 700 hertz. *Calculate the actual frequency of the whistle. Velocity of sound* = 350 m/s.

Here $v = 350$ m/s

$a = 36$ km/hr $= 10$ m/s

$n' =$ **700** hertz

$n = ?$

The source is moving towards a stationary observer

$$\therefore \qquad n' = \left(\frac{V}{V-a}\right)n$$

$$n = n'\left(\frac{V-a}{V}\right)$$

$$n = 700\left(\frac{350-10}{350}\right);$$

$$n = \mathbf{680\ \ hertz.}$$

Example 9.4. *Two aeroplanes A and B are approaching each other with a speed of* 360 km/hr. *The frequency of the whistle emitted by A is* 1000 hertz. *Calculate the apparent pitch of the whistle as heard by the passengers of aeroplane B. Velocity of sound in air* = 350 m/s.

Here $v = 350$ m/s

$a = 360$ km/hr $= 100$ m/s

$b = 360$ km/hr $= 100$ m/s

$n = 1000$ hertz.

The source and the observer are moving towards each other

$$\therefore \qquad n' = n\left[\frac{V+b}{V-a}\right]$$

$$\therefore \qquad n' = 1000\left(\frac{350+100}{350-100}\right]$$

$$n' = \mathbf{1350\ hertz.}$$

Example 9.5. *Two aeroplanes A and B are moving away from one another with a speed of* 720 km/hr. *The frequency of the whistle emitted by A is* 1100 hertz. *Calculate the apparent frequency of the whistle as heard by the passengers of the aeroplane B. Velocity of sound in air* = 350 m/s.

(Delhi, 1992)

Here $V = 350$ m/s

$a = 720$ km/hr $= 200$ m/s

$b = 720$ km/hr $= 200$ m/s

$n = 1100$ hertz.

The source and the observer are moving away from one another

$$\therefore \quad n' = n\left(\frac{V-b}{V+a}\right)$$

$$n' = 1100\left(\frac{350-200}{350+200}\right)$$

$n' =$ **300 hertz.**

Example 9.6. *Two aeroplanes A and B are approaching each other and their velocities are* 108 km/hr *and* 144 km/hr *respectively. The frequency of a note emitted by A as heard by the passengers in B is* 1170 hertz. *Calculate the frequency of the note heard by the passengers in A. Velocity of sound* = 350 m/s.

(Delhi, 1991)

Here $V = 350$ m/s

$a = 108$ km/hr $= 30$ m/s

$b = 144$ km/hr $= 40$ m/s

$n' = 1170$ hertz.

$n = ?$

The source and the observer are approaching each other.

$$\therefore \quad n' = n\left(\frac{V+b}{V-a}\right)$$

$$1170 = n\left(\frac{350+40}{350-30}\right)$$

$$n = \frac{1170 \times 320}{390}$$

$n =$ **960 hertz.**

Example 9.7. *Two aeroplanes pass each other in opposite directions and one of them is blowing a whistle of frequency 540* hertz. *Calculate the frequencies of the notes heard in the other aeroplane (i) before and (ii) after they have passed each other. Velocity of either of the aeroplanes is* 540 km-hr^{-1} *and the velocity of sound* = 350 ms^{-1} (Delhi, 1971)

Here
$$V = 350 \text{ m/s}$$
$$a = 540 \text{ km/hr} = 150 \text{ m/s}$$
$$b = 540 \text{ km/hr} = 150 \text{ m/s}$$
$$n = 540 \text{ hertz.}$$

(*i*) In the first case, the source and the observer are moving towards each other. Therefore,

$$n' = n\left[\frac{V+b}{V-a}\right]$$

$$n' = 540\left[\frac{350+150}{350-150}\right]; \; n' = \textbf{1350 hertz.}$$

(*ii*) In the second case, the source and the observer are moving away from each other.

$$n' = n\left[\frac{V-b}{V+a}\right]$$

$$n' = 540\left[\frac{350-150}{350+150}\right]; \; n' = \textbf{216 hertz.}$$

Example 9.8. *An observer on the railway platform observed that as a train passed through the station at* 90 km/hr, *the frequency of the whistle appeared to drop by* 400 hertz. *Find the frequency of the whistle. Velocity of sound in air* = 350 m/s. (Bombay, 1971)

Here
$$V = 350 \text{ m/s}$$
$$a = 90 \text{ km/hr} = 25 \text{ m/s}$$
$$n = ?$$
$$n_1 - n_2 = 400$$

Here n_1 is the apparent pitch when the train is approaching the observer and n_2 is the apparent pitch when the train is moving away from the observer.

$$\therefore \quad n_1 = n\left[\frac{V}{V-a}\right]$$

and
$$n_2 = n\left[\frac{V}{V+a}\right]$$

$$\therefore \qquad n_1 - n_2 = n\left[\frac{V}{V-a}\right] - n\left[\frac{V}{V+a}\right]$$

$$n_1 - n_2 = n\left[\left(\frac{V}{V-a}\right) - \left(\frac{V}{V+a}\right)\right]$$

$$\therefore \qquad 400 = n\left[\left(\frac{350}{350-25}\right) - \left(\frac{350}{350+25}\right)\right]$$

$$n = \frac{400}{\left(\frac{350}{325}\right) - \left(\frac{350}{375}\right)}$$

$$n = \mathbf{2785 \cdot 7 \text{ hertz.}}$$

Example 9.9. *A car emitting sound of frequency* 200 hertz *is moving away from a stationary observer and towards a rigid flat wall. The velocity of the car is 5 m/s. Calculate the number of beats heard per second by the observer. Velocity of sound in air* = 350 m/s.

Here $n = 200$ Hz

$v = 350$ m/s

Velocity of the car, $a = 5$ m/s

The observer hears sound of apparent frequency

$$n_1 = \left(\frac{v}{v+a}\right) n$$

$$n_1 = \left(\frac{350}{350+5}\right) 200$$

$$n_1 = \mathbf{197 \cdot 2 \text{ Hz.}} \qquad \ldots (1)$$

The observer also hears sound of apparent pitch n_2 due to the waves reflected by the wall. It means the source is moving towards the stationary observer.

$$n_2 = \left(\frac{v}{v-a}\right) n$$

or $$n_2 = \left(\frac{350}{350-5}\right) 200$$

or $$n_2 = 203$$

Therefore, the number of beats produced in one second is given by

$$n_2 - n_1 = 203 - 197{\cdot}2$$

$$n_2 - n_1 = \mathbf{5{\cdot}8.}$$

Example 9.10. *A motor car sounding a horn at a frequency of* 100 hertz *moves away from a stationary observer towards a rigid flat wall with a velocity of* 36 km/hr. *How many beats per second will be heard by the observer? The velocity of sound in air at room temperature* = 350 m/s.

Here
$$V = 350 \text{ m/s}$$
$$a = 36 \text{ km/hr} = 10 \text{ m/s}$$
$$n = 100 \text{ hertz}$$

Here the observer receives sound of apparent frequency n_1 from the moving sound directly and also sound of apparent frequency n_2 from the sound waves reflected from the wall.

$$n_1 = n\left[\frac{V}{V+a}\right]$$

$$n_1 = 100\left[\frac{350}{350+10}\right]$$

$$n_1 = 97{\cdot}23 \text{ hertz}$$

$$n_2 = n\left[\frac{V}{V-a}\right]$$

$$n_2 = 100\left[\frac{350}{350-10}\right]$$

$$n_2 = 102{\cdot}9 \text{ hertz.}$$

Number of beats/s

$$n_2 - n_1 = 102{\cdot}9 - 97{\cdot}23 = \mathbf{5{\cdot}67} \text{ hertz.}$$

Example 9.11. *The apparent frequency of the whistle of an engine changes in the ratio 6 : 5 as the engine passes a stationary observer. If the velocity of sound is* 352 m/s, *calculate the velocity of the engine.* (Delhi, 1974)

(*i*) When the source moves towards the stationary observer,

$$n_1 = \left(\frac{v}{v-a}\right)n \qquad \ldots (1)$$

(*ii*) When the source moves away from the stationary observer,

$$n_2 = \left(\frac{v}{v+a}\right)n \qquad \ldots (2)$$

Dividing (1) by (2)

$$\frac{n_1}{n_2} = \frac{v+a}{v-a}$$

Here $\frac{n_1}{n_2} = \frac{6}{5}$ (given)

$$v = 352 \text{ m/s}$$

$$a = ?$$

$\therefore$ $\frac{6}{5} = \frac{352+a}{352-a}$ or $a = \mathbf{32\ m/s.}$

Hence the velocity of the engine is 32 m/s.

Example 9.12. *Two trains travelling in opposite directions at* 100 km/hr *each, cross each other while one of them is whistling. If the frequency of the note is* 800 Hz *find the apparent pitch as heard by an observer in the other train:*

(*a*) *before the trains cross each other*

(*b*) *after the trains have crossed each other.*

Velocity of sound in air = 340 m/s. [Delhi (Sub.), 1976]

Here $V = 340$ m/s

$$a = 100 \text{ km/hr} = \frac{250}{9} \text{ m/s}$$

$$b = 100 \text{ km/hr} = \frac{250}{9} \text{ m/s}$$

$$v = \mathbf{800\ Hz.}$$

(*a*) In a first case, the source and the observer are moving towards each other. Therefore,

$$n' = n\left[\frac{V+b}{V-a}\right]$$

$$n' = 800\left[\frac{340+\left(\frac{250}{9}\right)}{340-\left(\frac{250}{9}\right)}\right]$$

$$n' = 800\left(\frac{3310}{2810}\right)$$

$$n' = \mathbf{942{\cdot}3}\text{ Hz.}$$

(*b*) In the second case, the source and the observer are moving away from each other.

$$n' = n\left[\frac{V-b}{V+a}\right]$$

$$n' = 800\left[\frac{340-\left(\frac{250}{9}\right)}{340+\left(\frac{250}{9}\right)}\right]$$

$$n' = 800\left(\frac{2810}{3310}\right)$$

$$n' = \mathbf{679{\cdot}2\ Hz.}$$

Example 9.13. *An observer on a railway platform noticed that when a train passed through the station, at a speed of* 72 km/hr, *the frequency of the whistle appeared to drop by* 500 Hz. *Calculate the actual frequency of the note given by the whistle. Velocity of sound in air* = 340 m/s.

Here
$$V = 340 \text{ m/s}$$
$$a = 72 \text{ km/hr} = 20 \text{ m/s}$$
$$n = ?$$
$$n_1 - n_2 = 500.$$

Here n_1 is the apparent pitch when the train is approaching the observer and n_2 is the apparent pitch when the train is moving away from the observer.

$$\therefore \quad n_1 = n\left[\frac{V}{V-a}\right]$$

and
$$n_2 = n\left[\frac{V}{V+a}\right]$$

$$\therefore \quad n_1 - n_2 = n\left[\frac{V}{V-a}\right] - n\left[\frac{V}{V+a}\right]$$

$$n_1 - n_2 = n\left[\left(\frac{V}{V-a}\right) - \left(\frac{V}{V+a}\right)\right]$$

$$500 = n\left[\left(\frac{340}{340-20}\right) - \left(\frac{340}{340+20}\right)\right]$$

or
$$n = \frac{500}{\left(\frac{340}{320}\right) - \left(\frac{340}{360}\right)}$$

$$n = \mathbf{4235 \cdot 3 \ Hz.}$$

Example 9.14. *The frequency of the horn of a car is observed to drop from* 272 hertz *to* 256 hertz *as the car passes a stationary observer. What is the speed of the car? Velocity of sound in air* = 346·5 m/s.

Here $v = 346 \cdot 5$ m/s

$$\frac{n_1}{n_2} = \frac{272}{256} = \frac{17}{16}$$

$$a = ?$$

(*i*) When the car approaches a stationary observer, the apparent pitch

$$n_1 = \left(\frac{v}{v-a}\right)n \qquad \text{... (1)}$$

(*ii*) When the car moves away from the stationary observer, the apparent pitch

$$n_2 = \left(\frac{v}{v+a}\right)n \qquad \text{... (2)}$$

Dividing (1) by (2)

$$\frac{n_1}{n_2} = \frac{v+a}{v-a}$$

or
$$\frac{17}{16} = \frac{346 \cdot 5 + a}{346 \cdot 5 - a}$$

$$a = \mathbf{10 \cdot 5 \ m/s.}$$

The speed of the car is 10·5 m/s.

Example 9.15. *Two observers A and B carry identical sound sources of frequency* 500 hertz. *If the observer A is stationary and B moves away from A at a speed of* 3·6 km/hr, *how many beats/s are heard by A and B respectively. Velocity of sound in air* = 350 m/s.

(1) In the first case, consider *A* as source and *B* as observer. It means the source is at rest and the observer is moving away from it. The apparent pitch

$$n_1 = \left(\frac{v-b}{v}\right)n$$

Here $v = 350$ m/s

$b = 3 \cdot 6$ km/hr = 1 m/s

$n = 500$

$$n_1 = \left(\frac{350-1}{350}\right)500$$

$$n_1 = \frac{349 \times 100}{350}$$

Number of beats/s heard by B

$$= n - n_1$$

$$= 500 - \frac{500 \times 349}{350} = \frac{500}{350} = 1{\cdot}43.$$

(2) In the second case, consider B as a source and A as an observer. It means the source is moving away from a stationary observer. The apparent pitch,

$$n_2 = \left(\frac{v}{v+a}\right) n$$

Here, $v = 350$ m/s

$a = 1$ m/s

$n = 500$ hertz

$$\therefore \quad n_2 = \left(\frac{350}{350+1}\right) 500$$

$$n_2 = \left(\frac{350}{351}\right) 500$$

Number of beats/s heard by A

$$= n - n_2$$

$$= 500 - \left(\frac{500 \times 350}{351}\right)$$

$$= \frac{500}{351} = \mathbf{1{\cdot}424.}$$

Example 9.16. *A motor car fitted with two sounding horns which have a difference in frequency by 320 vibrations per second is speeding at the rate of* 36 km/hr *towards a person standing. Calculate the difference in the frequencies of the notes heard by him. Velocity of sound in air* = 330 m/s.

(Madras, 1974)

Let n_1 and n_2 be the two frequencies

$$n_1 - n_2 = 320$$

$$n'_1 = \left(\frac{v}{v-a}\right) n_1 \quad \text{... (1)}$$

$$n'_2 = \left(\frac{v}{v-a}\right) n_2 \quad \text{... (2)}$$

Subtracting (2) from (1)

$$(n'_1 - n'_2) = \left(\frac{v}{v-a}\right)(n_1 - n_2)$$

Here $a = 36$ km/hr = 10 m/s

and $v = 330$ m/s

$$\therefore \quad n'_1 - n'_2 = \left(\frac{330}{330-10}\right)(320)$$

$$\therefore \quad n'_1 - n'_2 = \textbf{330 vibrations/s.}$$

Example 9.17. *At each of the two stations A and B, a siren is sounding with a constant frequency of* 250 cycles/s. *A cyclist from A proceeds straight towards B with a velocity of* 12 km/hr *and hears* 5 beats/s. *Calculate the velocity of sound.* (Punjab, 1993)

(1) In the first case, the cyclist is moving away from the source *A*.

$$\therefore \quad n_1 = \left(\frac{v-b}{v}\right)n$$

Here $b = 12 \text{ km/hr} = \frac{10}{3}$ m/s

$n = 250$

$$\therefore \quad n_1 = \left(\frac{v - \frac{10}{3}}{v}\right)250 \qquad \text{... (1)}$$

(2) In the second case, the cyclist is moving towards the source *B*.

$$\therefore \quad n_2 = \left(\frac{v+b}{v}\right)n$$

$$\therefore \quad n_2 = \left(\frac{v + \frac{10}{3}}{v}\right)250 \qquad \text{... (2)}$$

Subtracting (1) from (2)

$$n_2 - n_1 = \frac{250}{v} \times \frac{20}{3}$$

Here $n_2 - n_1 = 5$

$$\therefore \quad v = \frac{250 \times 20}{5 \times 3}$$

$$v = \textbf{333·33 m/s.}$$

Example 9.18. *The driver of a car moving towards a factory with a velocity of* 30 ms^{-1} *sounds the horn with a frequency of* 240 Hz. *Find the apparent frequency of sound heard by the watchman of the factory. Velocity of sound in air* = 350 m/s. (Osmania, 1992)

Here $v = 350$ m/s

$a = 30$ m/s

$n = 240$ Hz.

The source is moving towards a stationary observer,

$$\therefore \quad n' = \left(\frac{v}{v-a}\right)n$$

$$n' = \left(\frac{350}{350-30}\right)240$$

$$= \mathbf{262{\cdot}5\ Hz.}$$

Example 9.19. *Two express trains travelling at* 100 km/hr *are meeting each other while one of them is whistling. If the frequency of note is* 800, *find the apparent pitch as heard by an observer in the other train after they passed each other. Velocity of sound* = 340 m/s. (IAS, 1988)

Here $V = 340$ m/s

$$a = 100 \text{ km/hr} = \frac{100 \times 10^3}{3600} = 27{\cdot}78 \text{ m/s}$$

$$b = 100 \text{ km/hr} = \frac{100 \times 10^3}{3600} = 27{\cdot}78 \text{ m/s}$$

$n = 800$ Hz.

$n' = ?$

The source and observer are moving away from each other.

$$\therefore \quad n' = n\left[\frac{V-b}{V+a}\right]$$

$$= 800\left[\frac{340 - 27{\cdot}78}{340 + 27{\cdot}78}\right]$$

$$= 800\left[\frac{312{\cdot}22}{367{\cdot}78}\right] = \mathbf{679{\cdot}15\ Hz.}$$

EXERCISES

1. Explain Doppler's principle. Calculate the apparent pitch of a note due to the motion of source and the listener.
2. Discuss Doppler's effect in sound and obtain an expression for the apparent frequency of the note when the source and the listener are (*i*) moving towards each other, and (*ii*) moving away from each other.
3. Explain Doppler's effect. A source produces a note of frequency n and is moving towards a stationary observer with a uniform speed a. Show that the apparent pitch is

$$n' = n\left[\frac{v}{v-a}\right].$$

4. An observer on a railway platform observed a train passing through the station at a speed a. Show that the frequency of the whistle changes by

$$nv\left[\frac{2a}{v^2 - a^2}\right].$$

5. Explain Doppler's effect. Explain the apparent frequency of a note when both the source and the listener are in motion, the medium being at rest. *(Delhi, 1974)*
6. Explain Doppler's effect. Find an expression for the ratio of the apparent frequency to the real frequency of the notes emitted when there is relative movement between the source and the listener. *(Madras, 1974)*
7. Explain Doppler's principle. A source of sound produces a note of frequency n and is moving towards a stationary observer with a speed a. Show that the apparent pitch is given by

$$n' = n\left[\frac{v}{v-a}\right]$$

where v is the velocity of sound in air.

[Delhi (Addl. Physics), 1976]

8. State and explain Doppler's effect. Derive an expression for the change in frequency of a note when both the source of sound and the observer are in motion. *[Delhi (Sub.), 1976]*
9. Explain Doppler's effect. Obtain an expression for the frequency of a note heard by an observer, when both the source and the observer are in motion towards each other. *[Delhi (Supple.), 1976]*
10. What is Doppler's effect in sound? Obtain expressions for shifts in frequency separately for the cases of observer and source in motion. Are these expressions the same. Explain. *(I.A.S., 1975)*

11. Discuss Doppler's effect in light. Does the shift depend individually on the state of motion of the observer or that of the source? If not, why not?

(*I.A.S., 1975*)

12. A person is standing on a railway platform. An engine while approaching the platform blows a whistle of pitch 640 hertz. The speed of the engine is 108 km/hr. Velocity of sound is 350 m/s. Calculate the apparent pitch of the whistle as heard by the person. (**Ans.** 700 hertz)

13. A person is standing on a platform. A railway engine moving away from the person with a speed of 90 km/hr blows a whistle of pitch 730 hertz. Calculate the apparent pitch of the whistle as heard by the person. Velocity of sound in air = 340 m/s. (**Ans.** 680 hertz)

14. A person is standing near a railway track and a train moving with a speed of 72 km/hr is approaching him. The apparent pitch of the whistle as heard by the person is 680 hertz. Calculate the actual frequency of the whistle. Velocity of sound in air =340 m/s. (**Ans.** 640 hertz)

15. Two aeroplanes *A* and *B* are approaching each other with a speed of 540 km/hr. The frequency of the whistle emitted by *A* is 2000 hertz. Calculate the apparent pitch of the whistle as heard by the passengers of aeroplane *B*. Velocity of sound in air =350 m/s. (**Ans**. 500 hertz)

16. Two aeroplanes *A* and *B* are moving away from one another with a speed of 540 km/hr. The frequency of the whistle emitted by *A* is 1000 hertz. Calculate the apparent pitch of the whistle as heard by the passengers of aeroplane *B*. Velocity of sound in air = 350 m/s. (**Ans.** 400 hertz)

17. Two aeroplanes *A* and *B* are approaching each other and their velocities are 540 km/hr and 720 km/hr respectively. The frequency of the note emitted by *A* as heard by the passengers in *B* is 1100 hertz. Calculate the frequency of the note as heard by the passengers in *A*. Velocity of sound in air = 350 m/s. (**Ans**. 400 hertz)

18. Two aeroplanes pass each other in opposite directions and one of them is blowing a whistle of frequency 900 hertz. Calculate the frequencies of the notes heard in the other aeroplane (*i*) before and (*ii*) after they have crossed each other. Velocity of either of the aeroplane is 720 km/hr. Velocity of sound in air =350 m/s. [**Ans.** (*i*) 3300 hertz, (*ii*) 245.4 hertz]

19. What is Doppler effect? Derive an expression for the apparent frequency received by a stationary observer when the source of sound is in motion.

(*Osmania, 1992*)

CHAPTER 10

Fourier Analysis

10.1. Fourier Theorem

Fourier theorem deals with the summation of a number of simple harmonic vibrations, in which the vibrations are in the same straight line. The theorem also helps in the synthesis and analysis of complex forms of vibrations. This theorem was formulated by **J.B.T. Fourier** in 1828.

The theorem states that any single valued periodic function can be expressed as a sum of a number of simple harmonic terms which are multiples of the given function. The theorem is generally referred in relation to the study of transverse vibration of strings. However, the theorem has a wider scope.

Fourier's theorem deals primarily with the synthesis of a complex periodic vibration from simple harmonic terms and it also gives a method to analyse a complex vibration into its component vibrations.

Conditions. The theorem has the following two provisions :

(1) The displacement must be a single valued function and continuous. This condition is satisfied in all cases of mechanical vibrations because a single particle cannot actually have two different displacements simultaneously.

(2) The displacement must always have a finite value. This is true in the case of sound.

10.2. Fourier Series

The Fourier theorem can be expressed by the series

$$y = f(\omega t) = A_0 + A_1 \cos(\omega t + \alpha_1) + A_2 \cos(2\omega t + \alpha_2) + \ldots + A_m \cos(m\omega t + \alpha_m)$$

This can be written in the form

$$y = f(\omega t) = A_0 + \sum_{m=1}^{m=\infty} A_m \cos(m\omega t + \alpha_m) \qquad \ldots (1)$$

Here y represents the displacement of the complex periodic vibrations of frequency $\left(\frac{\omega}{2\pi}\right)$. A_1, A_2 . . , A_m are the amplitudes of the components of the simple harmonic vibrations and $\alpha_1, \alpha_2, \ldots, \alpha_m$ represent their respective initial phases.

It may be mentioned that sometimes it is convenient to represent y as a sum of sine and consine series in the form

$$y = f(\omega t) = \sum_{m-0}^{m=\infty} [A_m \sin m\omega t + B_m \cos m\omega t] \quad \ldots (2)$$

Here $\qquad A_0 = 0$

$$\therefore \qquad y = f(\omega t) = B_0 + \sum_{m=1}^{m=\infty} A_m \sin m\omega t + \sum_{m=1}^{m=\infty} B_m \cos m\omega t \quad \ldots (3)$$

10.3. Evaluation of Fourier Coefficients

The method of finding the amplitudes or Fourier coefficients (B_0, A_m and B_m) for all values of m is called Fourier analysis. In equation (3) the value of B_0 can be obtained by integrating this equation with respect to time t over a complete time period T

$$\therefore \qquad \int_0^T y\, dt = \int_0^T B_0\, dt + \int_0^T \left[\sum_{m=1}^{m=\infty} A_m \sin m\omega t\right] dt$$

$$+ \int_0^T \left[\sum_{m=1}^{m=\infty} B_m \cos m\omega t\right] dt \quad \ldots (4)$$

Here $\qquad \int_0^T \left[\sum_{m=1}^{m=\infty} A_m \sin m\omega t\right] dt = 0$

and $\qquad \int_0^T \left[\sum_{m=1}^{m=\infty} B_m \cos m\omega t\right] dt = 0$

$$\therefore \qquad \int_o^T y\, dt = B_0\, T$$

or $\qquad B_0 = \frac{1}{T}\int_0^T y\, dt \quad \ldots (5)$

To find the value of A_m, of the sine series, multiply all the terms of equation (3) by $\sin m\omega t$ and intergate the terms for a complete cycle 0 to T. Then,

$$A_m = \frac{2}{T}\int_0^T y \sin (m\omega t)\, dt \qquad \ldots (6)$$

Similarly the values of B_m of the cosine series can be obtained by multiplying all the terms of equation (3) by cos $m\omega t$ and integrating for a complete cycle 0 to T.

Thus we have

$$B_m = \frac{2}{T}\int_0^T y \cos (m\omega t)\, dt \qquad \ldots (7)$$

Results:

$$y = f(\omega t) = B_0 = + \sum_{m=1}^{m=\infty} A_m \sin m\omega t + \sum_{m=1}^{m=\infty} B_m \cos m\omega t$$

$$B_0 = \frac{1}{T}\int_0^T y\, dt \qquad \ldots (8)$$

$$A_m = \frac{2}{T}\int_0^T y \sin (m\omega t)\, dt$$

$$B_m = \frac{2}{T}\int_0^T y \cos (m\omega t)\, dt.$$

10.4. Square Wave

The analysis of a square wave according to Fourier theorem is an illustrative example. Let $y = f(\omega t) = 0$ at $t = 0$ and $t = \frac{T}{2}$. Let y be equal to $+a$ for $0 < t < \frac{T}{2}$ (Fig. 10.1). This function has a discontinuity at $t = 0$ and at $t = \frac{T}{2}$.

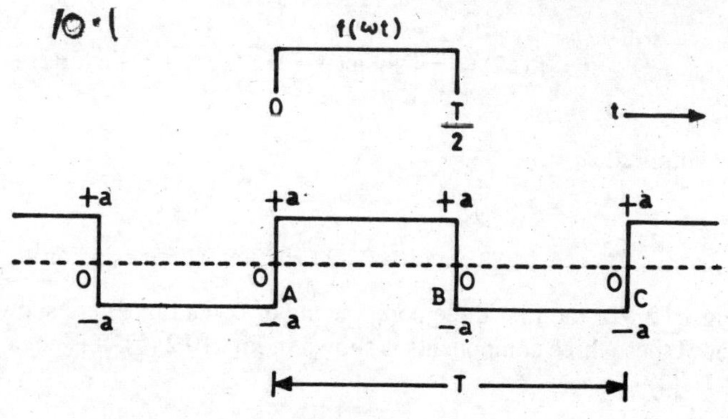

Fig. 10.1.

Consequently it does not satisfy the assumption that it is smooth everywhere. Therefore, a perfect representation cannot be reasonably expected from the Fourier series.

There is a **sharp overshoot** at $t = 0$ and at $t = \frac{T}{2}$ for every partial sum of the series. Addition of more and more terms makes the overshoot sharper but the displacement (y) does not become zero.

Here $$f(\omega t) = 0 \quad \text{for} \quad t = 0,$$

and $$f(\omega t) = +a \quad \text{for} \quad 0 < t < \frac{T}{2}$$

and $$f(\omega t) = 0 \quad \text{for} \quad t = \frac{T}{2}$$

and $$f(\omega t) = -a \quad \text{for} \quad \frac{T}{2} < t < T$$

Using the equations in (8), it is easy to obtain the values of

$$B_0 = 0$$

$$B_m = 0 \text{ for all values of } m$$

$$A_m = 0 \quad \text{for } m = 2, 4, 6 \qquad \text{(even integers)}$$

and $$A_m = \frac{4a}{m\pi} \text{ for } m = 1, 3, 5 \qquad \text{(odd integers)}$$

$$y = f(\omega t) = B_0 = + \sum_{m=1}^{m=\infty} A_m \sin m\omega t + \sum_{m=1}^{m=\infty} B_m \cos m\omega t$$

Substituting the values of B_0, A_m and B_m,

$$y = f(\omega t) = \frac{4a}{\pi}\left[\sin \omega t + \frac{1}{3}\sin(3\omega t) + \frac{1}{5}\sin(5\omega t) + \ldots\right]$$

If the amplitude $a = 1$

$$y = f(\omega t)$$

$$= 1{\cdot}273 \sin \omega t + 0{\cdot}424 \sin 3\omega t \; 0{\cdot}255 + \sin 5\omega t + \ \ldots \quad (9)$$

In Fig.10·2 (a), the first three components of equation (9) are shown. The resultant of these three components is shown in Fig. 10·2 (b). It is not a perfect square.

The resultant of the first fifteen terms of equation (9) is shown in Fig. 10.3.

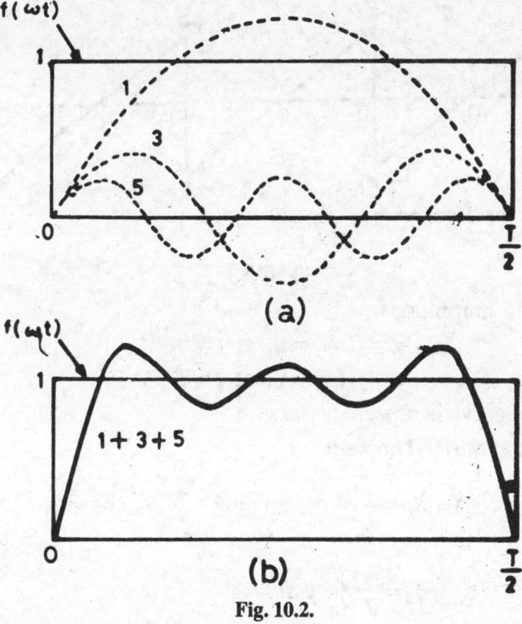

Fig. 10.2.

This figure is a near approach of a square wave form but not an exact square form. However, addition of more and more terms will give a resultant curve nearly approaching a square wave form.

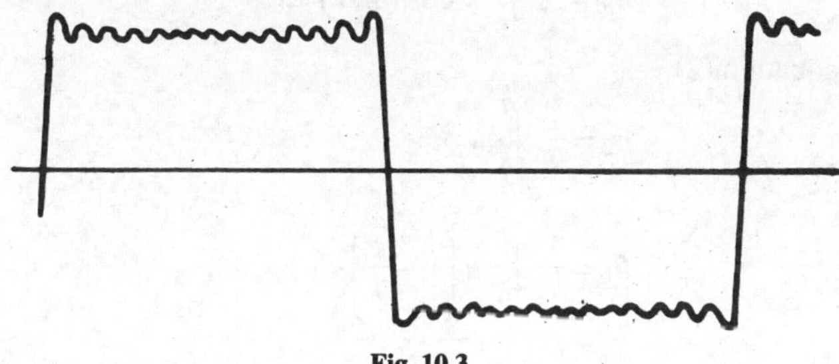

Fig. 10.3.

10.5. Saw-tooth Wave

The analysis of a saw-tooth wave according to Fourier Theorem can be illustrated as follows (Fig.10.4).

Saw-tooth wave can be represented by the equation

$$y = f(\omega t) = a\left(1 - \frac{t}{T}\right) \text{ for } 0 < t < T$$

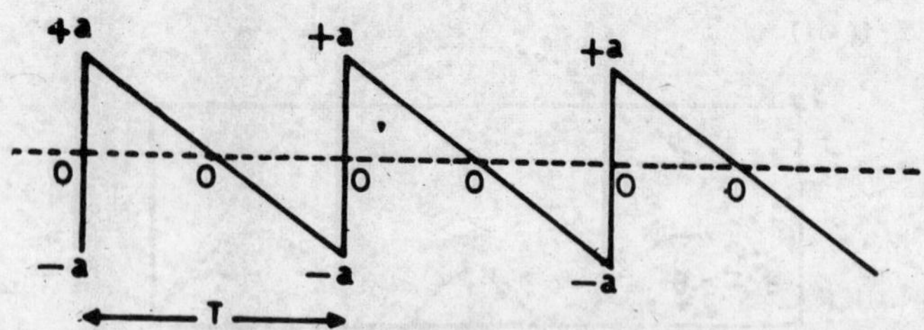

Fig. 10.4.

Here amplitude $= a$

$$y = f(\omega t) = a, \text{ at } t = 0$$

and

$$y = f(\omega t) = 0, \text{ at } t = T$$

After this the curve is repeated.

According to Fourier Theorem

$$y = f(\omega t) = B_0 = + \sum_{m=1}^{m=\infty} A_m \sin m\omega t + \sum_{m=1}^{m=\infty} B_m \cos m\omega t \qquad \ldots (1)$$

$$B_0 = \frac{1}{T} \int_0^T y \, dt \qquad \ldots . (2)$$

$$A_m = \frac{2}{T} \int_0^T y \sin (m\omega t) \, dt \qquad \ldots (3)$$

and

$$B_m = \frac{2}{T} \int_0^T y \cos (m\omega t) \, dt \qquad \ldots (4)$$

From equation (2)

$$B_0 = \frac{1}{T} \int_0^T y \, dt$$

$$B_0 = \frac{1}{T} \int_0^T a \left(1 - \frac{t}{T}\right) dt$$

$$B_0 = \frac{a}{T} \left[t - \frac{t^2}{2T} \right]_0^T$$

$$B_0 = \frac{a}{T} \left[T - \frac{T^2}{2T} \right]_0^T = \frac{a}{2}$$

$$B_0 = \frac{a}{2} \qquad \ldots (5)$$

From equation (3)

$$A_m = \frac{2}{T}\int_0^T y \sin(m\omega t)\, dt$$

$$= \frac{2}{T}\int_0^T a\left(1 - \frac{t}{T}\right)\sin(m\omega t)\, dt$$

$$= \frac{2a}{T}\int_0^T \sin(m\omega t)\, dt - \frac{2a}{T^2}\int_0^T t \sin(m\omega t)\, dt$$

$$A_m = \frac{2a}{T}\left[-\frac{\cos m\omega t}{m\omega}\right]_0^T - \frac{2a}{T^2}\left[\left(-\frac{t\cos m\omega t}{m\omega}\right)_0^T + \int_0^T \frac{\cos m\omega t}{m\omega}\, dt\right]$$

$$A_m = \frac{2a}{Tm\omega}[-\cos m\omega T + \cos(m\omega \times 0)]$$

$$+ \frac{2a}{T^2}\left[\frac{T\cos m\omega T}{m\omega}\right] - \frac{2a}{T^2}\left[\sin\frac{m\omega t}{m^2\omega^2}\right]_0^T$$

But $\quad -\cos(m\omega T) + \cos 0 = 0$

and $\quad \left[\sin(m\omega t\right]_0^T = 0 \qquad \left[\because\ T = \frac{2\pi}{\omega}\right]$

$$\therefore \qquad A_m = \frac{2a}{T}\left[\frac{\cos 2\pi m}{m\omega}\right] = \frac{2a}{Tm\omega} = \frac{a}{m\pi} \qquad \ldots (6)$$

From equation (4)

$$B_m = \frac{2}{T}\int_0^T y \cos(m\omega t)\, dt$$

$$B_m = \frac{2}{T}\int_0^T a\left(1 - \frac{t}{T}\right)\cos(m\omega t)\, dt$$

$$B_m = 0$$

Hence all the cosine terms of the Fourier series are absent.

In a saw-tooth wave, equation (1) of Fourier series can be written in the form

$$y = f(\omega t) = \frac{a}{2} + \sum_{m=1}^{m=\infty} \frac{a}{m\pi}\sin m\omega t \qquad \ldots (7)$$

This expression in the expanded form can be written as

$$y = \frac{a}{2} + \frac{a}{\pi}\left(\sin \omega t + \frac{1}{2}\sin 2\omega t + \frac{1}{3}\sin 3\omega t + \ldots + \frac{1}{m}\sin m\omega t \ldots\right) \qquad \ldots (8)$$

The addition of successive terms in the series given in equation (8) results in the saw-tooth wave form. Fig. 10.5 shows the resultant wave pattern for the

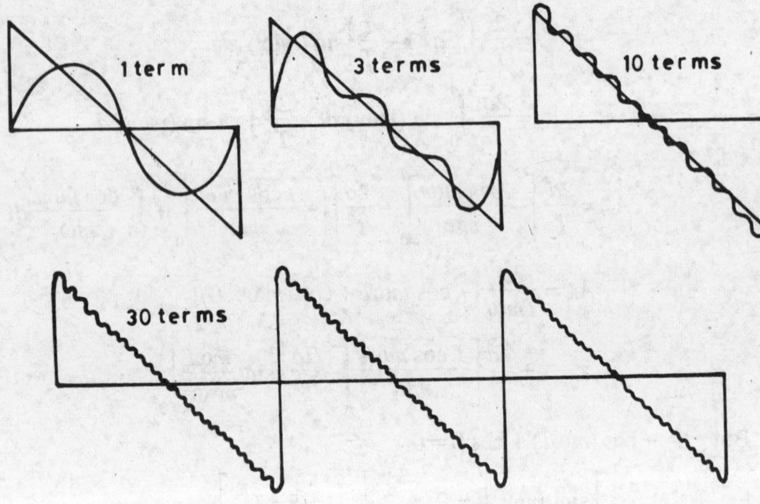

Fig. 10.5.

addition of two terms, three terms, 10 terms and 30 terms respectively. Taking only the first term, will give only a sine wave.

10.6. Application of Fourier Sine Series to Plucked String

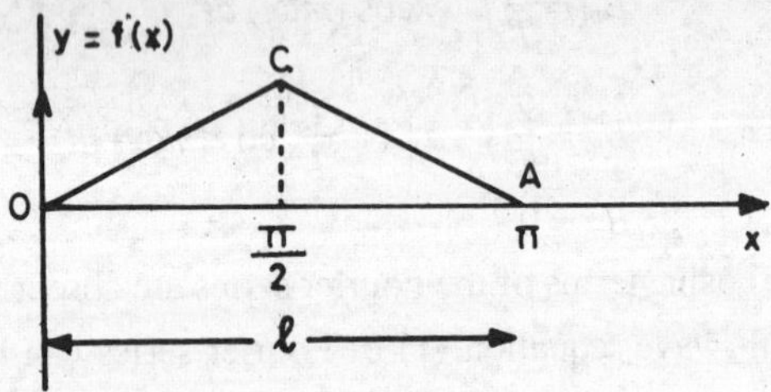

Fig. 10. 6.

Fig. 10.6 represents a function describing a sine series for interval 0 to π.

Let a string of length l be fixed between two points O and A. The midpoint of the string is pulled up by distance d. At $t = 0$, the velocity is zero at the point C and the displacement is d.

The function is given by

$$f(x) = x \quad \text{for} \quad (O < x < \pi/2)$$

$$f(x) = \pi - x \text{ for } (\pi/2 < x < \pi)$$

Taking $\quad f(x) = \Sigma A_m \sin mx$

$$A_m = \frac{2}{\pi}\int_0^{\pi/2} x \sin mx\, dx + \frac{2}{\pi}\int_{\pi/2}^{\pi} (\pi - x) \sin mx\, dx$$

$$\therefore \quad A_m = \left(\frac{4}{m^2 \pi}\right) \sin\left(\frac{m\pi}{2}\right)$$

When m is even

$$\sin \frac{m\pi}{2} = 0.$$

It means only those terms with odd values of m are present.

$$\therefore \quad f(x) = \frac{4}{\pi}\left[\frac{\sin x}{1^2} - \frac{\sin 3x}{3^2} + \frac{\sin 5x}{5^2} - \frac{\sin 7x}{7^2} + \ldots\right]$$

But at $\quad x = \pi/2,\ f(x) = \pi/2$

$$\therefore \quad \frac{\pi}{2} = \frac{4}{\pi}\left[\frac{\sin(\pi/2)}{1^2} - \frac{\sin(3\pi/2)}{3^2} + \frac{\sin(5\pi/2)}{5^2} - \frac{\sin(7\pi/2)}{7^2} + \ldots\right]$$

$$\frac{\pi}{2} = \frac{4}{\pi}\left[\frac{1}{1^2} + \frac{1}{3^2} + \frac{1}{5^2} + \frac{1}{7^2} + \ldots\right]$$

$$\frac{\pi^2}{8} = \left[\frac{1}{1^2} + \frac{1}{3^2} + \frac{1}{5^2} + \frac{1}{7^2} + \ldots\right]$$

$$= \sum_{m=0}^{\infty} \frac{1}{(2m+1)^2}$$

When $\quad m = 0, 1, 2, 3 \ldots\ldots$

When the string is fixed at its two ends and plucked at the centre through a distance d, the shape of the string is similar to the function discussed above.

The displacement at any point of the string is given by

$$\frac{d}{l/2} = \frac{y}{x} \text{ for } (0 < x < \pi/2)$$

and

$$\frac{d}{l/2} = \left(\frac{y}{l - x}\right) \text{ for } (l/2 < x < l)$$

Using the sine series expression

$$A_m = \frac{8d}{m^2 \pi^2} \sin \frac{m\pi}{2} = 0 \text{ for } m \textbf{ even.}$$

The displacement of the plucked string may be represented by adding all the permitted modes.

Hence,

$$y = \sum y_n = \sum_{n=1}^{n=\infty} A_n \sin\left(\frac{n\pi x}{l}\right)$$

where $n = 2m + 1$

and $$A_n = \left(\frac{8d}{n^2\pi^2}\right)\sin\left(\frac{n\pi}{2}\right).$$

Each term y_n in the above summation represents the space dependence of all allowed **eigen modes** when the string vibrates.

Let the string vibrate after it is released. The displacement in each **eigen mode** is represented in space and time as

$$y_n = A_n \begin{bmatrix}\sin\\ \cos\end{bmatrix}\phi \begin{bmatrix}\sin\\ \cos\end{bmatrix}\omega t$$

Here $$\phi = \frac{2\pi x}{\lambda}$$

For each mode

$$y_n = A \sin\phi \cos\omega t$$

When $\phi = n(\pi/2)$ wher n is even

$$\sin\phi = 0$$

It means all even modes of vibrations are missing.

It may also be noted that the amplitude of each mode is inversely proportional to n^2 (for odd values only).

It means for $n= 3$, the amplitude, is 1/9 of the first amplitude ($n = 1$).

$$\therefore \quad A_m \propto \frac{1}{n^2}$$

$$A_3 = \frac{A_1}{9}.$$

EXERCISES

1. State the Fourier's theorem and obtain expressions for Fourier's coefficients.
2. Discuss the application of Fourier's theorem for the analysis of a square wave.
3. Apply Fourier's series in analysing the saw-tooth wave form.

CHAPTER 11

Practical Applications

11.1. Siren

It consists of a wind chest. The upper disc D has holes near its circumference and is just above the disc A of the wind chest. The disc A has a number of holes equal to the number of holes in the disc. The holes in D are slanting and are just in opposite direction to the holes in the fixed disc A (Fig. 11.1). The disc D rotates about the spindle and the number of rotations made by it, is indicated by the counters C_1 and C_2.

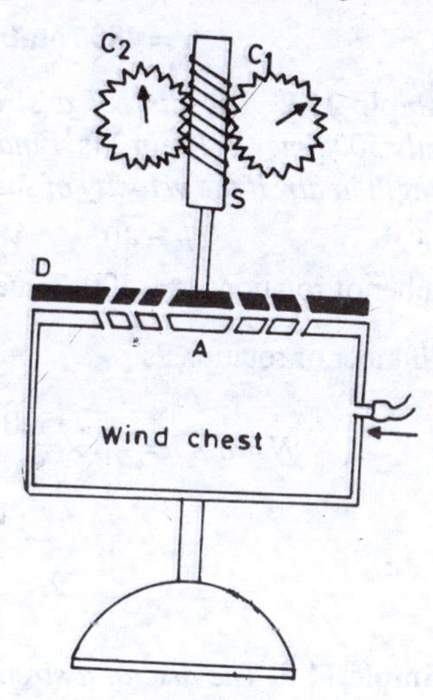

Fig. 11.1.

Working. The wind is pushed into the wind chest and each puff of air that escapes from the holes, rotates the upper disc D. When the holes in the disc D are just above the holes in A, air escapes. Suppose the upper disc has h holes and it rotates n times per second. The frequency of the sound produced by the air escaping out of the holes $= N = n \times h$.

Determination of the frequency of a note. The pressure of wind entering into the wind chest is adjusted such that the frequency of the sound produced by the siren is nearly equal to that of the given note. This can be done by increasing the speed of rotation of the disc D slowly. When the frequencies are nearly equal, count the number of beats produced per second. Let the number of beats produced per second be x.

The frequency of the note = $N + x$.

To avoid the counting of beats, the speed of rotation should be adjusted such that the two notes have the same pitch. In that case the frequency of the note = N.

Example 11.1. *The disc of a given siren has 32 holes. What must be the speed of rotation per minute of the siren disc so that the note emitted by the siren may be in unison that of a tuning fork which makes 256* vibrations per second.

Suppose, the number of rotations/minute = x

$\therefore$ Number of rotations/s $= \dfrac{x}{60}$

Here $\quad n = \dfrac{x}{60}$

$$h = 32, \quad N = 256$$

$$N = n \times h$$

$$256 = \frac{x}{60} \times 32$$

or $\quad$ $x =$ **480 rotations/minute.**

Example 11.2. *The disc of a siren having a circle of 40 holes rotates uniformly 500 times in 1* min 24s. *Find the frequency of the note emitted and its wavelength in air, if the velocity of sound in air is 34000* cm/s.

Here $\quad h = 40 \qquad V = 34000$ cm/s

Number of rotations = 500, Times = 1 min 24 s = 84 s

$\therefore$ Number of rotations/s $\quad n = \dfrac{500}{84}$

$$\therefore \quad N = n \times h = 40 \times \frac{500}{84} = 238{\cdot}1$$

But $\quad V = N\lambda$

$$\therefore \quad \lambda = \frac{V}{N} = \frac{34000}{238{\cdot}1} = \mathbf{142.8}$$

Example 11.3. *The disc of a siren having 60 holes makes 420* revolutions per minute. *Find the wavelength of its sound in air at* 0^0 C, *if the velocity of sound at* 0^0 C *is* 332 m/s. (IAS, 1989)

Number of holes, $h = 60$

Number of revolutions = 420 minute

$$n = \frac{420}{60} = 7 \text{ per second}$$

$$\nu = nh$$

$$\nu = 7 \times 60 = 420 \text{ Hz}$$

$$V = 332 \text{ m/s}$$

$$\lambda = \frac{V}{\nu}$$

$$= \frac{332}{420}$$

$$= \mathbf{0 \cdot 79 \text{ m.}}$$

11.2. Falling Plate Method

To the prong of a tuning fork a fine style is attached. The style touches a smoked paper fixed on a vertical plate. The vertical plate is fixed at the top. The tuning fork is set into vibration so that its prong having the style vibrates horizontally. The plate is allowed to fall under the action of gravity. The style traces a wave on the smoked paper as shown in Fig.11.2.

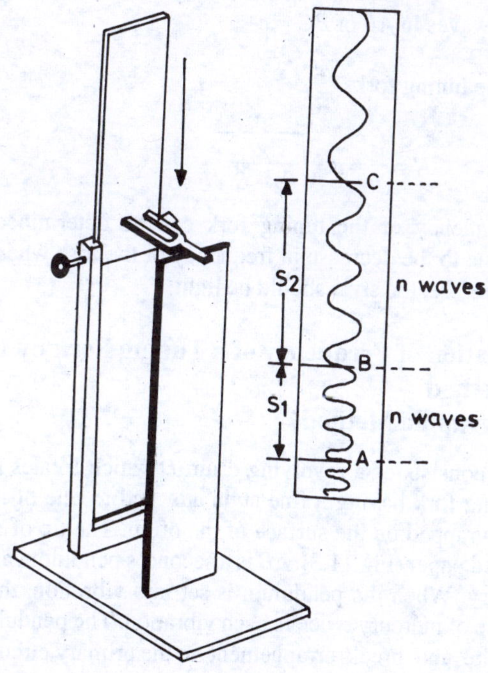

Fig. 11.2.

The wave trace near the lower edge of the plate is narrow and the wavelength increases with increase in distance from the lower end. Take a point A on the smoked paper and mark points B and C such that there are n complete waves in AB and also n complete waves in BC. Measure the distances AB and BC.

If u is the initial velocity of the plate at A and the time taken from A to B or B to C is t, then

$$S_1 = ut + \frac{1}{2} gt^2 \qquad \ldots (1)$$

$$(S_1 + S_2) = u\,(2t) + \frac{1}{2} g\,(2t)^2 \qquad \ldots (2)$$

Multiplying (1) by 2

$$2S_1 = 2ut + gt^2 \qquad \ldots (3)$$

Subtracting (3) from (2)

$$S_2 - S_1 = gt^2$$

$$t = \sqrt{\frac{S_2 - S_1}{g}} \qquad \ldots (4)$$

The number of waves in AB or $BC = n$

Frequency of the tuning fork $= \frac{n}{t}$

$$\therefore \qquad N = \frac{n}{t} = n\sqrt{\frac{g}{S_2 - S_1}} \qquad \ldots (5)$$

Hence, the frequency of the tuning fork can be determined. The only possible error is due to the decrease in frequency of the fork when the style is attached to it. Therefore, the style should be light.

11.3. Determination of Frequency of a Tuning Fork by Revolving Drum Method [Chronographic Method]

The apparatus consists of a revolving drum D which rotates at a uniform speed. T is a tuning fork having a fine style attached to one of its prongs. A smoked paper is wrapped on the surface of the drum. The tip of the style just touches the smoked paper (Fig.11.3). AB is a second's pendulum and C is a cup containing mercury. When the pendulum is set into vibration, the bob B just touches the surface of mercury twice in each vibration. The pendulum works as an automatic make and break arrangement in the primary circuit (Fig.11.3).

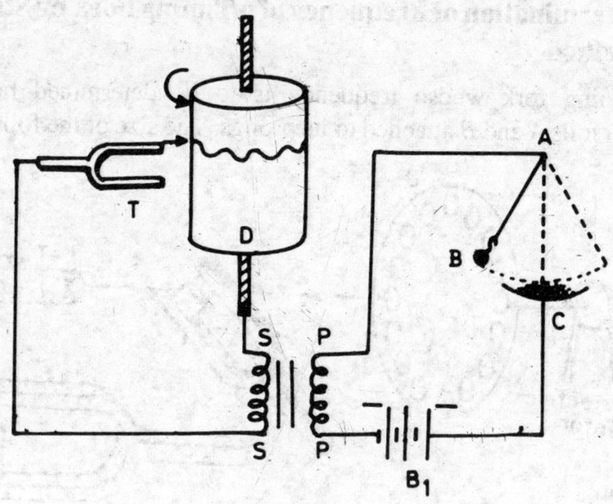

Fig. 11.3.

The secondary circuit is also shown in Fig.11.3. Whenever *B* makes contact with mercury, an induced *emf* is produced across *S, S.*

A spark is produced at the tip of the style and a mark is made on the smoked paper. These marks will be obtained on the smoked paper at equal intervals of time of one second.

The tuning fork and the pendulum are set into vibration. The style marks a wavy pattern on the smoked paper of the rotating drum. The marks are also obtained on the smoked paper. The smoked paper is removed and the number of complete waves between two consecutive marks is counted. The average number of waves between two marks gives the value of the frequency of the tuning fork.

Here the frequency of the tuning fork is determined with the style attached to one of its prongs. The experimental value is less than the actual frequency of the tuning fork. To determine this correction, a tuning fork of known frequency is taken with a style fixed to one of its prongs. The experiment is performed and the experimental value is found as discussed earlier. The difference between the actual frequency and the experimental value gives the value of the correction to be applied.

Suppose the actual frequency of the tuning fork is x, and the experimental value is y. The correction to be applied = + $(x - y)$

To determine the unknown value of the frequency of any tuning fork, the correction $(x-y)$ is added to the experimental value.

11.4. Determination of Frequency of a Tuning Fork by Stroboscopic Method

The tuning fork whose frequency is to be determined has two thin aluminium foils A and B attached to its prongs. The size of the foils is such that

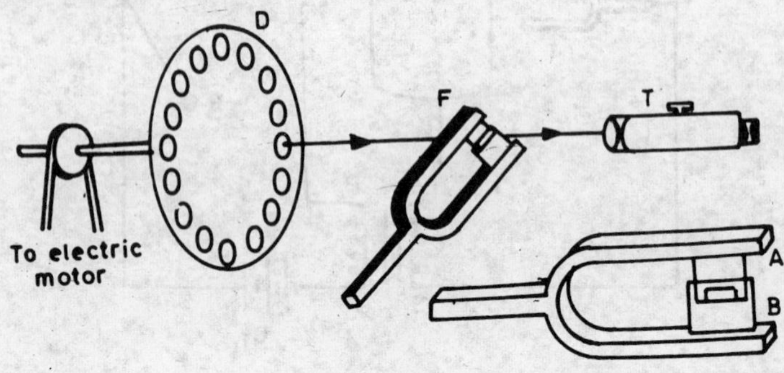

Fig. 11.4.

the window (slit) in one of the foils, opens when the prongs vibrate outwards and it closes when the prongs vibrate inward. The window opens only once in each vibration of the tuning fork (Fig.11.4).

D is a white circular disc having dots marked near the edge. These dots are equally spaced. The tuning fork F is placed in such a position that the dot lies in front of the window when it opens. The dots can be observed with the help of a telescope T. The disc D is rotated at a uniform speed with the help of an electric motor (Fig. 11.4).

The tuning fork is set into vibration and the speed of rotation of the disc is gradually increased. At a particular minimum speed on the disc the dot *appears stationary* in the field of view of the telescope. This means that during one complete vibration of the tuning fork, the position of one dot is exactly replaced by the next dot.

Let n be the minimum number of rotations of the disc per second when the dot appears stationary. Suppose the number of dots on the disc $= m$.

Time taken for one complete rotation of the disc $= 1/n$.

Time interval for one dot to be exactly replaced by the next dot.

$$t = \frac{1}{nm} \qquad \ldots (1)$$

This time t is also corresponds to the time taken by the tuning to complete one vibration.

Suppose the frequency of the tuning fork is N

$$\therefore \qquad t = \frac{1}{N} \qquad \ldots (2)$$

From equations (1) and (2)

$$\frac{1}{N} = \frac{1}{nm}$$

$$\therefore \qquad N = \mathbf{nm} \qquad \ldots (3)$$

If the speed of rotation of the disc is made double, the dot will again appear stationary. Similarly at the speeds $3n$, $4n$ etc. the dot will appear stationary.

Example 11.4. *A stroboscopic disc marked with 120 dots is rotated and the minimum speed adjusted till the dot appears stationary when looked through a slit attached to the prongs of a tuning fork. If the minimum speed of rotations of the disc is 150* rotations per minute, *calculate the frequency of the tuning fork.*

Here $\qquad m = 120$

$$n = \frac{150}{60} = 2{\cdot}5 \text{ rotations/s}$$

Frequency of the tuning fork,

$$N = nm$$

$$N = 2{\cdot}5 \times 120$$

or $\qquad N = \mathbf{300\ hertz.}$

11.5. Determination of Frequency of a Tuning Fork by Phonic Motor Method

In Fig.11.5, F is an electrically maintained tuning fork whose frequency is to be determined. W is a soft iron toothed wheel that can rotate but a horizontal axis. M_1 and M_2 are two electromagnets having their axes passing through the centre of the wheel. The electrical arrangement is shown in Fig.11.5.

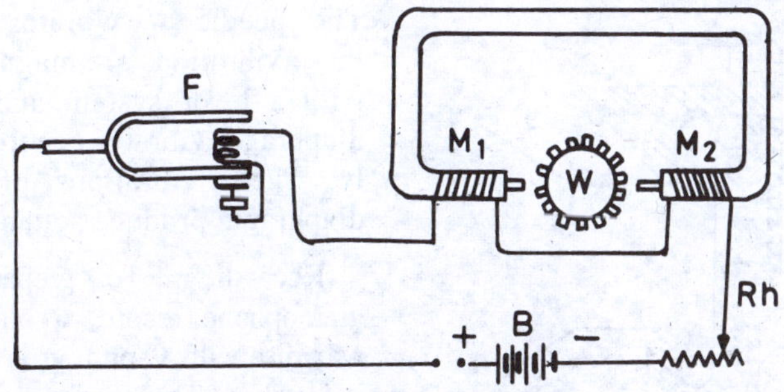

Fig. 11.5.

The tuning fork F is set into vibration. Initially the wheel W is rotated by hand. When any two diametrically opposite teeth of the wheel just pass the electromagnets at the instant of excitation, the teeth get impulses and move forward. The rotor will just advance through one tooth during one full vibration of the fork. The wheel will get an impluse during each excitation and will continue to rotate. The speed of rotation will depend on the frequency of the tuning fork. If the wheel is moving faster, the force of attraction between the tooth and the magnet, retards its motion. If the wheel is moving slower, the force of attraction accelerates it.

Suppose the wheel makes n rotations in one second and the number of teeth on the wheel is m.

Number of impulses received per second $= m \times n$.

If the frequency of the tuning fork is N, then

$$N = \mathbf{m} \times \mathbf{n}.$$

The speed of rotation of the wheel is determined by a mechanically geared counter. Knowing m and n the value of N can be calculated.

11.6. Gramophone

It consists of a turn table which is rotated with a clock work mechanism at a constant speed. It is provided with a sound box. The record is placed on the turn table and the needle of the sound box moves over the spiral groove. This groove has a wavy structure corresponding to the original sound recorded on it.

Sound Box. It consists of a diaphragm D and a horn H (Fig. 11.6). The diaphragm is held in position with the help of the rubber gaskets G, G. A lever L is fixed to the middle of the diaphragm. The lever is attached to the needle N. The needle N vibrates and these vibrations are magnified with a lever system and the diaphragm vibrates accordingly. These vibrations of the diaphragm produce sound.

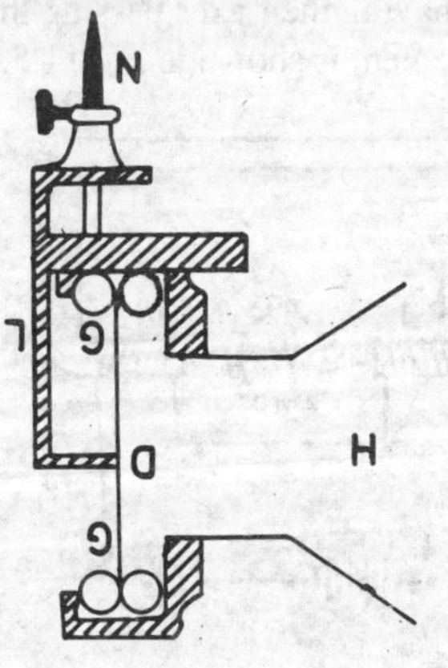

Fig. 11.6.

Recording. To prepare a gramophone record, a circular wax disc with a spiral groove on it is taken. The speaker speaks before the diaphragm and a sharp needle called a style vibrates along the path of the rotating wax record. A transverse wave is formed.

To prepare a large number of records, this wax disc is coated with graphite and electroplated uniformly with copper. The wax coating is removed by gently heating it. The copper disc remains with the grooves projecting outwards. This parent record is pressed against a plastic or shellac disc and the grooves are formed on it. These records prepared from the parent record are called sister records. Thus a large number of records can be obtained.

Sound reproduction. The record is kept on the turn table of the gramophone and the needle of the sound box is gently placed at the starting point of the groove, of the rotating record. When the needle movers along the groove, it vibrates and forces the diaphragm to vibrate accordingly. Thus the original recorded sound is reproduced. In a modern sound box, a pick up arrangement is used.

11.7. Microphone and Loud Speaker

Telephones are used to send and receive messages through electrical signals between two stations. As the electrical signals can flow round a closed circuit, two line wires are required. At both the ends of the line, a mouth piece and an ear piece are provided. The mouth piece works on the principle of microphone and the ear piece works as a small loudspeaker.

The microphone consists of a small ebonite box. Between the diaphragm *D* and the thin metal disc *P* the space is filled with carbon granules. [Fig.11.7 (*a*)].

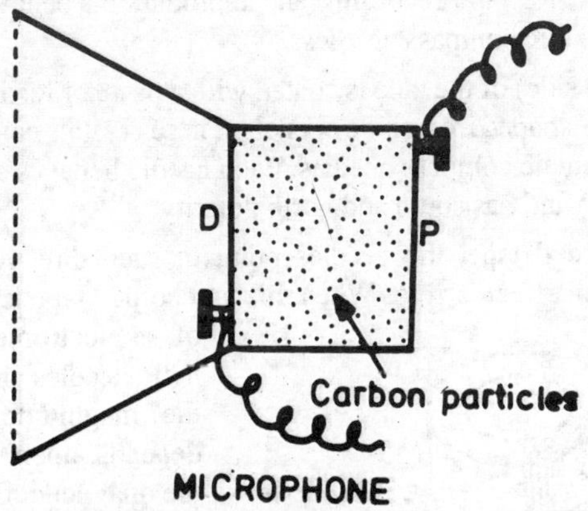

Fig. 11.7 (*a*) Microphone

The lead wires are taken from *P* and *D*. When a person speaks before the microphone, the diaphragm *D* vibrates. Due to the vibration of the diaphragm, the pressure on the carbon granules continuously changes. Consequently the resistance offered by the carbon granules in the circuit changes. When they are closely packed, the resistance decreases and the current in the circuit increases. When they are loosely packed, the resistance increases and the current in the circuit decreases. Thus the sound energy is converted into changing electric current.

The loud speaker consists of a diaphragm D and an electromagnet [Fig.11.7(*b*)]. When the changing current flows through the coil of the electromagnet, the magnetic flux changes. Due to the change in the magnetic flux, the diaphragm is set into vibration. Here electrical energy is converted into sound energy. The original sound is reproduced.

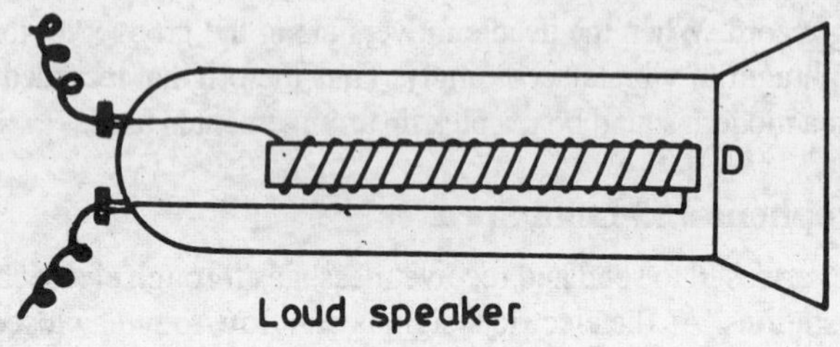

Fig. 11. 7 (*b*) Loud speaker.

11.8. Tape Recording

The most commonly used taperecorders are based on the principle of magnetism and are called magnetic taperecorders. Magnetic tapes of the form usually employed for tape recording and reproduction behave similar to an assembly of magnetic compass needles.

One side (dull side) of the tape is coated with a paste of ferric oxide (Fe_2O_3) containing needle shaped iron oxide crystals. These crystals can be imagined to be similar to magnetic compass needles. Each needle behaves as a tiny magnet (magnetic dipole) and has south and north polarity.

In an unrecorded tape, the needles point in such directions so that the resultant magnetic effect is zero. When this unrecorded tape is moved in front of an electromagnet, the magnetic needles are deflected and the magnitude of deflection depends upon the stength of the magnetic field. After this orientation, the needles no longer cancel their effect and the tape as a whole behaves as a magnet, magnetised to different extents at different

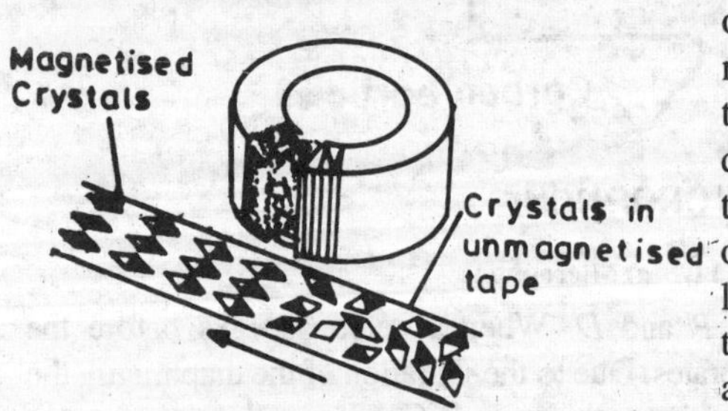

Fig. 11.8.

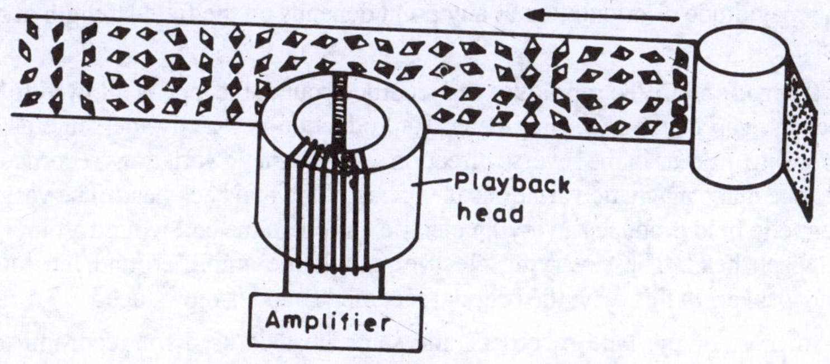

Fig. 11.9.

positions of the tape. The specific property of these iron oxide crystals is that they can retain their oriented positions for a very long time. Once recorded, the tape can serve as a permanent record unless it is erased.

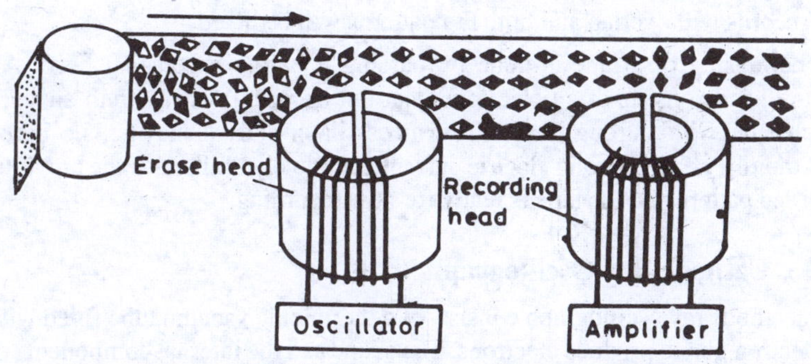

Fig. 11.10.

Recording. A magnetic tape has two sides. The dull side is coated with ferric oxide containing needle shaped crystals. The other side is shining. The tape is usually made of *cellulose acetate* or *poly vinyl chloride (PVC)* The tape is thin so that maximum length can be wound on a reel. The material of the tape must have flexibility but it should not stretch. The stretching of the tape can distort the recording. Sometimes **Mylar,** a kind of terylene is also used for the manufacture of the tape.

When a person speaks before a microphone, the sound vibrations are converted into varying electrical impulses. These electrical variations are fed into the amplifier. The amplified signal is fed to an electromagnet. (Fig. 11.10). The electromagnet is an iron ring having a very small air gap. This gap is not visible to the naked eye.

The dull side of the tape faces the gap and it is run, close to the gap. Varying electrical impulses produce corresponding variations in the magnetic field strength in the gap. The magnetic needles get oriented to different directions.

The magnitude of orientation at any point depends on the field strength at that instant.

Reproduction. To reproduce the recorded sound, the tape is allowed to run near the gap of the recording head. Reproduction is exactly the same as the recording process in the reverse direction. The magnetic variations recorded on the tape make magnetic variations in the gap of the playback head. The varying magnetic field produces a varying electric current in the coil wound around the playback head. These varying electric signals are amplified and fed into a loudspeaker. In this way, the original recorded sound is reproduced.

In most of the tape recorders, the same head is used for recording and reproduction of sound.

When the sound is to be recorded on a tape, it is initially passed in front of an erasing head, so that the previous record is erased.

Erasing the tape. Whenever new recording has to be made on a tape, it is essential that the previous recording is completely erased. If it is not done, the two records will overlap and both the patterns will be heard.

The erasing of a tape is done by passing it before an erasing head. An erasing head consists of an electromagnet (an iron ring with a small air gap) which is fed with a strong current from an oscillator of frequency 50 kilo-hertz. This rapidly varying strong electric current completely obliterates the previous recorded pattern and the tape is ready for new recording.

11.9. Cathode Ray Oscillograph

A cathode ray oscillograph consists of a thermionic vacuum tube fitted with an electron gun to produce electrons. The shape and the internal components of the cathode ray tube are shown in Fig. 11.11. The electrons transverse along the

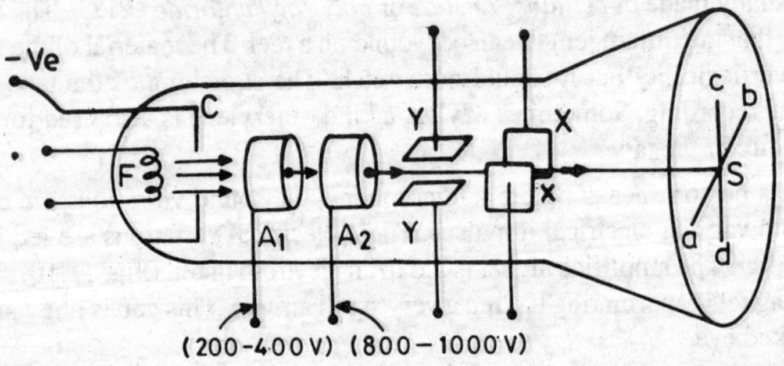

Fig. 11.11.

axis of symmetry of the tube, and in the absence of any deflecting electric or magnetic field, strike the centre of the screen and produce a fluorescent spot. The beam can be deflected by suitable electric or magnetic fields. The cathode ray oscillograph is a device useful for many purposes, such as, observing the wave form of an alternating current supply, of sound, studying the hysteresis of the ferro-magnetic materials and reception in television.

The cathode ray tube is evacuated to a pressure of 10^{-5} to 10^{-6} mm of Hg and electrons are produced by heating the tungsten filament *F* by low tension battery (not shown). *C* is a cylinder called the Wehnelt cylinder and it encloses the filament. This cylinder is maintained at a –ve potential so that a narrow beam of electrons is obtained through a narrow aperture in *C*. A_1 and A_2 are two cylindrical anodes with narrow openings. The potential of A_2 is higher than that of A_1. The two anodes function as a converging lens (electrostatic focussing). The combination of *F, C,* A_1 and A_2 is called the *electron gun*. The beam of electrons travel along the axis of symmetry of the tube and a fluorescent spot is obtained at *S*, the centre of the screen which is coated with a fluorescent material. The spot will appear yellowish green if coated with willemite (zinc ortho-silicate), blue with calcium or cadmium tungstate and white with zinc sulphide or zinc-cadmium sulphate.

The beam may be deflected horizontally or vertically by applying suitable electrostatic fields between the deflecting plates *XX* and *YY* respectively. The deflection of the beam is also possible by suitable magnetic fields produced by coils outside the tube.

The deflecting plates *XX* are connected to a '*time base*' voltage of known frequency of alternation. This voltage causes the spot to sweep across the screen (along the *X*-axis) at a uniform speed and then to quickly flash back to the starting point. The *PD* of which the wave form is to be investigated is applied between the plates *YY*.

If the voltage is applied only to the *XX* plates the spot sweeps a horizontal line passing through *S* and if the voltage is applied only to the *YY* plates the spot moves along a vertical line passing through *S*. If both the fields are applied, a wave form representing the voltage applied to the *YY* plates is traced and by synchronising this voltage with the time base voltage, *a fixed pattern of the wave form is obtained.* In case, two alternaing potentials with simple harmonic variation are applied between the plates *XX* and *YY*, the electronic beam traces *Lissajous'*

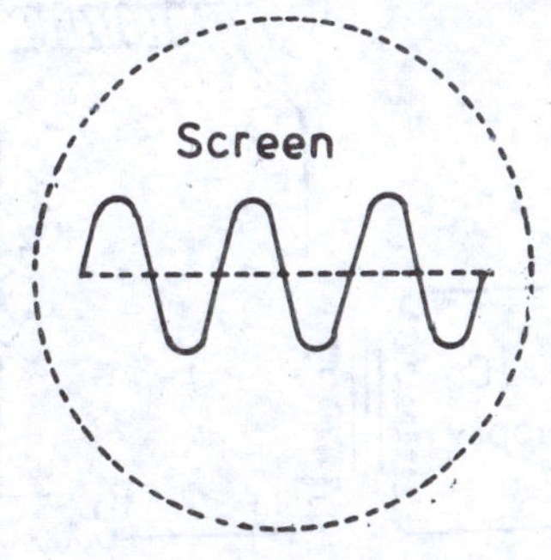

Fig. 11.12.

figures. A grid structure between the cathode and the anode controls the intensity of the beam. The extremely small inertia of the electrons enables the electron beam to follow the variations in the electric and magnetic fields which deflect it with practically no time lag. Persistence of vision and persistence of fluorescence give the illusion of an image. Fig. 11.12 shows a trace of the form on the screen.

Hysteresis curve with a cathode ray oscillograph. Fig.11.13 represents the diagram of connections to study the phenomenon of hysteresis in a bar of ferro-magnetic material. The bar *PQ* (say of soft iron) is supported horizontally near the cathode ray tube so that the length of the bar is parallel to the plates *YY* as shown. A solenoid would on this bar is connected to an alternating voltage from the secondary of a step down transformer and the plates *XX* are connected as shown. The strength of the current in the circuit is adjusted with the help of two variable non-inductive resistances R_1 and R_2. *S* is the screen. The axis of the tube is perpendicular to the plane of the paper.

The potential difference applied between the plates *XX* at any instant is equal to the potential difference between the points *L* and *M*. This potential difference will in turn depend on the strength of the current in the secondary circuit and hence is proportional to the magnetising field *(H)*. The displacement of the beam in the direction of the *X*-axis is proportional to *H* at any instant.

The plates *YY* are connected. When the iron bar is magnetised by the current passing through the solenoid, there is a magnetic field inside the tube parallel to the direction of the *X*-axis (the position of the rod *PQ* outside the tube corresponds to the function of the plates *YY*). The beam is travelling

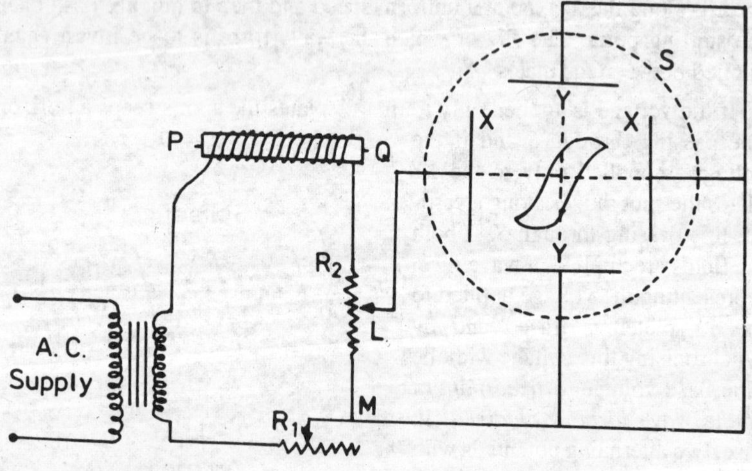

Fig. 11.13.

perpendicular to the plane of the paper and outwards. The electron beam will be deflected along the *Y*-axis according to Fleming's left hand rule (the conventional direction of current is opposite to the direction of motion of the electrons, *i.e.*, in this case perpendicular to the plane of the paper and inwards). As the current is alternating, the direction of the magnetic field also changes and hence the beam is deflected upwards or downwards depending on the magnetic polarity of *PQ* at that instant. The vertical shift of the beam corresponds to the intensity of magnetisation (*I*) of the specimen.

The force acting on an electron moving in an electric field is *Xe* where *X* is the intensity of the field and *e* is the charge on the electron. Due to the magnetic field the force is *Hev* where *H* is the intensity of the magnetic field and *v* is the velocity of the electron. These two forces *Xe* and *Hev* must be of the same order to obtain satisfactory results. Due to persistence of vision a hysteresis loop (*I-H* curve) is traced on the screen, the shift along the *X*-axis represents *H* and the shift along the *Y*-axis represents *I*.

Hysteresis curve for a specimen given in the form of a ring using a cathode ray oscillograph. To trace the hysteresis curve for a specimen given in the form of a ring, the circuit shown in Fig.10.14 is employed. *R* is the ring specimen. The alternating current supply is connected to the primary in series with the magnetic deflection coils *(MM)* which provide the time-base voltage. These coils are placed outside the tube. The direction of the magnetic field is along the plane of the paper and perpendicular to the axis of the tube. Due to these coils the beam is deflected along the *X*-axis (Fleming's left hand rule). Due to magnetic induction an alternating voltage is produced in the secondary and this voltage is connected to a condenser *C* through a high resistance R_1. The condenser helps to keep the induced *emf* in phase with the magnetising current.

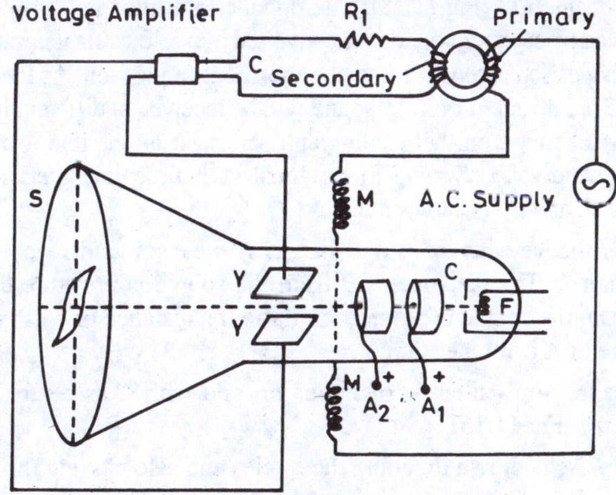

Fig. 11.14. B-H Curve with CRO.

The voltage across the condenser is amplified with a valve amplifier and fed to the plates *YY*. The deflection along the *X*-axis corresponds to the magnetising field *(H)* and the deflection along the *Y*-axis corresponds to the magnetic induction *(B)*. A hysteresis loop [*B-H* curve] is traced on the screen.

11.10. Applications of Cathode Ray Oscillograph

The cathode ray oscillograph [CRO] has a variety of applications in the laboratory, industry and research. It is commonly used (*i*) to study the wave form of *AC* voltages ; (*ii*) to study the wave form of sound from musical instruments ; (*iii*) to measure small *AC* and *DC* potentials ; (*iv*) to study the hysteresis loops of ferro-magnetic materials ; (*v*) to detect faults in radio sets and radio valves ; (*vi*) to study the operation of a radio valve during working ; (*vii*) to measure inductance, frequency, phase angle, power factor, dielectric losses etc., in *AC* circuits ; (*viii*) to study the mechanical stresses, indicator diagrams of internal combustion engines ; (*ix*) to study the heart's action [electro-cardiography], and (*x*) in Television receivers and Radar. A cathode ray oscillograph is very widely used in television receivers and radar equipments. Very small intervals of time of the order of micro-second can be measured with a cathode ray oscillograph.

11.11. Sound Ranging

The position of an enemy gun can be located by observing the time of the reception of sound at three different positions. This technique is called sound ranging.

Let *A, B* and *C* be three positions of observation. Suppose the gun is fired from some unknown position *O* (Fig.11.15). Microphones are fitted at each position. The sound received by each microphone is converted into an electrical impulse and fed into oscilloscopes separately at some common control position. As the distances of *A, B* and *C* are different from *Q*, sound will be received at different instants. From the signals obtained on the oscillograph screen, it is possible to find the difference in time intervals between various signals. Let these time intervals be *(i)* t_1 between *A* and *B*, and *(ii)* t_2 between *A* and *C*.

Suppose v is the velocity of sound in air. It means sound reaches *B*, t_1 seconds later than *A*. The distance of *B* from the source is vt_1 more than the distance of *A* from the source of sound. Similarly the distance of *C* is vt_2 more than the distance of *A* from the source.

With *B* as centre draw a circle of radius vt_1 and with *C* as centre draw a circle of radius vt_2 (Fig.11.15).

An imaginary arc is drawn touching the circles and also passing through *A*. The points *D, A* and *E* lie on this arc. The centre of the arc is found by geometrical construction. Here *O* is the centre of the circular arc. The enemy gun is situated at O.

The above geometrical construction is not simple. Usually the source of sound (position of the gun) is at a very large distance in comparison to the distances between A, B and B, C.

Let A and B be the two stations at which the observations are made (Fig. 11.16). O is the position of the enemy gun. The sound from O reaches B, t seconds later than A. The distance $OB - OA = vt$.

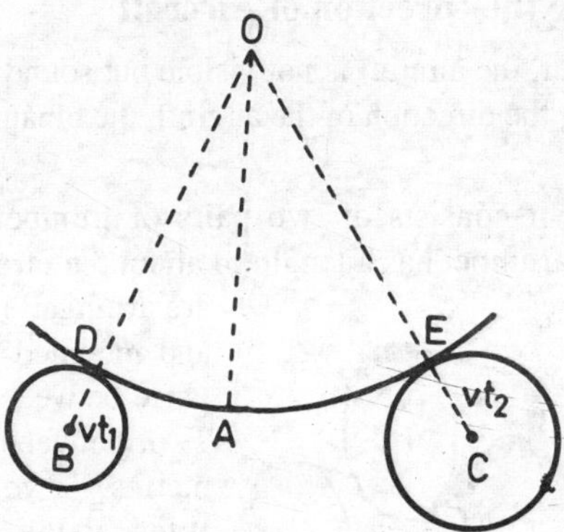

Fig 11.15.

The point O lies on a hyperbola drawn with A and B as foci. If O is at a large distance from G, (the centre of AB) and EOF and AD are perpendiculars on OG, then O lies on the asymtote OG such that

$$OB - OA = EB - FA$$

$$= BD$$

or $$BD = vt$$

Take $$AB = l$$

and $$\angle ABD = \angle\theta = \angle OGA$$

$$\cos\theta = \frac{BD}{AB} = \frac{vt}{l}$$

$\therefore$ $$\theta = \cos^{-1}\left[\frac{vt}{l}\right]$$

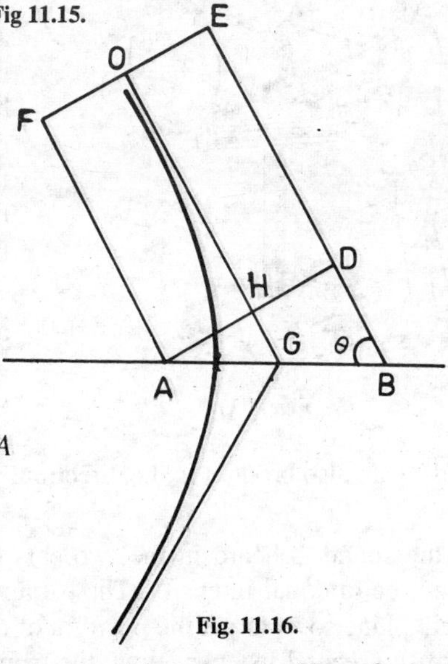

Fig. 11.16.

The value of θ can be calculated as v, t and l are known. Here, θ gives the direction of source of sound with respect to the positions A and B.

Similarly the angles θ corresponding to the positions of B and C, and also for C and A are found. The point of intersection of these three directions gives the position of the gun.

In actual practice, the three directions may not intersect at one point. In that case form a triangle from the three points of intersection. The source of sound lies within this triangle. The smaller the triangle, the higher will be the accuracy in locating the position of the gun.

11.12. Locating the Direction of Aircraft

In foggy weather, the aircraft is not visible but sound of the aircraft can be heard. For locating the direction of the aircraft, the binaural (two ears) method is employed.

The arrangement consists of two pairs of trumpets T_1, T_2 and T_3, T_4 (Fig.11.17). Each trumpet has a length of about 5 metres. The four trumpets are arranged in the form of a cross and mounted on a stand. The distance between the trumpets T_1 and T_2 or between T_3 and T_4 is about 4 metres. Two rubber tubes are fitted at the narrow ends of the trumpets T_1 and T_2 and these tubes are fixed in the ears of one of the listeners. This listener controls the horizontal motion. Similarly two rubber tubes from T_3 and T_4 are fitted in the ears of the second listener who controls the vertical motion.

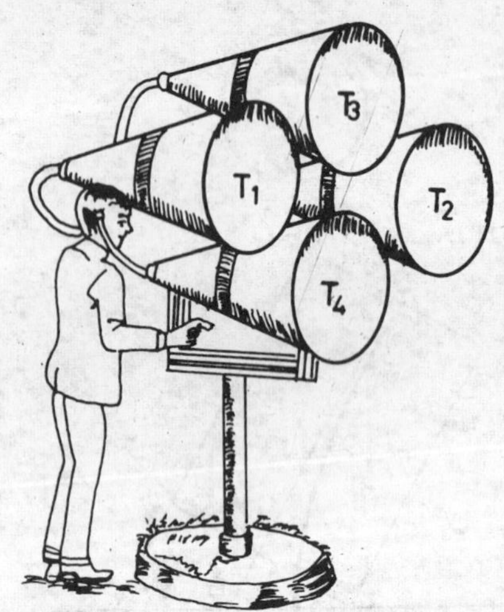

Fig. 11.17.

Initially the trumpets T_1 and T_2 are rotated in a horizontal plane. This is done to locate the position where the sound in the two ears is of equal intensity. Now, the trumpets T_3 and T_4 are also brought in the directions of T_1 and T_2. This fixes the vertical plane.

If the sound is heard in the two ears by the second listener, the sound may not be of equal intensity. The trumpets T_3 and T_4 are now rotated in a vertical plane so as to get the position of equal intensity of sound in the two ears of the second listener. Now, the trumpets T_1 and T_2 are brought in the directions of T_3 and T_4. Now all the four trumpets point in the direction of the aircraft.

11.13. Wavefront at Supersonic Speeds : Flight of the Bullet

Any body moving with a speed higher than that of sound is said to be moving with supersonic speed. In this case, the nature of the wavefront is conical.

Suppose a bullet is moving at supersonic speed V. The velocity of sound is v. At any instant, let P_1 be the position of the bullet. After time t, the bullet reaches the point P_2. In this time the longitudinal compressional waves of sound travel a distance vt. The wavefront due to the sound waves emitted at P_1 will be a sphere of radius vt. Here $P_1A_1 = vt$. The distance $P_1P_2 = Vt$. Here $P_1P_2 > P_1A_1$ (Fig. 11.18).

After a time $2t$ the bullet reaches the point P_3. From P_2 to P_3 the time taken is t. The sound waves emitted at P_1 will reach C_1 and D_1 and those emitted at

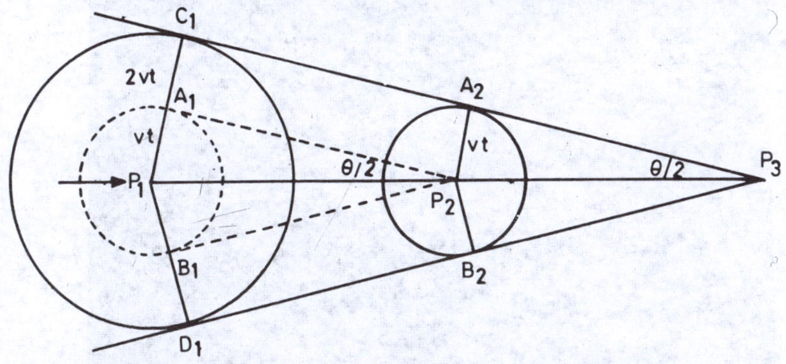

Fig. 11. 18.

P_2 will reach A_2 and B_2. $P_3 A_2 C_1$ and $P_3 B_2 D_1$ are tangents to the wavefronts at P_2 and P_1. Thus the wavefront is conical with the apex always at the position of the bullet. The cone angle is constant and depends upon the speed of sound and the speed of the bullet.

In Fig. 11.18, $\Delta P_3 P_1 C_1$ and $P_3 P_2 A_2$ are similar

$$\sin\frac{\theta}{2} = \frac{P_2 A_2}{P_3 P_2} = \frac{P_1 C_1}{P_3 P_1}$$

or
$$\sin\frac{\theta}{2} = \frac{vt}{Vt} = \frac{2vt}{2Vt}$$

or
$$\sin\frac{\theta}{2} = \frac{v}{V}$$

or
$$\frac{\theta}{2} = \sin^{-1}\left[\frac{v}{V}\right]$$

$$\theta = 2\left[\sin^{-1}\frac{v}{V}\right] \quad \ldots (1)$$

From equation (1), it is clear that the cone angle θ decreases with the increase in the speed of the bullet.

In Fig.11.19, the picture of the flight of the bullet at supersonic speed is shown. The eddies are seen clearly at the trail. When a bullet is fired at a supersonic speed, the observer at a distance hears three sounds :

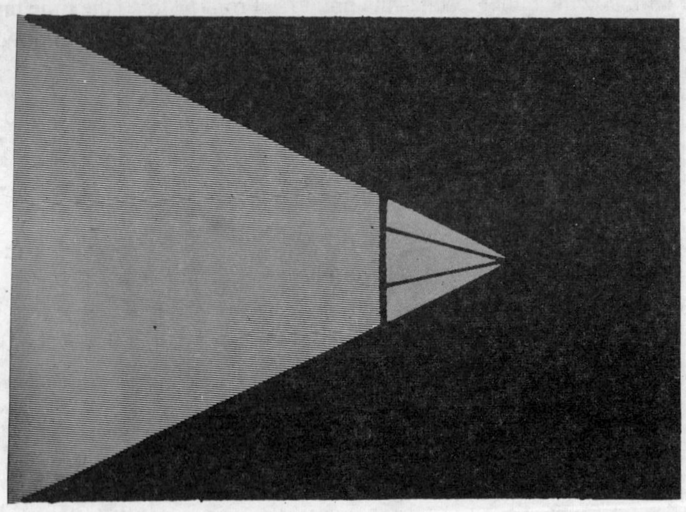

Fig. 11.19. Photograph Flight of Bullet.

(1) The click as the bullet moves past the observer. This is heard first.

(2) The buzzing sound due to the formation of eddies is heard by the observer.

(3) Lastly he hears the sound due to the explosion of the gun powder.

11.14. Acoustics

The branch of Physics that deals with the process of generation, reception and propagation of sound is called acoustics. This branch in fact covers many fields and is closely related to various branches of engineering. Some of the important fields of acoustics are (*i*) design of acoustical instruments, (*ii*) electro-acoustics *viz.* the branch relating to the methods of sound production and recording (microphones, amplifiers, loudspeakers etc.), (*iii*) architectural acoustics dealing with the design and construction of buildings, operas, music halls, recording rooms in radio and television broadcasting stations. In general architectural acoustics deals with the behaviour of sound waves in a closed space, and *(iv)* musical acoustics deals with the design of musical instruments.

11.15. Reverberation

It is observed that for a listener in a room or an auditorium, whenever a sound pulse is produced, he receives directly compressional sound waves from the source as well as sound waves from the walls, ceiling and other materials present in the room. The waves received by the listener are : *(i)* direct waves, and *(ii)* reflected waves due to multiple reflections at the various surfaces. The quality of the note received by the listener will be the combined effect of these two sets of waves. There is also a time gap between the direct wave received by the listener and the waves received by successive reflection. Due to this, the sound persists for sometime even after the source has stopped. This persistence of sound is termed as *reverberation.* The time gap between the initial direct note and the reflected note upto the minimum audibility level is called *reverberation time*. The reverberation time will depend on the size of the room or the auditorium, the nature of the reflecting material on the wall and the ceiling and the area of the reflecting surfaces.

In a good auditorium it is necessary to keep the reverberation time negligibly small. The intensity of sound as received by the listener, is shown graphically in Fig. 11.20. When a source emits sound, the waves spread out and the listener is aware of the commencement of sound when the direct waves reach his ears. Subsequently the listener receives sound energy due to reflected waves also. If the note is continuously sounded, the intensity of sound at the listener's ears gradually increases. After sometime a balance is reached between the energy emitted per second by the source and the energy lost or dissipated by walls or other materials. The resultant energy attains an average steady value, and to the listener the intensity of sound appears to be steady and constant. This is represented by the portion *BC* of the curve *ABCD*. If at *C*, the source stops emitting sound, the intensity of sound falls exponentially as shown by the curve *CD*. When the intensity of sound falls below the minimum audibility level, the listener will not hear the sound.

When a series of notes are produced in an auditorium (say speech or music) each note will give rise to its own intensity curve with respect time. The curves for these notes are shown in Fig 11.21.

For clear audibility of speech or music, it is necessary that *(i)* each separate note should give sufficient intensity of sound in every part of the auditorium, and *(ii)* each note should die down rapidly before the maximum average intensity due to the next note is heard by the listener. (Fig. 11.21). This is

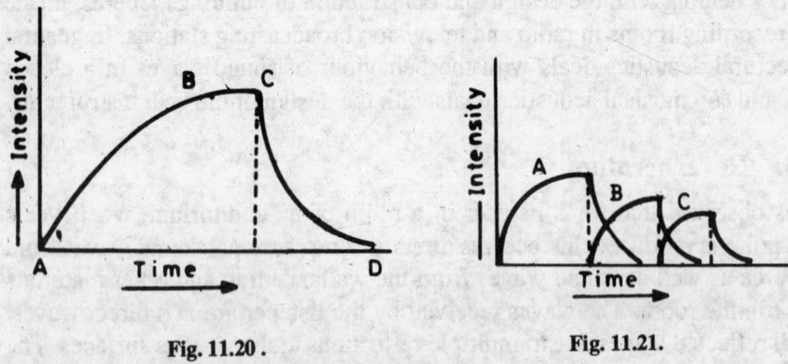

Fig. 11.20. Fig. 11.21.

particularly important with speech. In the case of music *comparatively more* reverberation can be tolerated.

11.16. Sabine's Reverberation Formula

Sabine developed the reverberation formula to express the rise and fall of sound in an auditorium. The main assumptions are :

(1) The average energy per unit volume is uniform. It is represented as σ.

(2) The energy is not lost in the auditorium. The energy lost is only due to the absorption of the material of the walls and ceiling and also due to escape through the windows and ventilators. Both these factors are included in the term **'absorption'** of energy.

Suppose a source is producing sound continuously. This sound energy is propagated in all directions. Let σ be the energy contained in a unit volume. The energy that is contained in a solid angle

$$d\phi = \frac{\sigma \, . \, d\phi}{4\pi}$$

Let this energy be incident on a unit surface area of the wall at an angle θ (Fig. 11.22). If the velocity of sound is v, then the total energy falling per second on a unit surface area of the wall.

$$= \left(\frac{\sigma \, . \, d\phi}{4\pi}\right)(\cos\theta) \, . \, v$$

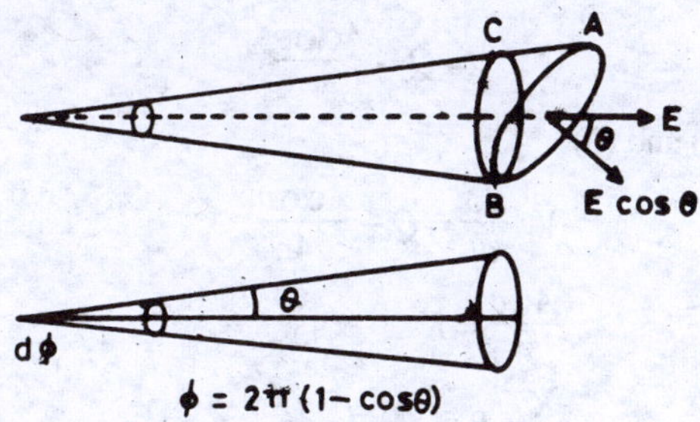

Fig. 11. 22.

The total energy falling per second within a hemisphere

$$= \frac{\sigma v}{4\pi} \cdot \int \cos\theta \, . \, d\phi$$

But $$\phi = 2\pi\,(1 - \cos\theta)$$

or $$d\phi = 2\pi \sin\theta \, . \, d\theta$$

Substituting this value of $d\phi$,

The total energy falling per second within a hemisphere

$$\frac{\sigma v}{4\pi} \int_0^{\pi/2} 2\pi \sin\theta \, . \cos\theta \, . \, d\theta$$

$$= \frac{\sigma v}{2} \left[-\frac{\cos^2\theta}{2} \right]_0^{\pi/2}$$

$$= \frac{\sigma v}{4}$$

Suppose α is the absorption coefficient of the walls that refers to the fraction of the incident energy not reflected from the walls. The amount of energy absorbed per second per unit area= $\frac{\alpha\sigma v}{4}$. If A is the area of the walls and the other absorbing materials including ceiling, windows and ventilators etc., the amount of energy absorbed per second

$$= \frac{A\alpha\sigma v}{4} \, .$$

Let V be the volume of the auditorium, the total energy = $V\sigma$. The rate of increase of energy $= \frac{d}{dt}(V\sigma)$

$$= V \frac{d\sigma}{dt} \qquad \ldots (1)$$

Suppose, the source supplies energy at the rate of Q units per second

Then, the rate of increase of energy

$$= Q - \frac{A\alpha\sigma v}{4} \quad \ldots (2)$$

Equating (1) and (2)

$$V \cdot \frac{d\sigma}{dt} = Q - \frac{A\alpha\sigma v}{4} \quad \ldots (3)$$

Take $\frac{A\alpha v}{4} = K$

and $\frac{K}{V} = \beta$ and $B = \frac{Q}{K} = \frac{4Q}{A\alpha v}$

From equation (3)

$$V \cdot \frac{d\sigma}{dt} = Q - K\sigma$$

$$\frac{d\sigma}{dt} = \frac{Q}{V} - \frac{K}{V} \cdot \sigma \quad \ldots (4)$$

The general solution of this equation is

$$\sigma = B + be^{-\beta t} \quad \ldots (5)$$

When $t = 0, \quad \sigma = 0$

$\therefore$ From equation (5)

$$0 = B + b$$

or $b = -B$

$$\sigma = B - Be^{-\beta t}$$

$$\sigma = B[1 - e^{-\beta t}]$$

Substituting the values of B and β

$$\sigma = \frac{4Q}{A\alpha v}\left[1 - e^{-\frac{(A\alpha v)\,.\,t}{4V}}\right] \quad \ldots (6)$$

Equation (6) represents the rise of average sound energy per unit time from the time the source commences to produce sound.

The maximum value of average energy per unit volume

$$\sigma max = \frac{4Q}{A\alpha v} \quad \ldots (7)$$

Similarly, after the source ceases to emit sound, the decay of the average energy per unit volume is given by

$$\sigma = \frac{4Q}{A\alpha v}\, e^{-\frac{(A\alpha v\,.\,t)}{4V}} \quad \ldots (8)$$

$$\sigma = \sigma\, e^{-\frac{(A\alpha v\,.\,t)}{4V}} \qquad \ldots (9)$$

The factor $\frac{A\alpha v}{4V}$ gives the reverberation time in the auditorium. If σ_o represents the minimum audible intensity after a time t_1, then from equation (9)

$$\sigma_0 = \sigma_{max}\, e^{-\frac{(A\alpha v)\,.\,t_1}{4V}} \qquad \ldots (10)$$

Here t_1 is the time interval between the cutting off the sound and the time at which intensity falls below the minimum audible level.

From equation (10)

$$\sigma_{max} = \sigma_0\, e^{+\frac{(A\alpha v\,.\,t_1)}{4V}}$$

Taking logarithms

$$\log_e\left(\frac{\sigma_{max}}{\sigma_0}\right) = \frac{A\alpha v}{4V}\,.\,t_1 \qquad \ldots (11)$$

Here α and σ_0 change with the frequency of sound.

For calculating the reverberation time, a standard steady intensity is required. Sabine took the value of $\frac{\sigma_{max}}{\sigma_0} = 10^6$.

From equation (11)

$$\log_e (10^6) = \frac{A\alpha v}{4V}\; t_1$$

$$2{\cdot}303 \times 6 = \frac{A\alpha v}{4V}\,.\,t_1$$

Taking velocity of sound approximately at room temperature as 350 m/s,

$$2{\cdot}303 \times 6 = \frac{A\alpha \times 350}{4V}\,.\,t_1$$

or

$$t_1 = \frac{2{\cdot}303 \times 24V}{350A\alpha}$$

$$t_1 = \frac{0{\cdot}158V}{A\alpha} \qquad \ldots (12)$$

In general

$$t_1 = \frac{0{\cdot}158V}{\Sigma A\alpha} \qquad \ldots (13)$$

In equation (12), the quantities are represented in M.K.S. units.

[**Note.** In F.P.S. units, taking v=1120 ft/s.]

$$t_1 = \frac{0{\cdot}05V}{A\alpha}$$

Equation (12) represents the Sabine's reverberation time formula.

According to equation (12), the reverberation time is (*i*) directly proportional to the volume of the auditorium, (*ii*) inversely proportional to the ceiling, area of the walls, etc., and (*iii*) inversely proportional to the total absorption plus transmission through open surfaces. It has been experimentally found that a reverberation time of 1·03 seconds is most suitable for all rooms having approximately a volume of less than 350 cubic metres.

To decrease the reverberation time, the walls of the auditorium are usually covered with materials having large absorption coefficients. The area of the surfaces of the walls is also increased in good cinema halls to decrease the reverberation time.

Example 11.5. *A hall of volume 5500* metre^3 *is found to have a reverberation time of* 2· 3 seconds. *The sound absorbing surface of the hall has an area of 750* metre^2. *Calculate the average absorption coefficient.*

(Bhagalpur, 1990)

Here $V = 5500\ \text{m}^3$

$t = 2{\cdot}3$ s

$A = 750\ \text{m}^2$

Absorption coefficient = α

Reverberation time,

$$t = \frac{0{\cdot}158V}{A\,\alpha}$$

$$\alpha = \frac{0{\cdot}158V}{At}$$

$$= \frac{0{\cdot}158 \times 5500}{750 \times 2{\cdot}3}$$

$$= \mathbf{0{\cdot}504.}$$

11.17. Determination of Absorption Coefficient

A source of frequency 512 hertz is taken. The time of reverberation in a hall is determined using a chronograph, (*i*) without the absorbing materials in the hall, and (*ii*) with the absorbing materials in the hall. Let these two times be t_1 and t_2 respectively. According to Sabine's formula

$$t_1 = \frac{0{\cdot}158V}{\Sigma\alpha_1 A_1} \quad \ldots (1)$$

and
$$t_2 = \frac{0{\cdot}158V}{\Sigma\alpha_1 A_1 + \alpha_2 A_2} \qquad \ldots(2)$$

Here α_2 is the absorption coefficient of the material of area A_2.

$$\therefore \qquad \frac{1}{t_1} = \frac{\Sigma\alpha_1 A_1}{0{\cdot}158V} \qquad \ldots(3)$$

$$\frac{1}{t_2} = \frac{(\Sigma\alpha_1 A_1) + \alpha_2 A_2}{0{\cdot}158V} \qquad \ldots(4)$$

$$\therefore \qquad \frac{1}{t_2} - \frac{1}{t_1} = \frac{\alpha_2 A_2}{0{\cdot}158V}$$

$$\therefore \qquad \alpha_2 = \frac{0{\cdot}158V}{A_2}\left[\frac{t_1 - t_2}{t_1 t_2}\right] \qquad \ldots(5)$$

From equation (5), knowing the values of t_1, t_2, A_2 and V, the value of α_2 can be calculated.

The absorption coefficients of some common materials calculated with a source of frequency 512 hertz are given in the following table.

The absorption coefficient of a given material is different at different frequencies. It is generally higher at higher frequencies.

Table 10.1

No.	*Material*	*Absorption Coefficient*
1.	Marble	0·01
2.	Common Plaster	0·03
3.	Glass	0·027
4.	Concrete	0·17
5.	Cork	0·23
6.	Asbestos	0·26
7.	Carpet	0·30
8.	Acoustic Plaster	0·30
9.	Acoustic Felt	0·45
10.	Fibre Board	0·50
11.	Heavy Curtains	0·50
12.	Hair or Felt	0·58
13.	Fibre Glass	0·75
14.	Perforated Cellulose Fibre Tiles	0·85

Example 11.6. *A room has dimensions $6 \times 4 \times 5$ metres. Calculate (i) the mean free path of the sound wave in the room, (ii) the number of reflections made per second by the sound wave with the walls of the room. Velocity of sound in air* = 350 m/s.

(*i*) The mean free path of the sound waves is defined as the average distance travelled by a sound wave through air between any two consecutive encounters with the walls of the room.

Mean free path,

$$L = \frac{4\text{ (Volume of the room)}}{\text{Total surface area}}$$

Here volume of the room $= 6 \times 4 \times 5 = 120 \text{ m}^3$

Total surface area $= 2[6 \times 4 + 4 \times 5 + 6 \times 5]$

$= 148 \text{ m}^2$

$$\therefore \qquad L = \frac{4 \times 120}{148}$$

or $\qquad$ $L = \mathbf{3{\cdot}243}$ **m.**

(*ii*) The number of reflections made per second

$$N = \frac{\text{Velocity of sound}}{\text{Mean free path}}$$

$$N = \frac{350}{3{\cdot}243}$$

$$N = \mathbf{107{\cdot}9}.$$

Example 11.7. *The volume of a room is* 600 m^3. *The wall area of the room is* 220 m^2, *the floor area is* 120 m^2 *and the ceiling area is* 120 m^2. *The average sound absorption coefficient, (i) for the walls is 0·03; (ii) for the ceiling is 0·8, and (iii) for the floor is 0·06. Calculate the average sound absorption coefficient and the reverberation time.*

The average sound absorption coefficient,

$$\alpha = \frac{\Sigma \alpha A}{\Sigma A}$$

or

$$\alpha = \frac{\alpha_1 A_1 + \alpha_2 A_2 + \alpha_3 A_3}{A_1 + A_2 + A_3}$$

Here $\qquad A_1 = 220 \text{ m}^2, \quad \alpha_1 = 0{\cdot}03$

$A_2 = 120 \text{ m}^2, \quad \alpha_2 = 0{\cdot}80$

$A_3 = 120 \text{ m}^2, \quad \alpha_3 = 0{\cdot}06$

$$\therefore \qquad \alpha = \frac{0{\cdot}03 \times 220 + 0{\cdot}8 \times 120 + 0{\cdot}06 \times 120}{220 + 120 + 120}$$

$$= \frac{109.8}{460}$$

$$= \mathbf{0.2389}$$

$$= 0.24 \text{ (approx)}$$

The total sound absorption of the room

$$= \alpha \Sigma A$$

$$= 0.24 \times 460$$

$$= \mathbf{110.40 \text{ metric sabines.}}$$

Reverberation time,

$$t = \frac{0.158V}{\alpha \Sigma A}$$

$$= \frac{0.158 \times 600}{110.4}$$

$$= 0.8588 \text{ s}$$

$$= \mathbf{0.86 \text{ s (app).}}$$

11.18. Acoustic Intensity

Acoustic intensity of a sound wave is defined as the average power transmitted per unit area in the direction of propagation of the wave.

Instantaneous power per unit area is equal to the product of instantaneous pressure (P) and instantaneous particle velocity (v). Average power per unit area measures the acoustic intensity.

∴ Acoustic intensity,

$$I = \frac{1}{T} \int_0^T Pv \, dt$$

Here $$P = -\rho C^2 \left(\frac{du}{dx} \right)$$

But $$u = A \cos(\omega t - kx)$$

and $$v = \frac{du}{dt} = \omega A \sin(\omega t - kx)$$

$$\frac{du}{dx} = -Ak \sin(\omega t - kx)$$

But $$kC = \omega$$

∴ $$P = -\rho CA \omega \sin(\omega t - kx)$$

Here C is the velocity of sound and ρ is the density of the medium.

$$\therefore I = \frac{1}{T} \int_0^T [-\rho CA\omega \sin(\omega t - kx)] \, [-\omega A \sin(\omega t - kx)] \, dt$$

$$I = \frac{\rho C\, \omega^2 A^2}{T} \int_0^T \sin(\omega t - kx)^2\, dt$$

$$= \frac{\rho C\, \omega^2 A^2}{T} \int_0^T (\cos^2 kx \sin^2 \omega t + \sin^2 kx \cos^2 \omega t - \tfrac{1}{2} \sin 2\omega t \sin 2kx)\, dt$$

$$= \frac{\rho C\, \omega^2 A^2}{T} \left(\frac{T}{2}\right)$$

or $$I = \frac{1}{2}\, \rho C\, \omega^2 A^2 \qquad \ldots (1)$$

Since $$P = -\rho C A \omega \sin(\omega t - kx)$$

$$P_{\max} = -\rho C A \omega$$

Root mean square value of pressure,

$$P_{RMS} = \frac{P_{\max}}{\sqrt{2}}$$

From equation (1)

$\therefore$ $$I = \frac{\rho^2 C^2 A^2 \omega^2}{2\, \rho C}$$

$$I = \frac{(P_{\max})^2}{2\, \rho C}$$

$$I = \frac{\left(\frac{P_{\max}}{\sqrt{2}}\right)^2}{\rho C}$$

or $$I = \frac{P^2_{RMS}}{\rho C} \qquad \ldots (2)$$

Example 11.8. *Compare the acoustic intensities in air and in water for the same acoustic pressure.*

Acoustic intensity,

$$I = \frac{P^2_{RMS}}{\rho C}$$

$$\frac{I_{air}}{I_{water}} = \frac{\left(\frac{P^2_{RMS}}{\rho C}\right)_{air}}{\left(\frac{P^2_{RMS}}{\rho C}\right)_{water}}$$

But $$(P_{RMS})_{air} = (P_{RMS})_{water}$$

$$\frac{I_{air}}{I_{water}} = \frac{(\rho C)_{water}}{(\rho C)_{air}}$$

$$\rho \text{ for air} = 1{\cdot}29 \text{ kg/m}^3$$

$$C \text{ for air} = 350 \text{ m/s}$$

$$\rho \text{ for water} = 1000 \text{ kg/m}^3$$

$$C \text{ for water} = 1470 \text{ m/s}$$

$$\therefore \quad \frac{I_{air}}{I_{water}} = \frac{1000 \times 1470}{1{\cdot}29 \times 350}$$

$$= \mathbf{3256.}$$

This shows that for the same acoustic pressure, the acoustic intensity in air is 3256 times that in water.

Example 11.9. *Compare the acoustic intensities in air and in water for the same frequency and amplitude.*

$$\frac{I_{air}}{I_{water}} = \frac{\frac{1}{2}(\rho C A^2 \omega^2)_{air}}{\frac{1}{2}(\rho C A^2 \omega^2)_{water}}$$

Here $(A\omega)_{air} = (A\omega)_{water}$

$$\therefore \quad \frac{I_{air}}{I_{water}} = \frac{(\rho C)_{air}}{(\rho C)_{water}}$$

$$\rho \text{ for air} = 1{\cdot}29 \text{ kg/m}^3$$

$$\rho \text{ for water} = 1000 \text{ kg/m}^3$$

$$C \text{ for air} = 350 \text{ m/s}$$

$$C \text{ for water} = 1470 \text{ m/s}$$

$$\therefore \quad \frac{I_{air}}{I_{water}} = \frac{1{\cdot}29 \times 350}{1000 \times 1470}$$

$$= \frac{1}{3256}$$

or $I_{water} = 3256\, I_{air}$

This shows that for the same frequency and amplitude, this acoustic intensity in water is 3256 times that in air.

Example 11.10. *A plane acoustic wave in air has an intensity of* 21 watts/m^2. *This wave strikes a wall at right angles to its surface. The area of the wall is* 50 m^2. *Calculate the force exerted on the wall. Velocity of the wave in air is* 350 m/s.

Acoustic intensity is the power per unit area.

$$I = \frac{\text{Power}}{\text{Area}} = \left(\frac{\text{Force}}{\text{Area}}\right) \times \text{Velocity}$$

or $I = PC$ watts/m^2

$$\therefore \qquad P = \frac{I}{C}$$

Here $\qquad I = 21$ watts/m^2

and $\qquad C = 350$ m/s

$$\therefore \qquad P = \frac{21}{350}$$

or $\qquad P = 0{\cdot}06$ newton/m^2

Area = 50 m^2

$\therefore$ Force = Pressure × Area

= 0·06 × 50

= **3 newtons.**

Example 11.11. *Calculate the amplitude of the displacement wave, given that for a frequency of 400* hertz, *the feeblest sound that could be heard corresponds to a pressure amplitude of* 8×10^{-5} newton/m^2. *The density of air is* 1·29 kg/m^3 *and velocity of sound in air is* 345 m/s.

Here $$A = \frac{P_{max}}{\rho C \omega}$$

Here $P_{max} = 8 \times 10^{-5}$ newton/m^2

$\rho = 1{\cdot}29$ kg/m^2

$C = 345$ m/s

$\omega = 2\pi f = 2\pi \times 400$

$= 800\,\pi$ radians/s

$$\therefore \qquad A = \frac{8 \times 10^{-5}}{1{\cdot}29 \times 345 \times 800\,\pi}$$

$$A = \mathbf{7{\cdot}15 \times 10^{-11}\ m.}$$

Note. It may be noted that this amplitude is of the order of molecular dimensions.

11.19. Acoustic Measurements

In acoustical measurements, it is necessary to use the logarithmic scale for measuring acoustic intensity, acoustic power and acoustic pressure. The logarithmic scale is called the decibel scale, abbreviated as db. The decibel scale always refers to the quantity to be measured logarithmically to some standard reference. The decibel is a dimensionless quantity because it refers to the ratio of two similar quantities.

The decibel value is equal to 10 times the logarithm of the ratio of the quantities measured, to the base 10.

1 bel = 10 decibels = 10 db

(1) Acoustic intensity level

Acoustic intensity level is written as

$$IL = 10 \log \left(\frac{I}{I_0} \right) \text{ db reference to } I_0 \text{ watts/m}^2$$

The standard acoustic intensity reference,

$$I_0 = 10^{-12} \text{ watt/m}^2$$

$$\therefore \quad IL = 10 \log \left(\frac{I}{10^{-12}} \right) \text{db reference to } 10^{-12} \text{ watt/m}^2$$

$$= 10 [\log I + \log 10^{12}] \text{ db}$$

$$= [10 \log I + 120] \text{ db}$$

(2) Acoustic pressure level

Acoustic pressure level is written as

$$PL = 10 \log_{10} \left(\frac{P}{P_0} \right)^2 \text{db}$$

reference to a pressure of P_0 newtons/m^2.

$$\therefore \quad PL = 20 \log_{10} \left(\frac{P}{P_0} \right) \text{db}$$

The standard acoustic pressure reference,

$$P_0 = 2 \times 10^{-5} \text{ newton/m}^2$$

$$\therefore \quad PL = 20 \log \left(\frac{P}{2 \times 10^{-5}} \right) \text{db}$$

reference to a pressure of 2×10^{-5} newton/m^2

$$\therefore \quad PL = 20 [\log P + \log 5 \times 10^4]$$

$$= [20 \log P + 20 \log 5 \times 10^4]$$

$$= [20 \log P + 20 \times 4{\cdot}6990]$$

$$= [20 \log P + 94] \text{ db (approx).}$$

Example 11.12. *The sound from a drill gives a noise level of* 90 decibels *at a point at a short distance from it. What is the noise level at this point if four such drills are working simultaneously at the same distance from the point.*

(Delhi, 1973)

Here, $\frac{I_2}{I_1} = 4$

and $IL_1 = 90$ db

$$IL = 10 \log\left(\frac{I}{I_0}\right) \text{db}$$

In the first case

$$IL_1 = 10 \log\left(\frac{I_1}{I_0}\right)$$

In the second case

$$IL_2 = 10 \log\left(\frac{I_2}{I_0}\right)$$

Increase in noise level

$$IL_2 - IL_1 = 10\left[\log\left(\frac{I_2}{I_0}\right) - \log\left(\frac{I_1}{I_0}\right)\right]$$

$$= 10 \log\left(\frac{I_2}{I_1}\right)$$

But $\frac{I_2}{I_1} = 4$

$\therefore$ $IL_2 - IL_1 = 10 \log_{10} 4$

$$= 10 \times 0{\cdot}6021 = 6{\cdot}021 \text{ db}$$

$\therefore$ $IL_2 = IL_1 + 6{\cdot}021 = 90 + 6{\cdot}021 =$ **96·021 db.**

Example 11.13. *Calculate the increase in the acoustic intensity level when the sound intensity is doubled.*

$$IL = 10 \log\left(\frac{I}{I_0}\right) \text{db}$$

In the first case

$$IL_1 = 10 \log\left(\frac{I_1}{I_0}\right)$$

In the second case

$$IL_2 = 10 \log\left(\frac{I_2}{I_0}\right)$$

Increase in acoustic intensity level,

$$IL_2 - IL_1 = 10\left[\log\left(\frac{I_2}{I_0}\right) - \log\left(\frac{I_1}{I_0}\right)\right]$$

$$= 10 \log\left(\frac{I_2}{I_1}\right)$$

But $\frac{I_2}{I_1} = 2$

$\therefore$ $IL_2 - IL_1 = 10 \log_{10} 2 = 10 \times 0{\cdot}3010 =$ **3·01 db.**

Example 11.14. *Calculate the increase in sound pressure level when the sound pressure is doubled.*

$$PL = 10 \log \left(\frac{P}{P_0} \right)^2$$

$$= 20 \log \left(\frac{P}{P_0} \right) \text{db}$$

In the first case

$$PL_1 = 20 \log \left(\frac{P_1}{P_0} \right) \text{db}$$

In the second case

$$PL_2 = 20 \log \left(\frac{P_2}{P_0} \right) \text{db}$$

Increase in sound pressure level,

$$PL_2 - PL_1 = 20 \left[\log \left(\frac{P_2}{P_0} \right) - \log \left(\frac{P_2}{P_1} \right) \right]$$

$$= 20 \log \left(\frac{P_2}{P_1} \right)$$

But $\frac{P_2}{P_1} = 2$

$\therefore$

$$PL_2 - PL_1 = 20 \log_{10} 2$$

$$= 20 \times 0{\cdot}3010$$

$$= \mathbf{6{\cdot}02 \text{ db.}}$$

Example 11.15. *Calculate the (i) acoustic intensity, (ii) acoustic pressure of a plane acoustic wave in air of intensity level of* 100 decibels *reference to* 10^{-12} watt/m^2.

(1)

$$IL = 10 \log \left(\frac{I}{I_0} \right) db$$

Here $L_0 = 10^{-12}$ watt/ m^2

$\therefore$

$$IL = 10 \log \left(\frac{I}{10^{-12}} \right) \text{db reference to } 10^{-12} \text{ watt/m}^2$$

$$= 10 [\log I + \log_{10} 10^{12}] \text{ db}$$

$$= 10 [\log I + 12] \text{ db}$$

$$IL = [10 \log I + 120] \text{ db} \quad \ldots (1)$$

But

$$IL = 100 \text{ db} \quad \ldots (2)$$

From (1) and (2)

$$\therefore \quad 100 = 10 \log I + 120$$

or

$$10 \log I = -20$$

$$\log I = -2$$

$$I = \mathbf{10^{-2} \text{ watt/m}^2}.$$

(2) Acoustic pressure.

Acoustic intensity, $I = \dfrac{P^2}{\rho C}$

or $P = \sqrt{I \rho C}$

Here $I = 10^{-2}$ watt/m^2

Density of air, $\rho = 1{\cdot}29$ kg/m^3

Velocity of sound in air,

$$C = 350 \text{ m/s}$$

$$\therefore \quad P = \sqrt{10^{-2} \times 1{\cdot}29 \times 350}$$

or $P = \mathbf{2{\cdot}124 \text{ newtons/m}^2}$.

Example 11.16. *Calculate the acoustic pressure of a plane acoustic wave of pressure level of* 100 decibels *reference to a standard acoustic pressure level of* 2×10^{-5} newton/m^2.

Here $PL = 10 \log \left(\dfrac{P}{P_0} \right)^2$ db

reference to a pressure P_0 newton/m^2

$$PL = 20 \log \left(\frac{P}{P_0} \right) \text{db}$$

Here $P_0 = 2 \times 10^{-5}$ newton/m^2

$$\therefore \quad PL = 20 \log \left(\frac{P}{2 \times 10^{-5}} \right) \text{db}$$

reference to a pressure of 2×10^{-5} newton/m^2

$$PL = 20 [\log P + \log 5 \times 10^4] \text{ db}$$

$$PL = 20 [\log P + 4{\cdot}6990] \text{ db}$$

or $PL = [20 \log P + 94]$ db

But $PL = 100$ db

$$\therefore \quad 100 = 20 \log P + 94$$

$$20 \log P = 6$$

$$\log P = 0{\cdot}3$$

$$P = \mathbf{1{\cdot}995 \text{ newtons/m}^2}$$

Example 11.17. *Calculate the acoustic intensity level in each case, at a distance of* 10 metres *from a source which radiates energy at the rate of* 3·14 watts. *Using reference intensities of (i)* 100 watts/m^2. *(ii)* 1 watt/m^2 *(iii)* 10^{-12} watt/m^2.

Power radiated $= W = 3{\cdot}14$ watts

Distance $= 10$ m

Intensity $I = \dfrac{W}{\text{Area}}$

$$= \frac{3{\cdot}14}{4\pi r^2}$$

$$I = \frac{1}{4r^2}$$

or $I = \dfrac{1}{400} = 0{\cdot}0025 \text{ watt/m}^2$

(*i*) $IL = 10 \log \left(\dfrac{I}{I_0} \right) \text{db}$

Here $I_0 = 100 \text{ watts/m}^2$

$I = 0{\cdot}0025 \text{ watt/m}^2$

∴ $IL = 10 \log \left(\dfrac{0{\cdot}0025}{100} \right)$

$IL =$ **−46·21 db reference to 100 watts/m²**

(*ii*) Here $I_0 = 1 \text{ watt/m}^2$

$I = 0{\cdot}0025 \text{ watt/m}^2$

∴ $IL = 10 \log \left(\dfrac{0{\cdot}0025}{1} \right) \text{db}$

$IL = 10 \log (25 \times 10^{-4}) \text{ db}$

$IL =$ **−26·021 db reference to 1 watt/m²**

(*iii*) Here $I_0 = 10^{-12} \text{ watt/m}^2$

$I = 0{\cdot}0025 \text{ watt/m}^2$

$$IL = 10 \log \left(\frac{I}{I_0} \right) \text{db}$$

$$= 10 \log \left(\frac{0{\cdot}0025}{10^{-12}} \right) \text{db}$$

$$= 10 \log (25 \times 10^{8}) \text{ db}$$

$IL =$ **93·979 db reference to 10^{-12} watt/m².**

Example 11.18. *An air-conditioning unit operates at a sound intensity level of* 75 db. *If it is operated in a room with an existing sound intensity level of* 70 db, *what will be the resultant intensity level?*

In the first case,

$$IL_1 = 10 \log \left(\frac{I_1}{I_0} \right) \text{db}$$

But $$IL_1 = 75 \text{ db}$$

$\therefore$ $$75 = 10 \log \left(\frac{I_1}{I_0} \right)$$

or $$\text{antilog } 7{\cdot}5 = \left(\frac{I_1}{I_0} \right)$$

or $$I_1 = I_0 \text{ (antilog } 7{\cdot}5)$$

$$I_1 = 6{\cdot}99 \times 10^7 \, I_0 \text{ watt/m}^2 \quad \ldots (1)$$

In the second case,

$$IL_2 = 10 \log \left(\frac{I_2}{I_0} \right) \text{db}$$

But $$IL_2 = 70 \text{ db}$$

$\therefore$ $$70 = 10 \log \left(\frac{I_2}{I_0} \right)$$

or $$I_2 = I_0 \text{ (antilog } 7)$$

$$I_2 = 10^7 \, I_0 \text{ watts/m}^2 \quad \ldots (2)$$

The resultant sound intensity,

$$I = I_1 + I_2$$

$$I = 6{\cdot}99 \times 10^7 \, I_0 + 10^7 \, I_0$$

$$I = (7{\cdot}99 \times 10^7)\, I_0 \text{ watts/m}^2$$

The resultant intensity level

$$IL = 10 \log \left(\frac{I}{I_0} \right) \text{db}$$

$$IL = 10 \log \left(\frac{7{\cdot}99 \times 10^7 \, I_0}{I_0} \right) \text{db}$$

$$IL = 10 \log 7{\cdot}99 \times 10^7$$

or $$\mathbf{IL = 79{\cdot}025 \text{ db}}$$

The resultant intensity level is **79·025 decibles.**

Example 11.19. *Two sources of sound A and B emit sound waves of different frequencies. The two sound pressure levels as recorded at a place C are* 80 db *and* 75 db *respectively. Calculate the resultant sound pressure level at C due to the combined effect.*

In the first case,

$$PL_1 = 20 \log \left(\frac{P_1}{P_0} \right) = 80 \text{ db}$$

or $$\log \left(\frac{P_1}{P_0} \right) = 4$$

$$P_1 = P_0 \text{ (antilog 4)}$$

$$P_1 = 10^4 \, P_0 \text{ newtons/m}^2 \qquad \ldots (1)$$

In the second case,

$$PL_2 = 20 \log \left(\frac{P_2}{P_0} \right) = 75 \text{ db}$$

or $$\log \left(\frac{P_2}{P_0} \right) = 3{\cdot}75$$

$$P_2 = P_0 \text{ (antilog 3·75)}$$

$$= 5{\cdot}6 \times 10^3 \, P_0 \text{ newtons/m}^2 \qquad \ldots (2)$$

The total sound pressure at C due to the combined effect,

$$P = P_1 + P_2$$

$$= (10^4 + 5{\cdot}6 \times 10^3) \, P_0$$

$$P = 15{\cdot}6 \times 10^3 \, P_0 \text{ newtons/m}^2 \qquad \ldots (3)$$

or $$\frac{P}{P_0} = 15{\cdot}6 \times 10^3 \qquad \ldots (4)$$

The combined sound pressure level,

$$PL = 20 \log \left(\frac{P}{P_0} \right) \text{db}$$

$$= 20 \log (15{\cdot}6 \times 10^3) \text{ db}$$

$$= 20 \, (4{\cdot}195) \text{ db}$$

$$= \mathbf{83{\cdot}90 \text{ db}}$$

Thus the total sound pressure level at C is **83·90 decibles**.

Example 11.20. *A light source 1* kWH, *is radiating energy uniformly. Determine the intensity of the electric field at a distance of one metre from the source.* (IAS, 1988)

$$E = 1 \text{ kWH}$$

It means energy radiated by the source in one hour = 1 kWH

Energy radiated in one second = 1kW-s

$$= 10^3 \text{ J/s}$$

Intensity, $$I = \frac{E}{4\pi R^2}$$

$$= \frac{10^3}{4\pi \times 1}$$

$$= \mathbf{79{\cdot}57} \text{ watts/m}^2 \text{ (or J/ m}^2 - \text{s)}$$

Example 11.21. *A source of sound emits energy in all directions at a rate of* 0·638 joule/second. *Find the intensity level at a distance of* 11·5 metres *reckoned above a datum of* $1{\cdot}2 \times 10^4$ watt/cm^2. (Bhagalpur, 1990)

Here

$$\text{Power radiated} = 0{\cdot}638 \text{ J/s}$$

$$P = 0{\cdot}638 \text{ watts}$$

$$\text{Distance } r = 11{\cdot}5 \text{ m}$$

$$\text{Intensity } I = \frac{P}{\text{Area}}$$

$$= \frac{0{\cdot}638}{4\pi r^2}$$

$$= \frac{0{\cdot}638}{4\pi\,(11{\cdot}5)^2}$$

$$= 3{\cdot}84 \times 10^{-4} \text{ watts/m}^2$$

Also $$I_0 = 1{\cdot}2 \times 10^4 \text{ watts/cm}^2$$

$$= 1{\cdot}2 \times 10^8 \text{ watts/m}^2$$

Intensity level, $$IL = 10 \log \left(\frac{I}{I_0} \right)$$

$$IL = 10 \log \left(\frac{3{\cdot}84 \times 10^{-4}}{1{\cdot}2 \times 10^8} \right) \text{db}$$

$$= 10 \log (3{\cdot}2 \times 10^{-12})$$

$$= \mathbf{-114{\cdot}95 \text{ db.}}$$

Example 11.22. *The noise from an aeroplane engine* 100 m *from an observer is* 40 db *in intensity. What will be the intensity when the aeroplane flies overhead at an altitude of* 2 km ? (IAS)

$$I_1 = \frac{P}{4\pi R_1^2}$$

$$I_2 = \frac{P}{4\pi R_2^2}$$

$$\frac{I_2}{I_1} = \frac{R_1^2}{R_2^2}$$

Here $R_1 = 100$ m, $R_2 = 2000$ m

$$\frac{I_2}{I_1} = \frac{(100)^2}{(2000)^2} = \frac{1}{400} \text{ or } \frac{I_1}{I_2} = 400$$

$$IL_1 = 10 \log \left(\frac{I_1}{I_0}\right)$$

$$IL_2 = 10 \log \left(\frac{I_2}{I_0}\right)$$

Decrease in intensity level

$$IL_1 - IL_2 = 10 \left[\log \left(\frac{I_1}{I_0}\right) - \log \left(\frac{I_2}{I_0}\right) \right]$$

$$= 10 \log \left(\frac{I_1}{I_2}\right)$$

$$= 10 \log_{10} (400)$$

$$= 10 \times 2{\cdot}6021 = 26{\cdot}021 \text{ db}$$

$$IL_2 = IL_1 - 26{\cdot}021 = 40 - 26{\cdot}021$$

$$\mathbf{IL_2 = 13{\cdot}979 \text{ db.}}$$

11.20. Factors Affecting the Acoustics of Buildings

Reverberation is one of the important single factors that affects the acoustics of a room or a hall. Besides reverberation, there are also other factors like loudness, focussing echelon effect, extraneous noise, resonance etc.

(1) **Loudness.** The speech of a person in a hall can be heard by an audience consisting of about 1000 persons. However, to ensure uniform distribution of sound intensity in the hall, electrically amplified loud speakers are used. These speakers are kept at different places in the auditorium and are located generally at a height higher than the speaker's head. Amplifiers, however, make the low

frequency tones more prominent and hence the amplification has to be kept low. The presence of low artificial ceilings improves the audibility in general.

(2) **Focussing**. The presence of cylindrical or spherical surfaces on the walls of the ceiling gives rise to undesirable focussing.

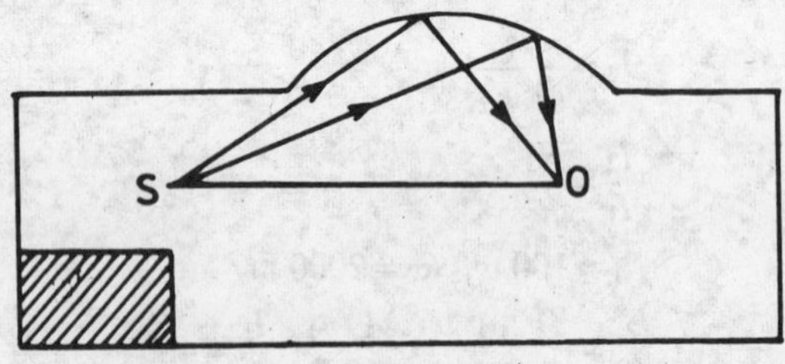

Fig. 11.23.

In Fig. 11.23 the observer at O receives sound from the speaker along the direct path SO. The observer also receives the sound waves after reflection from the ceiling. Thus the intensity of sound at O is comparatively higher than other positions in the auditorium. It may also happen that the direct and the reflected waves are in opposite phase. This results in minimum intensity of sound at O. Further, the direct and the reflected waves may form a stationary wave pattern. This causes uneven distribution of sound intensity.

(3) **Echelon Effect**. If there is regular structure similar to a flight of stairs or a set of railings in the hall, the sound produced in front of such a structure may produce a *musical note* due to regular successive echoes of sound reaching the observer. Such an effect is called *echelon effect*. If the frequency of this note is within the audible range, the listener will hear only this note prominently. To avoid echelon effect, the stair cases are covered with carpets to avoid reflection of sound.

(4) **Extraneous Noise**. The extraneous noise may be due to (*i*) sound received from outside the room and (*ii*) the sound produced by fans etc. inside the auditorium. The external sound cannot be completely eliminated but can be minimised by using double or triple windows and doors. Proper attention must also be paid to maximum permissible speed of fans and the rate of air circulation in the room. The air conditioning pipes should be covered with cork and insulated accoustically from the main building.

(5) **Resonance.** The acoustics of a building may also be affected by resonance. If there is resonance for any audio frequency note, the intensity of the note will be entirely different from the intensity desired. In halls of large size, the resonance frequency is much below the audible limit and harmful effects due to resonance will not be present.

11.21. Sound Distribution in an Auditorium

The design of the auditorium requires smooth decay and growth of sound. In an auditorium, the sound must be distributed or diffused over the whole area. To ensure these factors, acoustic treatment is given *viz.* scattering effect of

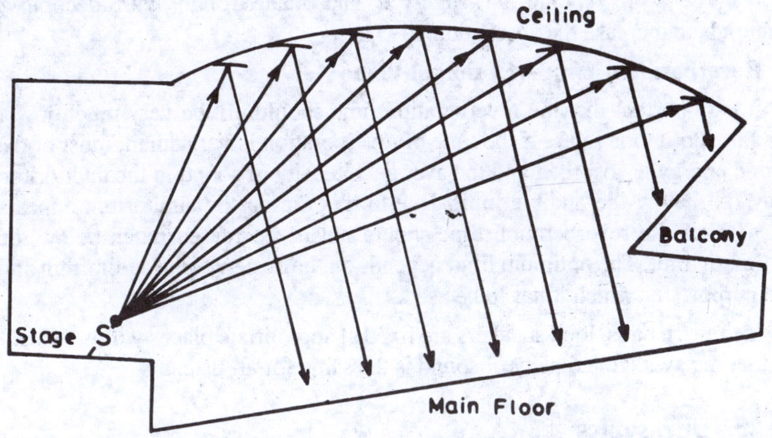

Fig. 11.24.

objects, irregularities on the wall surfaces, fixing absorptive material on the walls etc. In Fig. 11.24, the first reflection of sound waves at different positions of the ceiling is shown. It is clear from the figure that the reflected sound is distributed evenly in the whole auditorium *viz.* the main floor and the balcony. This design enables an even distribution of sound intensity.

11.22. Requisites for Good Acoustics

The reverberation of sound in an auditorium is due to multiple reflections taking place at various surfaces present within the auditorium. The acoustics of an auditorium can be improved by using the surfaces with high absorption coefficient. This will reduce the reverberation time below the optimum value. This can be achieved as follows :

(1) By hanging heavy curtains.

(2) By hanging pictures and maps.

(3) By having a few open windows.

(4) By having good audience. Each person is equivalent to about 0·50 sq metre area of an open window.

(5) The curved walls and corners bounded by two walls should be avoided. This is done to avoid (*i*) concentration of sound and (*ii*) dead spaces.

(6) Upholstered seats should be provided so that the absorption is approximately the same with or without the audience.

(7) The walls and the ceiling should be covered with the materials having high absorption coefficient *i.e.* with perforated card boards, felt, asbestos, fibre glass etc.

(8) The walls should be engraved and made rough with decorative materials to increase the absorption.

Reverberation Time—Optimum Value

It is important that the reverberation time should not be very much lower than the optimum value. If the time of reverberation is very small, most of the sound energy is absorbed and the average intensity of sound in the auditorium may fall below the audible limit. Due to this reason the auditorium appears dead. Thus the reverberation time should not be decreased much below the optimum limit. The optimum limit depends on the design of the auditorium and the purpose for which it has to be used.

In modern halls, loud speakers are fixed at appropriate places within the hall so that the average intensity of sound is the same for all listeners.

11.23. Ultrasonics

Sound waves of high frequency are called ultrasonics. The frequency of **ultrasonics** is higher than 20,000 hertz. These frequencies are beyond the audible limit. Their wavelength are small. For a frequency of 20,000 hertz, the wavelength at room temperature is $\frac{350}{20,000} = 0{\cdot}0175$ metre. These waves exhibit the properties of audible sound waves and also show some new phenomena.

The sound waves of frequency lower than the audible limit are called **infrasonics**. Supersonics refers to the velocities higher than the velocity of sound *i.e.* more than 1200 km/hour.

Ultrasonic waves have a large number of practical applications. Ultrasonic waves can be produced by the following methods : (*i*) Galton whistle, (*ii*) Magnetostriction oscillator and (*iii*) Piezo-electric oscillator.

11.24. Production of Ultrasonic Waves

1. Galton Whistle. Galton whistle works on the principle of organ pipe. It consists of a closed end air column A whose length can be adjusted with the help of a movable piston. The piston P can be moved to the desired position with the help of a screw S_1. The open end of the pipe A is fitted with a lip L. The position of the pipe C can be adjusted with the help of the screw S_2. The gap

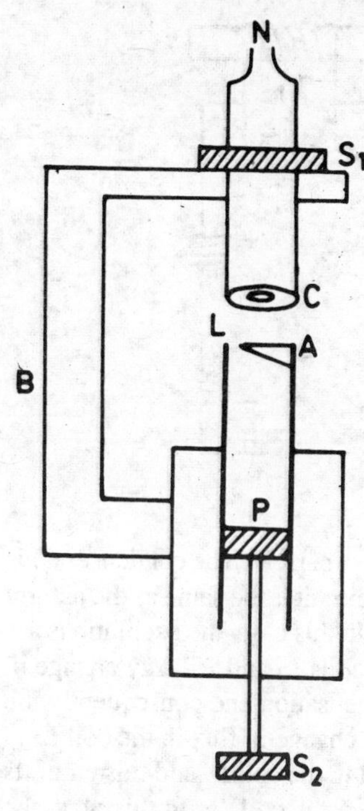

Fig. 11. 25.

between the ends of A and C can be adjusted with the help of the screw S_2. (Fig.11.25).

An air blast is blown through the nozzle N at the top. The blast of air coming out of C strikes against the lip L and the column of air in the pipe is set into vibration. By adjusting the length of the air column in A, it is brought to the resonant position. The resonant frequency will depend on the length and diameter of the pipe A. If l is the length of the air column in A, x the end correction, then the wavelength

$$\lambda = 4(l + x)$$

The frequency of sound is,

$$n = \frac{V}{\lambda}$$

$$n = \frac{V}{4(l + x)}$$

With the help of this whistle, frequencies of the order of 30,000 hertz can be produced. The micrometer screw S_1 can also be calibrated to give directly the frequency of the sound.

2. Magnetostriction Oscillator. Ultrasonic waves can be produced by using the principle of magnetostriction. According to this principle, if a ferromagnetic material in the form of a bar (like iron or nickel), is subjected to alternating magnetic field, the bar expands and contracts in length alternately. The frequency of contraction or expansion is twice the frequency of the alternating magnetic field. The alternating magnetic field is produced with the help of an oscillatory circuit. Due to the longitudinal contraction and expansion of the bar, longitudinal compressional waves are produced in the medium surrounding the bar. Frequencies ranging from a few hundred hertz to about 300,000 hertz can be produced with this arrangement.

The experimental arrangement is shown in Fig. 11.26. XY is a bar of ferromagnetic material say of iron or nickel. The bar is clamped in the middle. L_1 and L_2 are two coils surrounding the bar PQ. L_1 and C_1 are connected in parallel and the combination is connected in the plate circuit. L_2 is connected between the grid and the cathode. The milliammeter is connected in the plate circuit. The values of L_1 and C_1 determine the frequency of the oscillatory circuit.

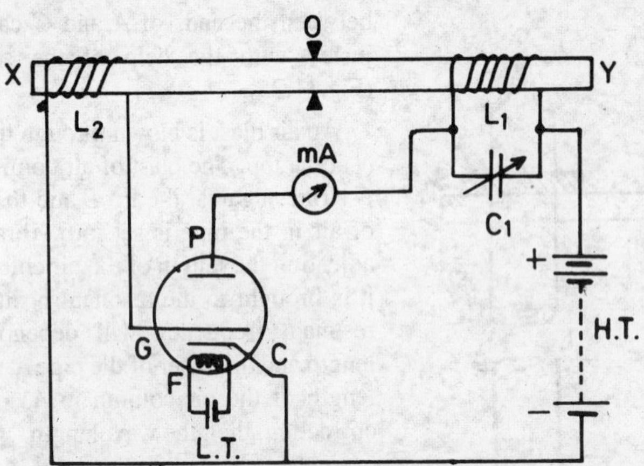

Fig. 11.26.

Initially the bar is magnetised by passing direct current. The condenser C_1 is adjusted so that the frequency of the oscillatory circuit is the same as the natural frequency of longitudinal vibrations of the bar. In this case, the oscillations are maintained by the coupling effect between the coils L_1 and L_2. Any change in the plate current brings about a change in magnetisation and consequently the length of the rod changes. This gives rise to the change in flux in the coil L_2 in the grid circuit. And induced e.m.f. is produced in the coil L_2 and this e.m.f. is amplified by the triode valve and reacts back on the coil L_1. In this way, the oscillations are maintained and the amplitude of oscillations will be large.

At the start, the milliameter shows alternation when the oscillations are set up. The condenser C_1 is adjusted to the position where these alternations are maximum. In this position, the vibrations are most intense.

The frequency of the ultrasonic waves produced by this method will depend on the length, density and elasticity of the material of the bar.

3. Piezo-electric Oscillator. In 1880, the Curie brothers, discovered that certain asymmetric crystals exhibit *Piezo electric effect.* In this effect, if one pair of opposite faces of a crystal is subjected to pressure, the other pair of opposite faces develop opposite electric charges. The sign of the charges changes when the faces are subjected to tension instead of pressure. The converse of Piezo electric effect is also true. According to this, if alternating voltages are applied to one pair of faces, the corresponding changes in the dimensions of the other pair of faces of the crystal are produced. These changes have been found to be more pronounced when thin slices of crystals of quartz, tourmaline and Rochellel salt are used. The direction of the cut of the crystal with reference to the optic axis is quite important.

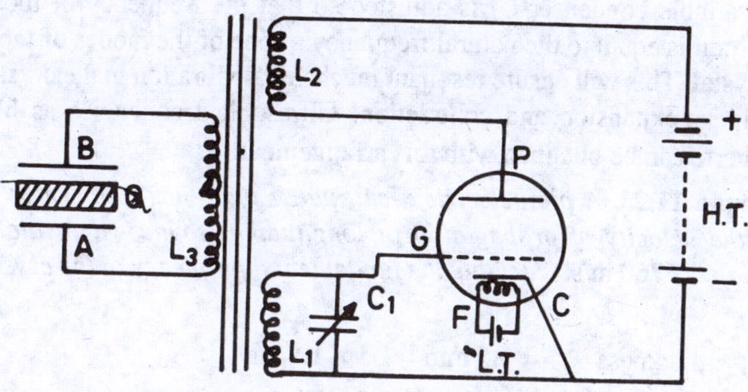

Fig . 11.27.

Thus, when the two opposite faces of a quartz crystal, their faces being cut perpendicular to the optic axis, are subjected to alternating voltage, the other pair of opposite faces experiences stresses and strains. The quartz crystal will continuously contract and expand. Elastic vibrations are set up in the crystal.

When the frequency of the alternating voltage is equal to the natural frequency of vibration of the crystal or its simple higher multiples, the crystal is thrown into resonant vibrations and the amplitude will be large. These vibrations are longitudinal in nature. The frequency of vibration is,

$$n = \frac{p}{2l}\sqrt{\frac{Y}{\rho}} \qquad \ldots (1)$$

Here $P = 1, 2, 3, \ldots$ etc.

Y is the elasticity and ρ is the density of the crystal.

The velocity of longitudinal waves in the crystal,

$$v = \sqrt{\frac{Y}{\rho}} \qquad \ldots (2)$$

The value of $v = 5{\cdot}5 \times 10^3$ m/s for quartz.

For a crystal of length 0·05 m, the frequency for the first mode of vibration will be

$$\frac{1 \times 5{\cdot}5 \times 10^3}{2 \times 0{\cdot}05} = 5{\cdot}5 \times 10^4 \text{ hertz}$$

The other modes of frequency are simple integral multiples of $5{\cdot}5 \times 10^4$ hertz.

For experimental arrangement, the circuit diagram is as shown in Fig. 11.27. Q is a thin slice of quartz crystal cut with its opposite faces perpendicular to the optic axis. The crystal is placed between two metal plates A and B. The plates A and B are connected to the coil L_3. Coils L_1, L_2 and L_3 are inductively coupled. Coil L_2 is connected in the plate circuit. The tank circuit L_1 and C_1 is connected between the grid and the cathode.

The variable condenser C_1 is adjusted so that the frequency of the oscillatory circuit is equal to the natural frequency of one of the modes of vibration of the crystal. This will ensure resonant mechanical vibration in the crystal due to the linear expansion and contraction. Ultrasonic frequencies as high as 5×10^8 hertz can be obtained with this arrangement.

Example 11.23. *A piezoelectric x-cut quartz plate has a thickness of* 1·5 mm. *If the velocity of propagation of longitudinal sound waves along the x-direction is* 5760 m/s, *calculate the fundamental frequency of the crystal.*

(IAS)

Here, thickness $= 1{\cdot}5 \text{ mm} = 1{\cdot}5 \times 10^{-3} \text{ m}$

$$v = 5760 \text{ m/s}$$

Thickness,
$$t = \frac{\lambda}{2}$$

$$\lambda = 2t = 2 \times 1{\cdot}5 \times 10^{-3} \text{ m}$$
$$= 3 \times 10^{-3} \text{ m}$$

Frequency,
$$n = \frac{v}{\lambda}$$

$$n = \frac{5760}{3 \times 10^{-3}}$$
$$= \mathbf{1920 \times 10^3 \ Hz}$$
$$= \textbf{1920 kilo–hertz.}$$

11.25 Detection of Ultrasonic Waves

Ultrasonic waves propagated through a medium can be detected in a number of ways. Some of the methods employed are as follows :

(1) **Kundt's tube method.** Ultrasonic waves can be detected with the help of Kundt's tube. At the nodes, lycopodium powder collects in the form of heaps. The average distance between two adjacent heaps is equal to half the wavelength. This method cannot be used if the wavelength of ultrasonic waves is very small *i.e.* less than a few millimeters. In the case of a liquid medium, instead of lycopodium powder, powdered coke is used to detect the position of nodes.

(2) **Sensitive flame method.** A narrow sensitive flame is moved along the medium. At the position of the antinode, the flame is steady. At the position of the node the flame flickers because there is change in pressure. In this way positions of nodes and antinodes can be found out in a medium. The average distance between two adjacent nodes is equal to half the wavelength. If the value of the frequency of the ultrasonic wave is known, the velocity of the ultrasonic wave through the medium can be calculated.

(3) **Thermal detectors.** This is the most commonly used method of detection of ultrasonic waves. In this method, a fine platinum wire is used. This wire is moved through the medium. At the position of nodes, due to alternate compressions and rarefactions, adiabatic changes in temperature take place. The resistance of the platinum wire changes with respect to time. This can be detected with the help of Callendar and Garrifith's bridge arrangement. At the position of the antinodes, the temperature remains constant and the resistance of the platinum wire remains constant. This will be indicated by the undisturbed balanced position of the bridge.

(4) **Quartz crystal method.** This method is based on the principal of Piezo-electric effect. When one pair of the opposite faces of a quartz crystal is exposed to the ultrasonic waves, the other pair of opposite faces developed opposite charges. These charges are amplified and detected using an electronic circuit.

11.26. Acoustic Grating

A quartz crystal Q is placed between two metal plates. These two plates are connected to an audio-frequency oscillator. The frequency of the oscillator is adjusted so that the crystal vibrates in resonance. Due to the longitudinal compression and expansion of the crystal, ultrasonic waves are produced.

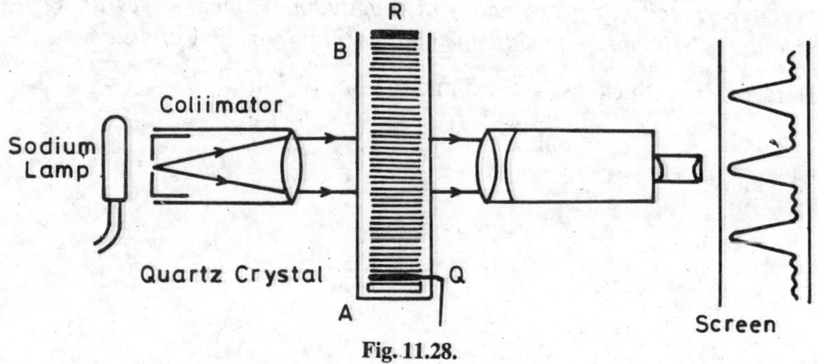

Fig. 11.28.

These waves are reflected by the reflector R (Fig. 11.28). Due to the superposition of the direct and the reflected waves, longitudinal stationary waves are produced in the medium. These waves give rise to fixed positions of nodal and antinodal planes. At the nodal planes, the density is maximum and at the antinodal planes, the density is minimum. This arrangement behaves like a diffraction grating and is called acoustic grating.

This arrangement is suitable for the measurement of wavelength and velocity of ultrasonic waves in a medium. In Fig. 11.28, the acoustic grating is mounted on the prism table of a spectrometer. A parallel beam of sodium light from the collimator is normally incident on the acoustic grating. The diffracted beam of light is viewed through the telescope. The diffraction pattern consists of a central maximum and principal maxima on either side. The directions of the principal maxima are given by the expression

$$d \sin \theta_n = n\lambda \qquad \ldots (1)$$

Here $n = 1, 2, 3$, etc. and θ_n is the angle of diffraction for the nth order. λ is the wavelength of sodium light and d is the distance between two adjacent nodal or antinodal planes. Knowing n, θ_n and λ, the value of d can be calculated from equation (*i*). If λ_a is the wavelength of the ultrasonic waves through the medium,

$$d = \frac{\lambda_a}{2}$$

or

$$\lambda_a = 2d \qquad \ldots (2)$$

If the resonant frequency of the Piezo electric oscillator is N, the velocity of the ultrasonic waves,

$$v = N\lambda_a$$

$$v = 2Nd \qquad \ldots (3)$$

This method is useful in measuring the velocity of ultrasonic waves through liquids and gases at various temperatures.

Example 11.24. *A piezoelectric x-cut quartz plate has a thickness of* 1·5 mm. *If the velocity of propagation of longitudinal sound waves along the x-direction is* 5760 m/s, *calculate the fundamental frequency of the crystal.*

Here, thickness $= 1{\cdot}5 \text{ mm} = 1{\cdot}5 \times 10^{-3}$ m

$$v = 5760 \text{ m/s}$$

Thickness $= \dfrac{\lambda}{2}$

$$\lambda = 2 \times \text{thickness}$$

$$= 2 \times 1{\cdot}5 \times 10^{-3}$$

$$= 3 \times 10^{-3} \text{ m}$$

Frequency, $n = \dfrac{v}{\lambda}$

$$n = \frac{5760}{3 \times 10^{-3}}$$

$$= 192 \times 10^4 \text{ Hz}$$

$$= \mathbf{1920 \text{ kilohertz.}}$$

Example 11.25. *An ultrasonic beam is used to determine the thickness of a steel plate. It was noticed that the difference in two adjacent harmonic frequencies is* 50 kilohertz. *The velocity of sound in steel is* 5000 m/s. *Calculate the thickness of the steel plates.*

The velocity of ultrasonic waves,

$$v = 2\,Nd$$

Here N is the fundamental frequency and d is the thickness.

$$N = \frac{v}{2d}$$

The harmonic frequencies are the multiples of the fundamental frequency.

$$N_{(n)} - N_{(n-1)} = N = \frac{v}{2d}$$

$$\therefore \quad d = \frac{v}{2\,(N_n - N_{n-1})}$$

Here $\quad N_n - N_{n-1} = 50$ kilo hertz

$$= 50{,}000 \text{ hertz}$$

$$v = 5{,}000 \text{ m/s}$$

$$\therefore \quad d = \frac{5{,}000}{2 \times 50{,}000}$$

$$d = \mathbf{0{\cdot}05\ m}$$

Hence the thickness of the steel plate is 0·05 metre.

Example 11.26. *A quartz crystal of thickness* 0·001m *is vibrating at resonance. Calculate the fundamental frequency. Given, Y for quartz =* $7{\cdot}9 \times 10^{10}$ newtons/m2 *and* ρ *for quartz* = 2650 kg/m³.

For longitudinal vibrations,

$$v = \sqrt{\frac{Y}{\rho}}$$

Here $\quad Y = 7{\cdot}9 \times 10^{10}$ newtons/m³

and $\quad \rho = 2650$ kg/m³

$$\therefore \quad v = \sqrt{\frac{7{\cdot}9 \times 10^{10}}{2650}}$$

$$v = 5460 \text{ m/s}$$

For the fundamental mode of vibration,

$$\text{Thickness} = \frac{\lambda}{2}$$

$$\therefore \quad \lambda = 2 \times \text{thickness}$$

$$= 2 \times 0{\cdot}001$$

$$= 0{\cdot}002 \text{ m}$$

Frequency, $$n = \frac{v}{\lambda}$$

$$n = \frac{5460}{0{\cdot}002}$$

$$= 2730{,}000 \text{ hertz}$$

$$= \mathbf{2730\ kilo\text{-}hertz.}$$

Example 11.27. *A quartz crystal of thickness* 0·001 m *radiates ultrasonic waves into water of frequency 20 kilo-hertz and intensity* 5×10^4 watts/m². *Calculate (i) maximum acceleration, and (ii) maximum displacement. Velocity of sound in water =* 1480 m/s *and density of water =* 1000 kg/m³. *Density of the crystal =* 2650 kg/m³.

Sound intensity,

$$I = \frac{P^2}{2\rho v} \text{ watts/m}^2$$

Here P is the sound pressure, ρ is the density and v is the velocity of sound.

∴ Sound pressure,

$$P = \sqrt{2I\rho v}$$

Here $$I = 5 \times 10^4 \text{ watts/m}^2$$

$$\rho = 1000 \text{ kg/m}^2$$

$$v = 1480 \text{ m/s}$$

∴ $$P = \sqrt{2 \times 5 \times 10^4 \times 1000 \times 1480}$$

$$= \mathbf{3{\cdot}85 \times 10^5\ newtons/m^2}$$

Force = Pressure × Area = Mass × Acceleration

∴ $$PA = Ma$$

Here $$M = \text{mass of the crystal}$$

$$= \text{Volume} \times \text{Density} = A \times t \times D$$

∴ $$PA = (AtD)\, a$$

(*i*) Maximum acceleration,

$$a = \frac{P}{tD}$$

$$a = \frac{3{\cdot}85 \times 10^5}{0{\cdot}001 \times 2650}$$

or $$a = \mathbf{1{\cdot}45 \times 10^5\ m/s^2}$$

(*ii*) Maximum displacement,

$$y = \frac{a}{\omega^2}$$

$$= \frac{1{\cdot}45 \times 10^5}{(2\pi f)^2}$$

Here $$f = 20 \text{ kilo-hertz} = 20{,}000 \text{ hertz}$$

$$y = \frac{1{\cdot}45 \times 10^5}{(2\pi \times 20{,}000)^2} = \mathbf{9{\cdot}4 \times 10^{-6} \text{ m.}}$$

11.27. Applications of Ultrasonic Waves

Ultrasonic waves have a large number of practical applications.

(1) **Depth of sea.** Ultrasonic waves of high frequency are used to determine the depth of the sea. A Piezo electric quartz oscillator is used for this purpose. The crystal is placed between two metal plates and the plates are connected to a spark oscillator, producing damped oscillations. The frequency of the damped oscillator is tuned to be the same as the natural frequency of the quartz crystal. The quartz crystal itself acts as a transmitter and a receiver of the ultrasonic waves. The ultrasonic waves transmitted by the crystal are directed towards the bed of the sea. These waves are reflected back from the bed and the echo is detected by the crystal itself. In this case the metal plates are automatically connected to an amplifier and a cathode ray oscillograph. The time interval between the emitted signal and the echo is determined with the help of the oscillograph. Knowing the velocity of sound through sea water and the time interval, the depth of the sea can be calculated. Suppose, t is the time interval between the transmission of the ultrasonic wave and receipt of the echo and v the velocity of sound waves through sea water, then depth of the sea,

$$h = \frac{v \times t}{2}.$$

This method is also suitable to detect the presence and depth of submarines, rocks etc. from the surface of sea water. The instrument directly calibrated to show the depth of sea is called a fathometer or echometer.

(2) **Signalling**. Ultrasonic waves are used for directional signalling. The frequency of ultrasonic waves is higher than the audible sound waves. Therefore, the wavelength is comparatively small. Due to the small wavelength, ultrasonic waves can be sent in the form of a short beam. If a quartz crystal, taken in the form of a disc of radius r, is used as a source of ultrasonic waves, the angle of the cone containing these waves is given by

$$\sin\theta = \frac{0{\cdot}61\lambda}{r}.$$

For small wavelengths, θ is small. Even for a small amplitude of the vibrating crystal, large amount of energy is radiated whereas it is not possible in the case of audio frequency waves.

Recently, ultrasonic microscope has been invented. It is used to detect concealed objects. The frequency is very high so that the wavelength is of the order of the wavelength of visible light.

(3) **Heating effects**. When a beam of ultrasonic waves is passed through a substance, it gets heated. If ultrasonic waves pass through water at 0°C, the water can be made to boil.

(4) **Mechanical effects.** Ultrasonic drills are used to bore holes in steel and other metals or their alloys. Here the drill oscillates with ultrasonic frequency and can bore any hard metal.

(5) **Crack in metals**. Ultrasonic waves can be used to detect cracks or discontinuity in metal structures. In this case, an emitter and detector of ultrasonic waves are used. Ultrasonic waves from the emitter are directed towards the metal. The reflected beam is detected by the detector. If there is a crack or discontinuity, there will be rise in energy received by the detector, if the emitter and the detector are on the same side. If the emitter and the detector are on the opposite sides of the metal, there will be fall in energy at the regions of cracks or discontinuity.

(6) **Formation of alloys.** Alloys of uniform composition are obtained by subjecting the constituents to an ultrasonic beam. The two constituents are well mixed by the ultrasonic waves, even though the constituents differ in density.

(7) **Chemical effect.** Ultrasonic waves act like catalytic agents and accelerate chemical reactions. They bring about a number of chemical changes. Some of the chemical applications are as follows :

(*a*) When potassium iodide is subjected to ultrasonic waves, it liberates iodine.

(*b*) Water is decomposed into hydrogen and hydroxyl ions, by the action of ultrasonic waves.

(*c*) Ultrasonic waves reduce mercuric chloride into mercurous chloride.

(*d*) Water and oil are immisicible. An emulsion of water and oil is obtained when the mixture is subjected to ultrasonic waves. Similarly an emulsion of water and mercury can be prepared.

(*e*) Ultrasonic waves accelerate crystallization.

(*f*) Ultrasonic waves explode nitrogen iodide.

(8) **Soldering**. Aluminium cannot be soldered by ordinary soldering method. To solder aluminium, ultrasonic waves are used in addition to the electrical soldering iron. The ultrasonic waves remove the oxide film and facilitates soldering.

(9) **Medical applications.** Ultrasonic waves have a large number of applications in the field of medicine. Some of the important applications are as follows :

(*a*) *Neuralgic pain.* Ultrasonic waves are useful for relieving neuralgic and rheumatic pains. The affected portion of the body is exposed to ultrasonic waves. The waves produce a soothing massage action and relieves pain.

(*b*) *Arthritis.* Ultrasonic waves are used to relieve pain due to arthritis. Here a small metal head, vibrating with a frequency of more than 10^6 hertz is moved over the skin of the patient. These vibrations after passing through the tissues, produce a deep massage action. The patient is relieved of the pain.

(*c*) *Contracted fingers.* Ultrasonic waves are used to restore the contracted fingers. They are also used to loosen up the scar tissues in various parts of the human body.

(*d*) *Broken teeth.* Ultrasonic waves are used by dentists for the proper extraction of broken teeth.

(*e*) *Bloodless surgery.* Ultrasonic waves are used in bloodless surgery. Here the ultrasonic waves are focussed on a sharp instrument and the tissues are destroyed without any loss of blood.

American doctors have used such instruments for conducting bloodless brain operations.

(10) **Sterilization.** Ultrasonic waves can destroy unicellular organisms. Bacteria perish under the action of ultrasonic waves. Ultrasonic waves are used in the sterilization of water and milk.

(11) **Enemy of lower life**. When some lower animals like rats, frogs, fish, etc., are exposed to ultrasonic waves, they become lame.

Ultrasonic waves are having more and more practical applications in all fields. Active research work is still in progress to study the effect of ultrasonic waves in mechanical, biological, chemical, physical and industrial fields.

EXERCISES

1. Describe the working of a siren. Explain how it can be used to determine the frequency of a note.
2. Describe the falling plate method for determining the frequency of a tuning fork.
3. Discuss the revolving drum method (choronographic method) for determining the frequency of a tuning fork.
4. Discuss the stroboscopic method for determining the frequency of a tuning fork.
5. Explain with details the method of determining the frequency of a tuning fork by using a phonic motor.
6. Explain briefly the principles involved in recording and reproduction of sound in a magnetic tape recorder. *(Nagpur, 1974)*
7. Give the principle, construction and working of a cathode ray oscillograph. Explain briefly the applications of a cathode ray oscillograph.
8. Explain clearly how sound ranging is done.
9. Discuss the nature of wave front formed during the flight of a bullet at supersonic speed.

10. What is reverberation and on what factors does it depend?
11. Derive Sabine's reverberation formula and explain its significance.
12. What are ultrasonics? Describe their production.
13. Discuss briefly the applications of ultrasonics. *(Behrampur, 1972)*
14. How are ultrasonics detected? Explain the principle and working of a thermal detector.
15. Discuss briefly the applications of ultrasonics in *(i)* depth finding and *(ii)* signalling.
16. Discuss briefly the application of ultrasonics in medicine and surgery.
17. Describe with experimental details and full explanation the stroboscopic method of determining the frequency of a tuning fork. *(Delhi, 1972)*
18. What is a phonic wheel ? How does it work? *(Delhi, 1972)*
19. Describe the falling plate method for the determination of the frequency of a tuning fork giving the relevant theory.
20. Discuss the phonic wheel method for the determination of the frequency of a tuning fork.
21. Enumerate the features that an auditorium should have for good acoustics. How can these features be incorporated in the design of an auditorium? What remedial measures would you suggest for the following defects in an existing auditorium?

    (i) Uneven distribution of sound.

    (ii) Unintelligible audibility of speech. *(Nagpur, 1974)*
22. Discuss briefly the applications of ultrasonics. *[Delhi (Add. Physics), 1976]*
23. Describe the stroboscopic method for finding the frequency of the tuning fork. *(Delhi, 1976)*
24. What are ultrasonic waves? Describe a method of producing ultrasonic waves using piezoelectric crystals. Mention some of the uses of ultrasonic waves. *(IAS, 1976)*
25. What are ultrasonic waves? How are they produced and to what uses are they put? *(Delhi, 1976)*
26. Describe in detail the falling plate method for determining the frequency of a tuning fork. *[Delhi (Supple), 1976]*
27. Explain what causes reverberation in a hall and how it can be reduced? *(Kanpur, 1975)*
28. What are ultrasonics? How are these produced and detected?

29. Describe, with a diagram, the working of an electrically maintained tuning fork. *(Gauhati, 1992)*

30. Define intensity and loudness of sound. Also define bel and phon. *(Bhagalpur, 1990).*

31. Write short notes on :

(i) Gramophone;
(ii) Tape recording;
(iii) Microphone;
(iv) Loud Speaker;
(v) Cathode ray oscillograph;
(vi) Sound ranging;
(vii) Flight of a bullet;
(viii) Reverberation;
(ix) Absorption coefficient;
(x) Acoustics of buildings;
(xi) Galton's whistle;
(xii) Magnetostriction oscillator;
(xiii) Piezo-electric oscillator;
(xiv) Acoustic grating;
(xv) Ultrasonics and its use;
(xvi) Recording and reproduction of sound;
(xvii) Echo and its use in depth ranging;
(xviii) Frequency of sound waves and their measurement;
(xix) Sabbine's formula.

CHAPTER 12

Allied Phenomena

12.1. Wave Velocity and Group Velocity

The equation for a plane progressive wave is given by

$$y = a \sin 2\pi \left(\frac{t}{T} - \frac{x}{\lambda} \right)$$

$$y = a \sin (\omega t - kx)$$

where $$\omega = \frac{2\pi}{T} = 2\pi n$$

and $$k = \frac{2\pi}{\lambda}$$

Here ω is called the angular frequency of the wave and k is called the wave vector.

Also,

$$\frac{\omega}{k} = \frac{2\pi n}{2\pi/\lambda} = n\lambda = v_p$$

Here v_p is called the *phase velocity* or the *wave velocity*. In other words, $v_p = \frac{\omega}{k}$ is the velocity with which a plane progressive wave front travels forward. It has a constant phase = $(\omega t - kx)$

$$\omega t - kx = \text{constant}$$

Differentiating this equation with respect to t and x,

$$\omega - k\frac{dx}{dt} = 0$$

or $$\frac{dx}{dt} = \frac{\omega}{k} = v_p$$

Group velocity. Consider a group of progressive waves with angular frequencies in the range ω and $\omega + \Delta\omega$ and wave vectors in the range k and $k + \Delta k$ superimposing on one another. The resultant wave pattern consists points of maximum amplitude and points of minimum amplitude. In between any two consecutive maximum and minimum, the amplitude gradually decreases. Between any two consecutive minima, say A and C, there is a position of maximum amplitude midway between A and C *i.e.*, at B. The dotted loop represents a group of waves. This group of waves is called a packet and with time the packet moves forward in the medium with a velocity, called the *group velocity.* (Fig. 12.1).

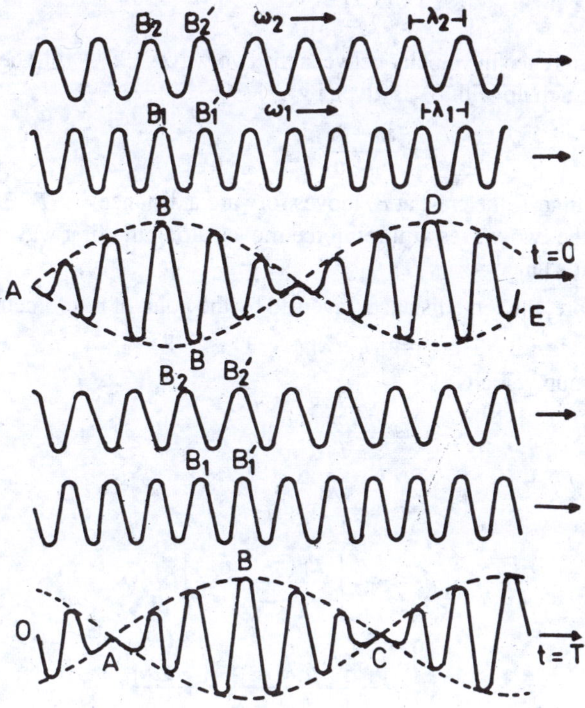

Fig. 12.1.

Consider two plane progressive waves travelling in the positive direction of the X-axis. The equations of the two waves at any instant are,

$$y_1 = a_1 \sin(\omega_1 t - k_1 x) \qquad \ldots (1)$$

$$y_2 = a_2 \sin(\omega_2 t - k_2 x) \qquad \ldots (2)$$

Here $$k_1 = \frac{2\pi}{\lambda_1} \qquad \ldots (3)$$

and $$k_2 = \frac{2\pi}{\lambda_2} \qquad \ldots (4)$$

The wave velocities for the two waves are

$$v_1 = \frac{\omega_1}{k_1}$$

and $$v_2 = \frac{\omega_2}{k_2}$$

Initially the two peaks B_1 and B_2 are at the same position. Here $\lambda_2 > \lambda_1$ and $v_1 > v_2$. By the time the peak at B_2 travels to B_2', the peak at B_1 travels to B_1'. The difference in distance between the positions of B_2' and B_1'

$$= (\lambda_2 - \lambda_1)$$

The difference in velocity between the two waves $= v_1 - v_2$. The time taken by B_1' to catch up with B_2' is given by

$$T = \frac{\lambda_2 - \lambda_1}{v_1 - v_2} \qquad \ldots (5)$$

In this time T, the crest at B_1 moves forward a distance $= v_1 T$. But the point at which the two waves will reinforce moves along the first wave from B_1 to B_1' i.e., a distance $= \lambda_1$.

Therefore, the total distance traversed by the point of reinforcement

$$= v_1 T + \lambda_1$$

$\therefore$ Group velocity,

$$v_g = \frac{v_1 T + \lambda_1}{T} \qquad \ldots (6)$$

or $$v_g = v_1 + \frac{\lambda_1}{T}$$

or $$v_g = v_1 + \left[\frac{\lambda_1 (v_1 - v_2)}{\lambda_2 - \lambda_1} \right]$$

$$v_g = \frac{v_1\lambda_2 - v_1\lambda_1 + v_1\lambda_1 - v_2\lambda_1}{(\lambda_2 - \lambda_1)}$$

or $$v_g = \frac{v_1\lambda_2 - v_2\lambda_1}{(\lambda_2 - \lambda_1)}$$

Dividing the numerator and denominator by $\lambda_1 \lambda_2$,

$$v_g = \frac{\frac{v_1}{\lambda_1} - \frac{v_2}{\lambda_2}}{\frac{1}{\lambda_1} - \frac{1}{\lambda_2}}$$

Multiplying the numerator and denominator by 2π

$$v_g = \frac{\left(\frac{2\pi}{\lambda_1}\right)v_1 - \left(\frac{2\pi}{\lambda_2}\right)v_2}{\left(\frac{2\pi}{\lambda_1}\right)v_1 - \left(\frac{2\pi}{\lambda_2}\right)}$$

or $$v_g = \frac{k_1 v_1 - k_2 v_2}{k_1 - k_2} \qquad \ldots (7)$$

But $$v_1 = \frac{\omega_1}{k_1} \quad \text{and} \quad v_2 = \frac{\omega_2}{k_2}$$

$\therefore$ $$v_g' = \frac{\omega_1 - \omega_2}{k_1 - k_2} \qquad \ldots (8)$$

If the difference in angular frequency and difference in wave vector is extremely small,

$$v_g = \frac{d\omega}{dk} \qquad \ldots (9)$$

The group velocity v_g is the speed with which the point of reinforcement of the resultant wave travels. The group velocity is different from the individual wave velocity. v_g can be higher or lower than v_p depending upon the relative directions of the individual waves. The modulated wave due to the super-imposition will be in the form of wave packets as shown in Fig. 12.2. This group of waves or wave packet in each dotted envelope travels with the group velocity and not with wave velocity.

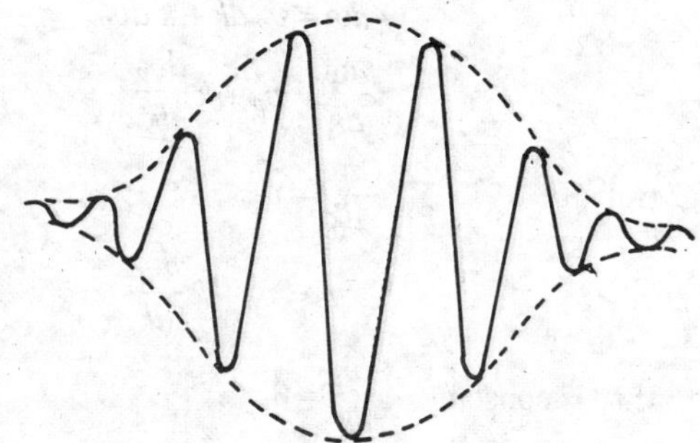

Fig . 12.2.

Electromagnetic Radiation in Vacuum

In the case of a non-dispersive medium (vacuum), the phase velocity and group velocity of electromagnetic radiations are equal.

For an electromagnetic wave, the phase velocity,

$$v_p = \frac{\omega}{k} = c \qquad \ldots (1)$$

Here c is the velocity of electromagnetic radiations,

Also $$\omega = ck$$

$$\frac{d\omega}{dk} = c \qquad \ldots (2)$$

But $$\frac{d\omega}{dk} = v_g$$

$\therefore$ $$v_g = \frac{d\omega}{dk} = c \qquad \ldots (3)$$

Thus the phase velocity and the group velocity are equal to c for electromagnetic radiation (light) in vacuum. It means modulation propagates with velocity c.

Non-dispersive Waves

Light waves in vacuum are non-dispersive, i.e. the phase velocity does not depend upon frequency.

In general,

$$v_p = \frac{\omega}{k} \qquad \ldots (1)$$

or $$\omega = v_p k \qquad \ldots (2)$$

Differentiating

$$d\omega = v_p \, dk + k \, dv_p$$

$\therefore$ $$\frac{d\omega}{dk} = v_p + k \frac{dv_p}{dk} \qquad \ldots (3)$$

But $$\frac{d\omega}{dk} = v_g$$

$\therefore$ $$v_g = v_p + k \frac{dv_p}{dk} \qquad \ldots (4)$$

If v_p is constant, $$\frac{dv_p}{dk} = 0$$

$\therefore$ $$v_g = v_p \qquad \ldots (5)$$

Thus the group velocity and the phase velocity are equal if $\frac{dv_p}{dk} = 0$.

The other examples of non-dispersive waves are :

(*i*) **Audible sound waves**

Here, $$v_p = \sqrt{\frac{\gamma P}{\rho}}$$

But $$v_p = \frac{\omega}{k}$$

$\therefore$ $$\omega = k \sqrt{\frac{\gamma P}{\rho}} \qquad \ldots (6)$$

(*ii*) **Transverse waves on a continuous string**

Here $$v_p = \sqrt{\frac{T}{m}}$$

But $$v_p = \frac{\omega}{k}$$

$\therefore$ $$\omega = k\sqrt{\frac{T}{m}} \qquad \ldots (7)$$

12.2 Group Velocity and Phase Velocity in a Dispersive Medium

The energy transfer in a wave takes place with a group velocity. The phase velocity,

$$v_p = \frac{\omega}{k}$$

and group velocity

$$v_g = \frac{d\omega}{dk} \quad \text{where} \quad k = \left(\frac{2\pi}{\lambda}\right)$$

The phase velocity v_p is frequency dependent *i.e.* radiation of different wavelengths travel with different velocities in a medium. Such a phenomenon is known as dispersion.

Here $$\omega = k\, v_p$$

Group velocity

$$v_g = \frac{d\omega}{dk}$$

$$v_g = \frac{d}{dk}\,[k\, v_p\,]$$

$$v_g = v_p + k\left(\frac{dv_p}{dk}\right) \qquad \ldots (i)$$

Equation (*i*) gives a relation between group velocity and phase velocity in a dispersive medium.

Also $$k = \frac{2\pi}{\lambda}$$

or $$\lambda = \frac{2\pi}{k}$$

$$\frac{d\lambda}{dk} = -\left(\frac{2\pi}{k^2}\right) = \frac{-\lambda^2}{2\pi} \qquad \ldots (ii)$$

From equations (*i*) and (*ii*)

$$v_g = v_p + \left(\frac{2\pi}{\lambda}\right)\left(\frac{dv_p}{d\lambda} \times \frac{d\lambda}{dk}\right) \quad \ldots (iii)$$

$$v_g = v_p + \frac{2\pi}{\lambda}\left(\frac{dv_p}{d\lambda}\right) \times \left(\frac{-\lambda^2}{2\pi}\right)$$

$$v_g = v_p - \lambda\left(\frac{dv_p}{d\lambda}\right) \quad \ldots (iv)$$

We can also write

$$\frac{1}{v_g} = \frac{dk}{d\omega}$$

$$\frac{1}{v_g} = \frac{d}{d\omega}\left(\frac{\omega}{v_p}\right)$$

$$\frac{1}{v_g} = \frac{1}{v_p} - \frac{\omega}{v_p^2}\frac{dv_p}{d\omega} \quad \ldots (v)$$

Since $\omega = 2\pi\nu$

$$\frac{1}{v_g} = \frac{1}{v_p} - \left(\frac{\nu}{v_p^2}\right)\left(\frac{dv_p}{d\nu}\right) \quad \ldots (vi)$$

Refractive index, $n = \dfrac{c}{v_p}$

$$c = n\, v_p$$

Differentiating

$$0 = n\frac{dv_p}{dv} + v_p\frac{dv}{d\nu}$$

$$\frac{dv_p}{d\nu} = -\left(\frac{v_p}{n}\right)\frac{dn}{d\nu}$$

Substituting this value in equation (*vi*)

$$\frac{1}{v_g} = \frac{1}{v_p} - \left(\frac{\nu}{v_p^2}\right)\left(-\frac{v_p}{n}\right)\left(\frac{dn}{d\nu}\right)$$

$$= \frac{1}{v_p} + \left(\frac{\nu}{n\, v_p}\right)\left(\frac{dn}{d\nu}\right)$$

But $$n\, v_p = c$$

$\therefore$
$$\frac{1}{v_g} = \frac{1}{v_p} + \left(\frac{v}{C}\right)\left(\frac{dn}{dv}\right) \qquad \ldots (vii)$$

Also
$$c = v\,\lambda$$

Differentiating
$$0 = v d\lambda + \lambda dv$$

$$\frac{d\lambda}{dv} = -\left(\frac{\lambda}{v}\right)$$

$$\frac{dn}{dv} = \frac{dn}{d\lambda} \times \frac{d\lambda}{dv}$$

$$= -\left(\frac{\lambda}{v}\right)\left(\frac{dn}{d\lambda}\right)$$

Substituting this value in equation (*vii*), we get

$$\frac{1}{v_g} = \frac{1}{v_p} - \left(\frac{\lambda}{c}\right)\left(\frac{dn}{d\lambda}\right) \qquad \ldots (viii)$$

Here λ is the wave length in vacuum.

Multiplying equation (*viii*) by C

$$\left(\frac{c}{v_g}\right) = \left(\frac{c}{v_p}\right) - \lambda\left(\frac{dn}{d\lambda}\right)$$

$\left(\frac{c}{v_g}\right) = n_g,$ group refractive index of the medium.

$\left(\frac{c}{v_p}\right) = n,$ refractive index of the medium.

$$n_g = n - \lambda\left(\frac{dn}{d\lambda}\right) \qquad (ix)$$

It means group refractive index is less than the refractive index of the medium.

12.3. Velocity of a Wave on the Surface of a Liquid

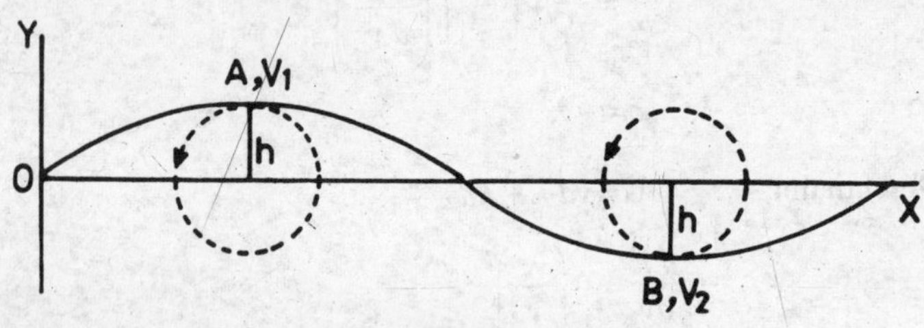

Fig. 12.3.

When a stone is dropped on the surface of water, transverse waves are produced and the disturbance travels radially outwards.The molecules of the liquid vibrate transversely. The velocity of propagation of a transverse wave on the surface of a liquid depends on (*i*) gravity (*ii*) hydrostatic pressure and (*iii*) surface tension. As the liquids are incompressible, transverse vibration of the molecules is also accompanied by longitudinal vibrations, otherwise, there should be a hollow space under a crest and high compression at the position of a trough. The continuity of the liquid can only be maintained by the longitudinal vibration of the molecules in the direction of propagation of the wave. Thus the molecules possess (*i*) transverse vibration and (*ii*) longitudinal vibration. The resultant of these two rectangular vibrations makes the molecule to move in a circle in the anticlockwise direction if the wave propagates in the positive direction of *X*-axis. During the wave propagation, it appears that the molecules roll along a circular path (Fig. 12.3).

Further, at the positions of crest or trough (points *A* and *B*), the particle velocity is zero for the vertical vibration. Similarly, for a molecule at *C*, the horizontal velocity is zero and the vertical velocity is maximum. The vertical and horizontal velocities differ in phase by $\frac{\pi}{2}$.

Effect of Gravity

Consider a wave on a liquid surface moving in the positive direction of *X*-axis with a velocity v. Let V_1, V_2 and V_3 be the resultant velocities of the molecules of the liquid at *A, B* and *C* respectively. The resultant velocity is the resultant of wave velocity and particle velocity.

Applying Bernoulli's theorem for points *A* and *B*

$$\frac{P}{\rho} + hg + \frac{1}{2} V_1^2 = \frac{P}{\rho} - hg + \frac{1}{2} V_2^2$$

As air pressure *P* is the same at all points on the liquid surface

$$hg + \frac{1}{2} V_1^2 = -hg + \frac{1}{2} V_2^2$$

$$V_2^2 - V_1^2 = 4hg \qquad \dots (i)$$

Let $\frac{dx}{dt}$ represent the horizontal velocity at A and B. At A and B the vertical velocity is zero.

$\therefore$
$$V_1 = v - \left(\frac{dx}{dt}\right) \qquad \dots (ii)$$

and
$$V_2 = v + \left(\frac{dx}{dt}\right) \qquad \dots (iii)$$

At the point C, the vertical velocity is $\frac{dy}{dt}$ and the wave velocity is v.

$$V_3 = \sqrt{v^2 + \left(\frac{dy}{dt}\right)^2}$$

Applying Bernoulli's theorem for the points A and C,

$$hg + \frac{1}{2} V_1^2 = \frac{1}{2} V_3^2 \qquad \dots (v)$$

For the points B and C

$$-hg + \frac{1}{2} V_2^2 = \frac{1}{2} V_3^2 \qquad \dots (vi)$$

Adding (v) and (vi)

$$V_3^2 = \frac{V_1^2 + V_2^2}{2} \qquad \dots (vii)$$

Substituting the values of V_1, V_2 and V_3,

$$v^2 + \left(\frac{dy}{dt}\right)^2 = \frac{\left(v - \frac{dx}{dt}\right)^2 + \left(v + \frac{dx}{dt}\right)^2}{2}$$

$$v^2 + \left(\frac{dy}{dt}\right)^2 = v^2 + \left(\frac{dx}{dt}\right)^2$$

$$\frac{dy}{dt} = \frac{dx}{dt} \qquad \dots (viii)$$

It means that the maximum vertical velocity is equal to the maximum velocity and differs in phase by $\pi/2$. The velocity amplitudes are equal and the resultant motion of the molecule is circular.

As the wave progresses in the positive direction of X-axis, the motion of the molecules of the liquid is circular in the anticlockwise direction in a vertical plane. If T is the time period, the speed of the molecule along the circular path

$$= \left(\frac{2\pi a}{T}\right)$$

$$\therefore \qquad \frac{dx}{dt} = \frac{dy}{dt} = \frac{2\pi a}{T}$$

Therefore

$$V_1 = v - \left(\frac{2\pi a}{T}\right)$$

and

$$V_2 = v + \frac{2\pi a}{T}$$

Substituting these values in equation (i), we get,

$$\left[v + \left(\frac{2\pi a}{T}\right)\right]^2 - \left[v - \left(\frac{2\pi a}{T}\right)\right]^2 = 4hg$$

or

$$\frac{8\pi a v}{T} = 4hg$$

$$v = \frac{gT}{2\pi}$$

If λ is the wave length of the wave,

$$T = \frac{\lambda}{v}$$

$$\therefore \qquad v = \frac{g\left(\frac{\lambda}{v}\right)}{2\pi}$$

or

$$v^2 = \frac{\lambda g}{2\pi}$$

$$v = \sqrt{\frac{\lambda g}{2\pi}} \qquad \text{... (ix)}$$

This equation represents the velocity of the gravity waves. It is only due to the effect of gravity.

12.4. Surface Tension Effect

To determine the effect of surface tension on the velocity of transverse wave on the surface of a liquid, consider a SHM wave (Fig. 12.4).

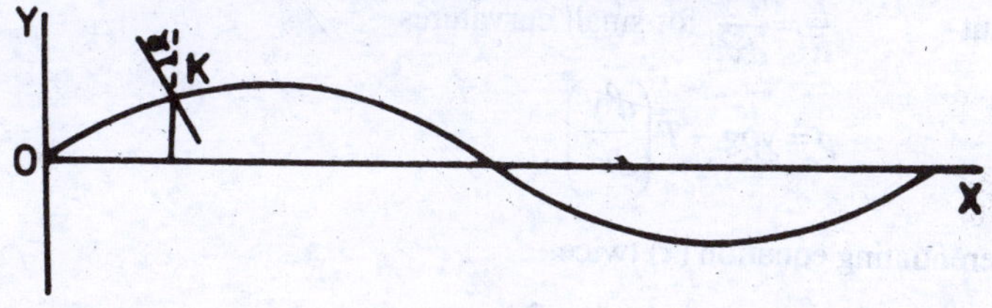

Fig. 12.4.

Let the displacement at any instant be y.

$$y = a \sin \frac{2\pi}{\lambda}(vt - x) \qquad \ldots (x)$$

$$y = a \sin \left[\frac{2\pi vt}{\lambda} - \frac{2\pi x}{\lambda}\right]$$

$$y = a \sin \left[\phi - \frac{2\pi x}{\lambda}\right]$$

Here ø is the initial phase at the origin.

At the point L, pressure due to gravity alone

$$p_1 = y\rho g$$

Excess of pressure at K due to surface tension,

$$p_2 = T\left[\frac{1}{R} + \frac{1}{R'}\right]$$

Here $R' = \alpha$ (for the cylindrical wave)

$\therefore$ $p_2 = \dfrac{T}{R}$ acting radially outwards at K

Therefore outward pressure due to surface tension

$$= \frac{T}{R}\cos\alpha$$

Resultant downward pressure,

$$p = p_1 - p_2$$

$$p = y\rho g - \frac{T}{R}\cos\alpha$$

As the amplitude of vibration is small, α is small.

$\therefore \qquad \cos \alpha = 1$

and $$p = y\rho g - \frac{T}{R} \qquad \ldots (xi)$$

But $\frac{1}{R} = \frac{d^2y}{dx^2}$ for small curvatures

$$\therefore \qquad p = y\rho g - T\left(\frac{d^2y}{dx^2}\right) \qquad \ldots (xii)$$

Differentiating equation (*x*) twice,

$$\frac{d^2 y}{dx^2} = -\frac{4\pi^2 a}{\lambda^2} \sin \frac{2\pi}{\lambda}(vt-x)$$

$$\frac{d^2y}{dx^2} = -\left(\frac{4\pi^2}{\lambda^2}\right)y$$

$$p = y\rho g + \frac{4\pi^2 TY}{\lambda^2}$$

$$p = y\rho\left[g + \frac{4\pi^2 T}{\rho\lambda^2}\right] \qquad \ldots (xiii)$$

It shows that the effect of surface tension is to increase the effective value of acceleration due to gravity by $\frac{4\pi^2 T}{\rho\lambda^2}$

Modifying equation (*ix*), we get,

$$v = \sqrt{\left(\frac{\lambda}{2\pi}\right)\left(g + \frac{4\pi^2 T}{\rho\lambda^2}\right)}$$

$$v = \sqrt{\frac{\lambda g}{2\pi} + \frac{2\pi T}{\rho\lambda}} \qquad \ldots (xiv)$$

It shows that the velocity depends upon (*i*) wave length λ, (*ii*) density ρ (*iii*) acceleration due to gravity g and (*iv*) surface tension T.

When $\qquad \lambda = 0 \qquad v = \infty$

Also, when $\qquad \lambda = \infty \qquad v = \infty$

Critical wavelength. The wavelength for which the velocity of the wave is minimum is called the critical wavelength.

From equation (*xiv*)

$$v^2 = \frac{\lambda g}{2\pi} + \frac{2\pi T}{\rho\lambda}$$

Differentiating

$$2\,v\frac{dv}{d\lambda}=\frac{g}{2\pi}+\frac{2\pi T}{\rho}\left(-\frac{1}{\lambda^2}\right)$$

For v to be minimum

$$\frac{dv}{d\lambda}=0 \text{ and } \lambda=\lambda_c$$

$$\frac{1}{\lambda_c^2}=\frac{g\rho}{4\pi^2 T}$$

or

$$\lambda_c=2\pi\sqrt{\frac{T}{g\rho}} \qquad \ldots (xv)$$

Minimum velocity. Substituting the value of

$$\lambda_c=2\pi\sqrt{\frac{T}{g\rho}} \text{ in equation (xiv)}$$

$$v_{min}=\left[2\left(\frac{gT}{\rho}\right)^{1/2}\right]^{1/2}$$

$$v_{min}=1{\cdot}414\left[\frac{gT}{\rho}\right]^{1/4} \qquad \ldots (xvi)$$

Example 12.1. *(A) Calculate the minimum velocity of the capillary waves on the surface of water. Surface tension of water* = $7{\cdot}2\times10^{-2}$ N/m.

[Delhi (Hons) 1980]

$$v_{min}=1{\cdot}414\left[\frac{gT}{\rho}\right]^{1/4}$$

Here

$$g=9{\cdot}8 \text{ m/s}^2$$

$$T=7{\cdot}2\times10^{-2} \text{ N/m}$$

$$\rho=10^3 \text{ kg/m}^3$$

$$\therefore \quad v_{min}=1{\cdot}414\left[\frac{9{\cdot}8\times7{\cdot}2\times10^{-2}}{10^3}\right]^{1/4}$$

$$= \mathbf{0{\cdot}2304 \text{ m/s.}}$$

12.5. Special Cases

(1) Gravity Waves

When $\lambda > \lambda_c$, in equation (*xiv*)

$\frac{2\pi T}{\rho\lambda}$ is negligibly small and

$$v = \sqrt{\frac{\lambda g}{2\pi}}$$

These waves are called gravity waves. The amplitude of the waves is extremely small as compared to wave length.

(2) Ripples

When in $\lambda < \lambda_c$, in equation (*xiv*) the term $\frac{\lambda g}{2\pi}$ can be neglected and

$$v = \sqrt{\frac{2\pi T}{\rho\lambda}}.$$

The propagation of the wave is due to surface tension and these waves are called **ripples** or **capillary waves**.

The wave length is short as compared to the amplitude.

If ν is the frequency and λ the wave length,

$$v = \nu\lambda$$

$$\therefore \qquad \nu\lambda = \sqrt{\frac{\lambda g}{2\pi} + \frac{2\pi T}{\rho\lambda}}$$

$$\nu^2\lambda^2 = \frac{\lambda g}{2\pi} + \frac{2\pi T}{\rho\lambda}$$

or
$$T = \frac{\lambda^2\rho}{4\pi^2}\,[2\pi\nu^2\lambda - g] \qquad \ldots (xvi)$$

With the help of the ripple tank experiment, using a tuning fork of known frequency ν and measuring λ by experiment, the value of surface tension T can be calculated.

12.6. Propagation of Electro-Magnetic Waves (Oscillatory Circuit)

Electromagnetic radiations can be produced by rapid variation of current in a conductor or by the acceleration of charged particles. The earliest and the most classic experiment to produce significant amount of radiation by electrical means was performed by Hertz in 1889. Maxwell asserted, mainly on theoretical considerations, that light is a form of electromagnetic radiation. The only difference between visible light and the radiations produced by electrical methods is in regard to the frequency of the waves. Radiation of sufficient energy can be obtained using alternating currents of high frequency.

If a charged condenser of capacity C discharges through an inductance L and a resistance R, the nature of the discharge depends on the value of $\left(\frac{R^2}{4L^2}-\frac{1}{LC}\right)$. If $\left(\frac{R^2}{4L^2}-\frac{1}{LC}\right)$ is + ve, the discharge is a periodic and dead beat.

If, $\left(\frac{R^2}{4L^2}-\frac{1}{LC}\right)=0$, the discharge is critically damped and if $\left(\frac{R^2}{4L^2}-\frac{1}{LC}\right)$ is – ve the discharge is damped and oscillatory.

The frequency of the oscillations in the last case is given by

$$f=\frac{1}{2\pi}\sqrt{\frac{1}{LC}-\frac{R^2}{4L^2}}$$

If the value of R is negligible then, $f=\frac{1}{2\pi\sqrt{LC}}$. Thus, the lower the value of $\sqrt{LC}$, the higher will be frequency of the oscillations. These high frequency currents produce an accompanying electromagnetic field. An oscillatory discharge circuit is of fundamental importance in producing electromagnetic waves.

Hertz's Experiment

In Hertz's experiment, the apparatus consisted of two large square plates A and B of brass or zinc. The plates are connected through thick copper wires to two *highly polished* brass knobs K_1 and K_2 (Fig.12.5). The distance between the centres of the plates is about 60 cm and these two plates form a condenser of low capacity. The secondary terminals S, S of an induction coil are connected to the wires p and q as shown. The distance between the knobs is 2 or 3 cm.

The high potential difference during break, across the terminals of the secondary of the induction coil, charges the condenser and when the potential difference between the plates is sufficient enough, the air gap between the knobs becomes conducting. Consequently, the condenser discharges producing damped electromagnetic waves. Due to the discharge of the condenser the potential difference between the plates falls and the air gap ceases to be conducting. However, the condenser is charged again to a high potential during the break of the primary of the induction coil and hence it discharges again through the air gap producing damped oscillatory waves. This process of charging and discharging of the condenser is continuous as long as the primary circuit of the induction coil is closed. The number of complete oscillations that

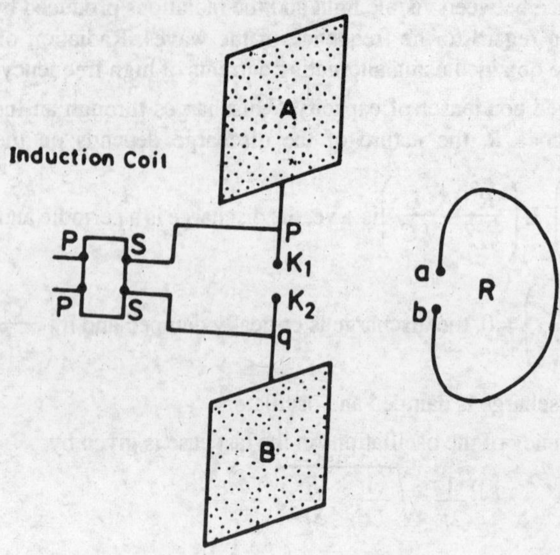

Fig. 12.5. Hertz's experiment.

occur during discharge is small and the time period of the oscillations is not constant because of the variations in the resistance of the air gap. With this type of arrangement, Hertz could obtain radiations of wavelength about 5·4 metres corresponding to a frequency 5·55 megacycles per second approximately.

For detecting these radiations, Hertz used an *almost closed metallic ring R* terminating in two small *well polished spheres a* and *b* (Fig. 12.5). This circular ring is placed near the knobs K_1 and K_2 in such a way, that the magnetic lines of force cut it. Consequently, an induced EMF is produced in the ring and the two knobs *a* and *b* cause a spark to pass through, provided the induced EMF is large. The detector has a natural frequency of its own depending on its values of *L* and *C.* By suitably adjusting the size of the ring and the distance between *a* and *b,* the frequency of the ring can be made equal to the frequency of the electro-magnetic radiations. In this resonant condition, maximum effect can be produced. Further, the effect will be maximum, if the plane of the detector ring is prependicular to the magnetic lines of force *i.e.,* the plane of the ring must be vertical and at right angles to the plane of the condenser plates. In the horizontal direction the effect observed will be zero.

12.7. Marconi's Experiments

In Hertz experiment the radiations could not be detected over long distances. The commercial use of electromagnetic waves came into existence with an important discovery by Marconi in 1895. He observed that if one of the terminals of the induction coil is earth connected and the other terminal connected to a wire hung high in the air (*aerial*) the radiations could be detected

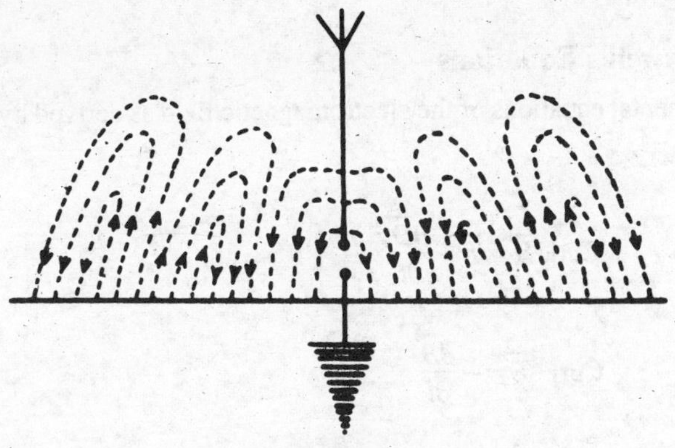

Fig. 12.6.

over long distances. Fig. 12.6 represents the upper half of the radiated electric field and the magnetic field is at right angles to the loops at every point. Application of a high frequency alternating potential to a wire hung in the air, forces electrons to move up and down the wire and the upper end of the wire is charged alternately positive and negative. The wire (called the *antenna*) and the earth act as the two plates of a condenser and an electrostatic field is produced round the antenna. The distribution of the electrostatic lines of force is shown in Fig. 12.6. The motion of electrons constitutes an electric current which produces a magnetic field round it. When the direction of the current in the antenna is reversed the electric and magnetic fields are also reversed. The electric and magnetic fields are at right angles to one another and they travel with the velocity of light outwards and constitute the electromagnetic waves.

Figure 12.7 represents the variation in the electric and magnetic field intensities. Indroduction of a Morse key in the primary circuit of the induction coil

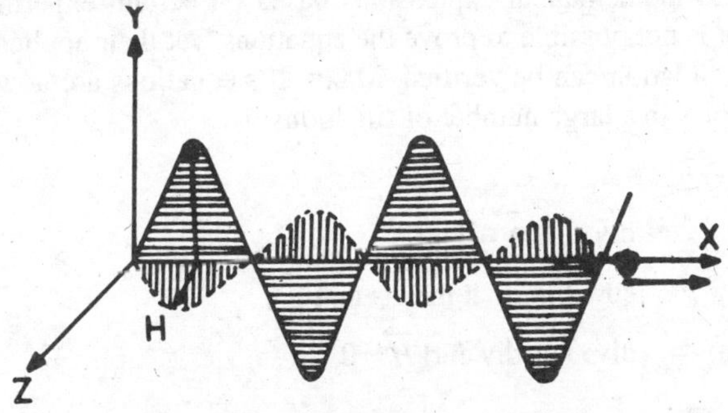

Fig. 12.7.

enables the transmission of coded messages. The number of damped wave trains radiated in space per second will depend on the time for which the Morse key is kept pressed.

12.8. Maxwell's Equations

The fundamental equations of the electromagnetic field as derived by Maxwell are :

$$\text{Curl } \vec{H} = \vec{J} + \frac{\partial \vec{D}}{\partial t} \quad \ldots (1)$$

$$\text{Curl } \vec{E} = -\frac{\partial \vec{B}}{\partial t} \quad \ldots (2)$$

$$\text{div } \vec{D} = \rho \quad \ldots (3)$$

$$\text{div } \vec{B} = 0 \quad \ldots (4)$$

Here $\vec{D}$ is the electric displacement, ρ is the electric charge density, $\vec{B}$ is the magnetic flux density, J is the current density and $\vec{E}$ is the electrical intensity. Equation (1) is an important contribution by Maxwell. The term $\frac{\partial \vec{D}}{\partial t}$ represents the displacement current density. In the absence of this term, equation (1) represents Ampere's law. Equation (2) represents Maxwell's generalisation of Neumann's law of electromagnetic induction. Equation (3) represents Gauss's flux law in electrostatics and equation (4) represents Gauss's flux law in magnetism.

This establishes the fact that a single magnetic pole cannot exist. Maxwell's equations relate to mathematical expressions based on certain experimental results. Though it is not possible to prove the equations, yet their applicability to any particular situation can be verified. Maxwell's equations are now used as guiding principles in a large number of situations.

For any vector $\vec{A}$,

$$\text{div curl } \vec{A} \equiv 0$$

If $\quad \text{curl } \vec{H} = \vec{J}$, it means that

$$\text{div } \vec{J} = \text{div curl } \vec{H} = 0$$

But from the equation of continuity of charge (equation 3)

$$\text{div } \vec{J} = -\frac{\partial \rho}{\partial t} = -\frac{\partial}{\partial t}\left[\text{ div } \vec{D}\right] \qquad \ldots (5)$$

i.e. div $\vec{J}$ is zero for steady state only.

From equation (5), $\text{div}\left(J + \frac{\partial D}{\partial t}\right) = 0$ ... (6)

This equation will reduced to div $J = 0$ in the case of steady state.

According to equation (6), the current density $\vec{J}$ should be written as

$$\vec{J} + \frac{\partial \vec{D}}{\partial t} \text{ in the non-study state.}$$

However $\vec{J}$ is the conduction current density and $\frac{\partial \vec{D}}{\partial t}$ is the displacement current density.

In equation (1), if $\vec{J}$ is omitted, it will be identical with equation (2). It means that a change in magnetic flux induces an electric field and a change in the electric flux induces a magnetic field. However, displacement currents will be significant only at rapidly varying fields *i.e.* only at high frequencies.

12.9. Propagation of Electromagnetic Waves in Isotropic Medium

The conduction current density $\vec{J} = 0$ for a non-conducting medium. In an isotropic medium, the electric displacement at any point has the same direction as the electrical intensity. Also magnetic induction has the same direction as the magnetic intensity. Here,

$$\vec{D} = \varepsilon \vec{E} \qquad \ldots (1)$$

and

$$\vec{B} = \mu \vec{H} \qquad \ldots (2)$$

Here ε is the permittivity of the medium and μ is the permeability of the medium. As the medium is isotropic, the values ε and μ will remain the same throughout the medium. Further, it is also assumed that

(*i*) ε and μ are independent of $\vec{E}$ and $\vec{H}$

(*ii*) there is no volume distribution of charge *i.e.*, $\rho = 0$

With these assumptions, Maxwell's equations can be rewritten as

$$\text{curl } \vec{H} = \varepsilon \frac{\partial \vec{E}}{\partial t}$$

or $$\text{curl } \vec{B} = \mu\,\varepsilon\,\frac{\partial E}{\partial t} \qquad \ldots(3)$$

$$\text{curl } \vec{E} = -\mu\,\frac{\partial \vec{H}}{\partial t}$$

or $$\text{curl } \vec{E} = -\,\frac{\partial \vec{B}}{\partial t} \qquad \ldots(4)$$

$$\text{div } \vec{E} = 0 \qquad \ldots(5)$$

$$\text{div } \vec{H} = 0$$

or $$\text{div } \vec{B} = 0 \qquad \ldots(6)$$

From equation (4)

$$\text{curl curl } \vec{E} = -\,\text{curl}\,\frac{\partial \vec{B}}{\partial t}$$

$$= -\frac{\partial}{\partial t}\left(\text{curl } \vec{B}\right) \qquad \ldots(7)$$

In any vector $\vec{A}$,

$$\text{curl curl } \vec{A} \equiv \text{grad div } \vec{A} - \nabla^2 A \qquad \ldots(8)$$

∴ Equation (7) can be written as

$$\text{grad div } \vec{E} - \nabla^2 \vec{E} = -\frac{\partial}{\partial t^2}\left(\text{curl } \vec{B}\right)$$

Substituting the value of curl $\vec{B}$ from equation (3)

$$\text{grad div } \vec{E} - \nabla^2 \vec{E} = -\,\mu\varepsilon\,\frac{\partial^2 \vec{E}}{\partial t^2} \qquad \ldots(9)$$

But $\text{div } \vec{E} = 0$

∴ $$\nabla^2 \vec{E} = \mu\varepsilon\,\frac{\partial^2 \vec{E}}{\partial t^2} \qquad \ldots(10)$$

Similarly it can be shown that

$$\nabla^2 \vec{H} = \mu\,\varepsilon\,\frac{\partial^2 \vec{H}}{\partial t^2}$$

$$\nabla^2 \vec{B} = \mu\,\varepsilon\,\frac{\partial^2 \vec{B}}{\partial t^2} \qquad \ldots(11)$$

Equations (10) and (11) have the same form as the general equation of wave motion. These equations also show that $\vec{E}$ and $\vec{B}$ are propagated with the same speed $\frac{1}{\sqrt{\mu\varepsilon}}$. These equations are linear and hence the sum of any number of solutions is also a solution. Therefore, electromagnetic waves also interfere according to the principle of superposition.

For the electromagnetic wave travelling through free space,

$$\varepsilon = \varepsilon_0 \text{ and } \mu = \mu_0$$

The velocity of the electromagnetic wave

$$\therefore \quad c = \frac{1}{\sqrt{\mu_0 \varepsilon_0}}$$

Here $\varepsilon_0 = 8{\cdot}85 \times 10^{-12}$ coulomb^2/N^2 - m^2

and $\mu_0 = 4\pi \times 10^{-7}$ henry/m

$$\therefore \quad c = \frac{1}{\sqrt{4\pi \times 10^{-7} \times 8{\cdot}85 \times 10^{-12}}}$$

$$c = 3 \times 10^8 \text{ m/s}$$

Hence Maxwell's equations for the fields in free space in SI system are :

$$\text{div } \vec{E} = 0 \quad \ldots (12)$$

$$\text{div } \vec{B} = 0 \quad \ldots (13)$$

$$\text{curl } \vec{E} = -\frac{\partial \vec{B}}{\partial t} \quad \ldots (14)$$

$$\text{curl } \vec{B} = \mu_0 \varepsilon_0 \frac{\partial \vec{E}}{\partial t} + \mu_0 \vec{J} \quad \ldots (15)$$

When $J = 0$

$$\text{curl } B = \mu_0 \varepsilon_0 \frac{\partial \vec{E}}{\partial t} \quad \ldots (16)$$

Thus Maxwell's theory of electromagnetic waves enables us to find the speed of electromagnetic waves purely from theoretical consideration. Light is also a form of electromagnetic wave. The experimental value of speed of light in free space agrees with the value derived on the basis of electromagnetic theory.

Example 12.2. *A certain radio station broadcasts programmes in a wavelength of* 200 m. *What is the frequency of this radiowave? Is it the frequency or wavelength that will change when the wave passes from air to water of refractive index 4/3. Find the frequency, wavelength and the velocity of the wave in water.* (IAS, 1977)

Here, $\lambda = 200$ m

$c = 3 \times 10^8$ m

(*i*) Frequency $\nu = \dfrac{c}{\lambda}$

$$= \frac{3 \times 10^8}{200}$$

$$= \mathbf{1{\cdot}5 \times 10^6 \text{ Hz}}$$

(*ii*) When the wave passes through water, frequency does not change but wavelength changes.

$$\mu = \frac{C_a}{C_\omega}$$

$$C_\omega = \frac{C_a}{\mu} = \frac{3 \times 10^8}{4/3} = \mathbf{2{\cdot}25 \times 10^8\ m/s}$$

$$v_\omega = \mathbf{1{\cdot}5 \times 10^6\ Hz}$$

$$\lambda_\omega = \frac{C_\omega}{v_\omega} = \frac{2{\cdot}25 \times 10^8}{1{\cdot}5 \times 10^6} = \mathbf{150\ m.}$$

12.10. Poynting Vectro

One main feature of an electromagnetic wave is that it can transport energy from point to point. The rate of energy flow per unit area in a plane electromagnetic wave is defined by a vector $\vec{S}$ called the **Poynting vector.**

$$\vec{S} = \frac{1}{\mu_0} \vec{E} \times \vec{B} \quad \ldots (1)$$

or

$$\vec{S} = \vec{E} \times \vec{H} \quad \ldots (2)$$

The unit of S is watts/m ^{2}, the direction of S gives the direction in which the energy is transferred. $\vec{E}$ and $\vec{B}$ refer to the instantaneous values at the point under consideration.

12.11. Thomson Model of the Atom

The experiments by J.J. Thomson showed the existence of electrons. Radioactivity proved the fact that atom consists of positively and negatively charged particles. As the atom is electrically neutral, the positive charge must be equal to the negative charge. The question arose as to *how the electrons are arranged in the atom.* Thomson suggested that the atom was spherical in shape. The whole mass of the atom was evenly distributed and the positive charge was distributed all over the mass. The electrons with negative charge were embedded within the atom (Fig. 12.8).

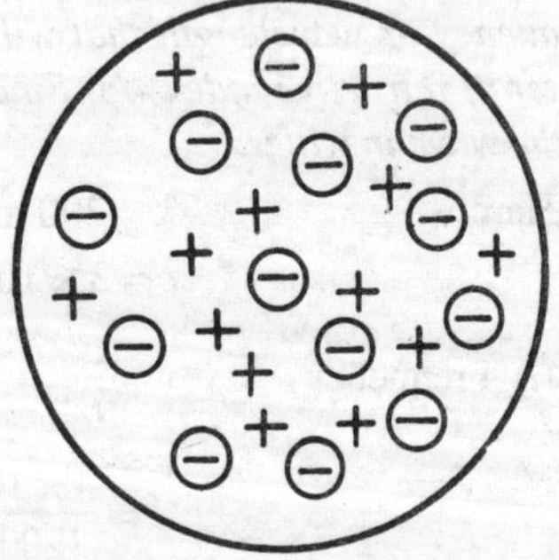

Fig. 12.8. Thomson Model

The whole positive charge of the atom was equal to the total charge of the electrons. The atom was like a plum *pudding*. The electrons oscillate or vibrate about their mean positions. The Thomson model could be successfully applied in simple gas laws and in simple physical and chemical problems.

The characteristics of Thomson model were :

(1) It explained the emission of electrons by heating as thermions and in photoelectric effect.

(2) It provided a mechanism for the emission of the electromagnetic waves. The electromagnetic waves were supposed to be emitted by oscillating bound electrons in the atom.

(3) It provided an elastic hard model required for the kinetic theory of gases.

(4) The dimensions of the atom were assumed as given by the mean free path and Van der Waal's considerations.

(5) It explained the existence of positively charged and negatively charged ions. A gain in electron by the neutral atom produces a negative ion and the loss of an electron by the neutral atom produces a positive ion.

Failure. The scattering of α-particles by the heavy atoms studied by Rutherford led him to the idea that the positive charge could not be distributed evenly in the whole mass of the atom.

12.12. Rutherford's Experiment–Scattering of α-Particles and Rutherford Model of the Atom

The scattering of α-particles was investigated by Rutherford and his co-workers. The experimental arrangement is shown in Fig. 12.9.

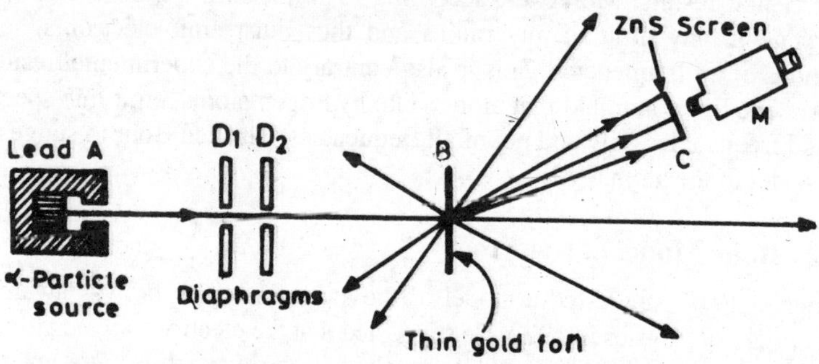

Fig. 12.9.

α-particles are emitted from the radioactive source A. After passing through the diaphragms D_1, D_2, a narrow beam of α-particles is incident on a thin gold foil B. While passing through the gold foil, the α-particles are scattered through different angles. The scattered α-particles in a particular direction are allowed to strike a screen C coated with zinc sulphide. When an α-particle is incident on zinc sulphide, it produces fluorescence and it is detected with the help of the microscope M. In some cases the α-particles are scattered through large angles and some are even reversed.

Rutherford came to the conclusion that positively charged α-particles must be encountering extremely dense concentrations of the positive charge at the centre of the atom. If a number of bullets are fired through a sand-bag containing a stone, some of the bullets striking the stone will be scattered. Thus, Rutherford considered the atom to be consisting of a nucleus containing a positive charge at its centre (Fig. 12.10). The electrons were considered to be distributed around the nucleus in the empty space of the atom. Thus the discovery of the nucleus of the atom is due to Rutherford. The nucleus is of great importance to explain the modern experimental and theoretical facts.

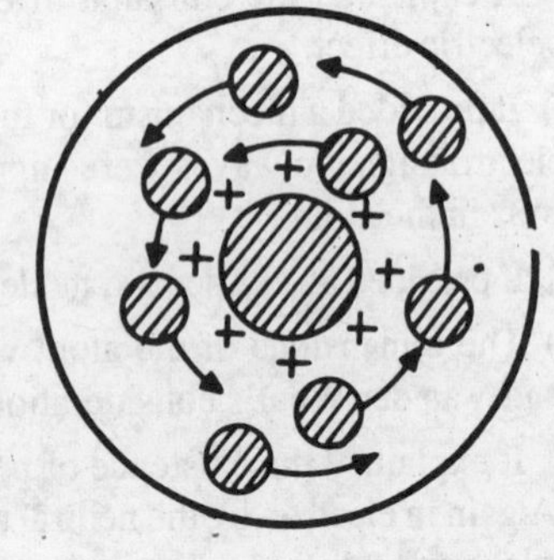

Fig. 12.10. Rutheford Model of the Atom.

Drawback. The planetary electron is constantly attracted towards the nucleus. The accelerated electron must emit the electromagnetic radiations. If the electron emits electromagnetic radiations as it rotates, it will finally be drawn nearer the nucleus. As the electron spirals down to the nucleus, it must emit electromagnetic waves of all frequencies. This fact is contradictory to the experimental results. Moreover, according to Rutherford's model, electrons can rotate in any orbit of any radius and thus must emit electromagnetic radiations of all frequencies. This is also contrary to the experimental results. Experimentally, it is found that atoms, like hydrogen atoms, emit line spectra of fixed frequencies only and not of all frequencies. This led Bohr to suggest a new model of the atom.

12.13. Bohr Model of the Atom

Bohr in 1913, suggested a model of the atom for which he was awarded Nobel Prize for Physics in 1922. He suggested that the electrons are negatively charged particles and move round the nucleus in various orbits. *The orbit of an electron is fixed.* He gave the name *energy levels* to these orbits. The electron cannot emit any energy when it moves in its own fixed orbit known as

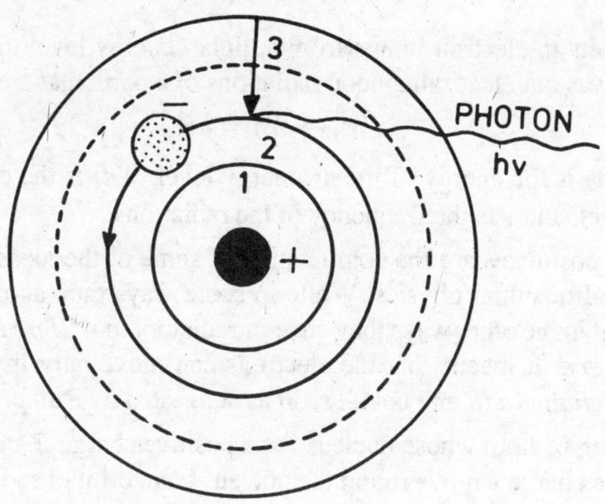

Fig. 12.11. Bohr's Model of the Atom.

stationary energy level. In Fig. 12.11 the dotted line orbit is ruled out in Bohr's model and only the thick line orbits are present. It means that the orbits of definite radii are present but orbits of all radii are ruled out. The electron gives out energy in the form of electromagnetic radiations or definite frequency when it jumps from higher energy level to lower level Fig. 12.11. The Bohr model opened the way to new concepts in the twentieth century.

12.14. Bohr's Theory of the Hydrogen Spectrum

Rutherford's experiment on the scattering of α-particles led Bohr to the conclusion that the atom consists of a positively charged nucleus at its centre. Moreover, Bohr applied the quantum theory of radiation as developed by Planck and Einstein to the Rutherford's model. His theory is mainly based on the following postulates.

Postulates. (1) The atom consists of a positively charged nucleus at its centre.

(2) The negatively charged particles known as electrons move round the nucleus in various orbits known as stationary energy levels. The electrons cannot emit radiation when moving in their own stationary levels.

(3) The Coulombian and Newtonian forces are applicable in the domain of the atom.

(4) The electrons revolve round the nucleus in various circular orbits and the angular momentum $mvr = \frac{nh}{2\pi}$

where n (= 1, 2, 3, 4, etc.,) is called the quantum number, and h is the Planck's constant.

(5) When an electron jumps from a higher energy level to a lower energy level, it gives out electromagnetic radiations of a particular frequency

$$Wn_2 - Wn_1 = h\nu$$

where Wn_2 is the energy of the n_2 energy level. Wn_1 is the energy of the n_1 energy level, and ν is the frequency of the radiations.

Bohr's postulates are the combination of some of the ideas of the classical physics and quantum physics. While in some ways they agree with classical physics but in the other ways they are contradictory to it. *Here the energy levels are quantized.* It means that the electrons can move only in some particular orbits of *definite radii* and not all radii as suggested by Rutherford's model.

Consider an atom whose nucleus has a positive charge E and let an electron of negative charge e move round the nucleus in an orbit of radius r.

$$E = Ze, \quad \text{where} \quad Z = 1 \quad \text{for hydrogen.}$$

The force of attraction between the nucleus and the electron in CGS esu,

$$= \frac{E\,e}{r^2} = \frac{Z\,e^2}{r^2}$$

As the electrons move along the circular orbit, the centripetal force $\left(= \frac{mv^2}{r} \right)$ is provided by the attractive force on the electron

$$\therefore \quad \frac{m\,v^2}{r} = \frac{Z\,e^2}{r^2} \quad \ldots (1)$$

$$\therefore \quad mv^2 = \frac{Z\,e^2}{r} \quad \ldots (2)$$

The angular momentum,

$$mvr = \frac{nh}{2\pi} \quad \ldots (3)$$

$$\therefore \quad v = \frac{nh}{2\pi mr} \quad \ldots (4)$$

Substituting the value of v in equation (2)

$$m\left(\frac{nh}{2\pi mr}\right)^2 = \frac{Z\,e^2}{r}$$

$$\therefore \quad r = \frac{n^2h^2}{4\pi^2 mZe^2} \quad \ldots (5)$$

Substituting this value of r in equation (4)

$$v = \frac{2\pi Ze^2}{nh} \qquad \ldots(6)$$

From equation (5), we find that $r \propto n^2$. The radii of the orbits are in the ratio of 1 : 4 : 9 : 16 : 25 : etc. The radius of the second orbit is four times the radius of the first orbit. For hydrogen atom, the radius of the first orbit

$$r = \frac{n^2h^2}{4\pi^2 Ze^2}. \text{ Here, } n = 1 \text{ and } Z = 1$$

$$r = 5{\cdot}29 \times 10^{-9} \text{ cm}$$

Equation (6) shows that the velocity of the electron decreases with the increase in the order of the orbit. The velocity of the electron in the fist orbit,

$$v = \frac{2\pi\, Ze^2}{nh}.$$

Substituting the values of n, h, Z and e

$$v = 2{\cdot}2 \times 10^8 \text{ cm/s}$$

Let us consider the energy of an electron in the orbit. The energy is partly potential and partly kinetic. From definition, the potential energy of an electron

$$= \int_\infty^r \frac{Ze^2}{x^2}\, dx = \frac{-Ze^2}{r}$$

$\therefore$ Potential energy of the electron $= \dfrac{-Ze^2}{r}$

Kinetic energy $= \frac{1}{2} mv^2$

From equation (2),

$$mv^2 = \frac{Ze^2}{r}$$

Kinetic energy $= \dfrac{Ze^2}{2r}$

The potential energy is negative because energy must be given to the electron to bring it far away from the nucleus to zero energy level.

Total energy = potential energy + kinetic energy

$$= -\frac{Ze^2}{r} + \frac{Ze^2}{2r} = -\frac{Ze^2}{2r}$$

Substituting the value of r, we get

$$W_n = -\frac{2\pi^2 me^4 Z^2}{n^2h^2} \text{ in CGS esu} \qquad \ldots(7)$$

where W_n represents the energy of the electron in the nth energy level. Here n may have values, 1, 2, 3, 4, etc.

Note. In *rationalised MKS units* or *SI units* the energy of an electron in the nth orbit

$$W_n = \frac{-2\pi^2 me^4 Z^2}{(4\pi\varepsilon_0)^2 n^2 h^2}$$

or
$$W_n = \frac{-me^4 Z^2}{8\varepsilon_0^2 n^2 h^2} \quad \text{joules} \qquad \ldots (8)$$

Here
$$m = 9{\cdot}1 \times 10^{-31} \text{ kg}, \quad e = 1{\cdot}6 \times 10^{-19} \text{ coulomb}$$

$$\varepsilon_0 = 8{\cdot}85 \times 10^{-12} \frac{\text{coulomb}^2}{\text{newton–m}^2}$$

$$h = 6{\cdot}624 \times 10^{-34} \quad \text{joule second.}$$

For an electron to be removed from its first orbit, in hydrogen atom, energy must be supplied. Therefore, ionization potential for hydrogen,

$$\phi = \frac{me^4 Z^2}{8\varepsilon_0^2 n^2 h^2}$$

For hydrogen, $Z = 1, \ n = 1$

$$\therefore \quad \phi = \frac{me^4}{8\varepsilon_0^2 h^2} \text{ joules} \qquad \ldots (9)$$

$$\phi = \frac{me^4}{8\varepsilon_0^2 h^2 \times 1{\cdot}6 \times 10^{-19}} \quad \text{electorn volts} \qquad \ldots (10)$$

When an electron jumps from energy levels n_2 to energy level n_1.

But
$$W_{n2} - W_{n1} = \frac{-2\pi^3 me^4 Z^2}{n_2^2 h^2} + \frac{2\pi^2 me^4 Z^2}{n_1^2 h^2}$$

$$W_{n2} - W_{n1} = \frac{2\pi^2 me^4 Z^2}{h^2}\left(\frac{1}{n_1^2} - \frac{1}{n_2^2}\right)$$

But; $W_{n2} - W_{n1} = h\nu$

$$\therefore \quad h\nu = \frac{2\pi^2 me^4 Z^2}{h^2}\left(\frac{1}{n_1^2} - \frac{1}{n_2^2}\right)$$

and
$$\nu = \frac{2\pi^2 me^4 Z^2}{h^3}\left(\frac{1}{n_1^2} - \frac{1}{n_2^2}\right)$$

In the case of hydrogen atom $Z = 1$

$$\therefore \quad \nu = \frac{2\pi^2 me^4}{h^3}\left(\frac{1}{n_1^2} - \frac{1}{n_2^2}\right)$$

But $$c = \nu \lambda, \quad \therefore \quad \nu = \frac{c}{\lambda}$$

$$\frac{c}{\lambda} = \frac{2\pi^2 me^4}{h^3}\left(\frac{1}{n_1^2} - \frac{1}{n_2^2}\right)$$

$$\frac{1}{\lambda} = \frac{2\pi^2 me^4}{ch^3}\left(\frac{1}{n_1^2} - \frac{1}{n_2^2}\right)$$

$\frac{1}{\lambda} = \nu$, known as the wave number

$$\nu = \frac{2\pi^2 me^1}{ch^3}\left(\frac{1}{n_1^2} - \frac{1}{n_2^2}\right)$$

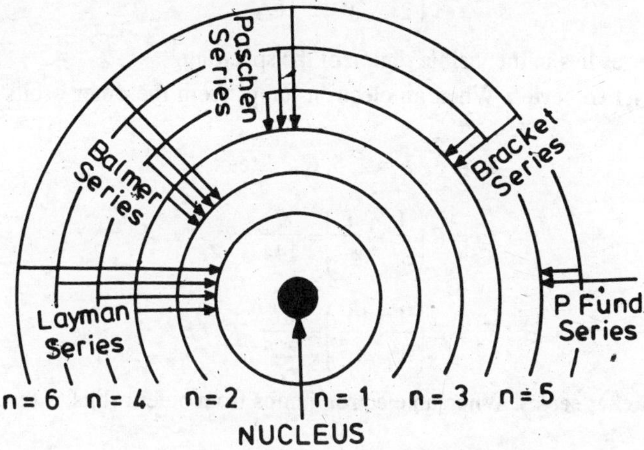

Fig. 12.12. Spectral Series of Hydrogen.

$\frac{2\pi^2 me^4}{ch^3} = R$ and is known as Rydberg constant.

Substituting the values, $R = 109678 \text{ cm}^{-1}$

$$\therefore \quad \overline{\nu}_1 = R\left(\frac{1}{n_1^2} - \frac{1}{n_2^2}\right) \qquad \ldots (11)$$

Spectral series of hydrogen atom. (1) Lyman series. When an electron jumps from the outer to the first orbit, the spectral lines are in the ultra-violet region.

Here, $$n_1 = 1, n_2 = 2, 3, 4, 5 \ldots$$

$$\nu_1 = R\left(\frac{1}{1} - \frac{1}{2^2}\right) = \frac{3}{4} R$$

$$\bar{\nu}_2 = R\left(\frac{1}{1} - \frac{1}{3^2}\right) = \frac{8}{9}\ R$$

$$\bar{\nu}_3 = R\left(\frac{1}{1} - \frac{1}{4^2}\right) = \frac{15}{16} R$$

(2) Balmer series. When an electron jumps from outer orbits to the second orbit.

$$n_1 = 2, n_2 = 3, 4, 5 \ldots, \text{etc.,}$$

$$\therefore \qquad \bar{\nu}_1 = R\left(\frac{1}{2^2} - \frac{1}{3^2}\right) = \frac{5}{36}\ R$$

$$\bar{\nu}_2 = R\left(\frac{1}{2^2} - \frac{1}{4^2}\right) = \frac{3}{16}\ R$$

This series lies in the visible region of the spectrum.

(3) Paschen series. When an electron jumps from the outer orbits to the third orbit.

$$n_1 = 3, n_2 = 4, 5, 6 \ldots, \text{etc.,}$$

$$\therefore \qquad \bar{\nu}_1 = R\left(\frac{1}{3^2} - \frac{1}{4^2}\right) = \frac{7}{144}\ R$$

$$\bar{\nu}_2 = R\left(\frac{1}{3^2} - \frac{1}{5^2}\right) = \frac{16}{225}\ R$$

(4) Bracket series. When an electron jumbs from outer orbits to the fourth orbit

$$n_1 = 4, n_2 = 5, 6, 7 \ldots, \text{etc.,}$$

$$\therefore \qquad \bar{\nu}_1 = R\left(\frac{1}{4^2} - \frac{1}{5^2}\right) = \frac{9}{400}\ R$$

$$\bar{\nu}_2 = R\left(\frac{1}{4^2} - \frac{1}{6^2}\right) = \frac{5}{144}\ R$$

(5) P-fund series. When an electron jumps from outer orbits to the fifth orbit.

$$n_1 = 5, n_2 = 6, 7, 8 \ldots, \text{etc.,}$$

$$\therefore \qquad \bar{\nu}_1 = R\left(\frac{1}{5^2} - \frac{1}{6^2}\right) = \frac{11}{900}\ R$$

$$\bar{v}_2 = R\left(\frac{1}{5^2} - \frac{1}{7^2}\right) = \frac{24}{1225}\ R$$

The last three series are in the infra-red region. When the Bohr's theory was suggested, only the Balmer and Paschen series for hydrogen atom were known. The other series as suggested theoretically according to equation (11) were verified experimentally also. The Lyman series (1916), the Bracket series (1922) and P-fund series (1929) studied experimentally were found to show the same wave numbers as suggested theoretically. When an electron absorbs energy, it moves from lower energy level to higher level as in the absorption spectrum. When an electron jumps from higher energy level to the lower energy level, it emits radiations as in the emission spectrum.

Example 12.3. *The Rydberg constant for hydrogen is* 109678 cm^{-1} *and for ionized helium* 109722 cm^{-1}. *Calculate the ratio of the mass of the proton to that of the electron, assuming the mass of the helium nucleus to be four times the mass of proton.*

The nucleus of the atom has motion though it is too heavy as compared to the electron. Due to this reason the reduced mass of the electron is

$$= m\left[\frac{M}{M+m}\right] = \frac{m}{\left[1 + \frac{m}{M_H}\right]}$$

Here m is the mass of the electron and M is the mass of the nucleus. For hydrogen, Rydberg constant,

$$R_H = \frac{Rm}{\left[1 + \frac{m}{M_H}\right]} \qquad \ldots (1)$$

For ionized helium, Rydberg constant

$$R_{He} = \frac{Rm}{\left[1 + \frac{m}{M_{He}}\right]} \qquad \ldots (2)$$

Here $$R = \frac{2\pi^2 me^4}{ch^3}\ \text{cm}^{-1}$$

Dividing (2) by (1)

$$\frac{R_{He}}{R_H} = \frac{\left[1 + \frac{m}{M_H}\right]}{\left[1 + \frac{m}{M_{He}}\right]}$$

Taking $M_{H_e} = 4M_H$ and simplifying,

$$\frac{M_H}{m} = \frac{R_H - \frac{1}{4} R_{He}}{R_{He} - R_H}$$

Here $R_H = 109678 \text{ cm}^{-1}$ and $R_{He} = 109722 \text{ cm}^{-1}$

$$\frac{\mathbf{M_H}}{\mathbf{m}} = \frac{109678 - \frac{1}{4} \times 109722}{109722 - 109678} = \mathbf{1869.}$$

Example 12.4. *The first member of Balmer series of hydrogen has a wave length of* 6563 Å. *Calculate the wave length of its second member.*

For Balmer series of the hydrogen atom

$$\frac{1}{\lambda_1} = \frac{5R}{36}$$

$$\frac{1}{\lambda_2} = \frac{3R}{16}$$

$$\therefore \quad \frac{\lambda_2}{\lambda_1} = \frac{20}{27} \quad \text{or} \quad \lambda_2 = \frac{20 \times \lambda_1}{27}$$

$$\lambda_2 = \frac{20 \times 6563}{27} = \mathbf{4861 \text{ Å}.}$$

Example 12.5. *Calculate the ionization potential, in electron volts, for hydrogen atom, given that the charge and mass of the electron are* $4{\cdot}8 \times 10^{-10}$ esu *and* 9×10^{-28} gram *respectively*

$$h = 6{\cdot}6 \times 10^{-27} \text{ erg second.}$$ (Delhi, 1968)

The energy required to remove an electron from the first orbit to infinity is called the ionization potential

$$\therefore \quad \phi = \frac{2\pi\, me^4\, Z^2}{n^2 h^2} \text{ ergs}$$

$$\phi = \frac{2\pi\, me^4\, Z^2}{n^2 h^2 \times 1{\cdot}6 \times 10^{-12}} \text{ electron volts}$$

Here $m = 9 \times 10^{-28}$ g, $\quad e = 4{\cdot}8 \times 10^{-10}$ esu

$h = 6{\cdot}6 \times 10^{-27}$ ergs, $\quad n = 1, \quad Z = 1$

$$\therefore \quad \phi = \frac{2 \times 22 \times 9 \times 10^{-28} \times (4{\cdot}8 \times 10^{-10})^4 \times 1}{7 \times 1 \times (6{\cdot}6 \times 10^{-27})^2 \times 1{\cdot}6 \times 10^{-12}}$$

$$\boldsymbol{\phi = 13{\cdot}51 \text{ eV}.}$$

Example 12.6. *Calculate the ionization potential, in electron volts, for hydrogen atom, given that*

$$e = 1{\cdot}6 \times 10^{-19} \text{ coulomb}, \quad m = 9 \times 10^{-31} \text{ kg}$$

$$h = 6{\cdot}6 \times 10^{-34} \text{ joule second}$$

$$\varepsilon_0 = 8{\cdot}85 \times 10^{-12} \frac{\text{coulomb}^2}{\text{newton-m}^2}.$$

In *rationalised MKS units* or *SI units,*

$$\phi = \frac{me^4}{8\varepsilon_0^2 h^2} \text{ joules}$$

$$\phi = \frac{m\,e^4}{8\,\varepsilon_0^2\,h^2 \times 1{\cdot}6 \times 10^{-19}} \text{ electron volts}$$

Here $m = 9 \times 10^{-31}$ kg, $e = 1{\cdot}6 \times 10^{-19}$ coulomb

$$\varepsilon_0 = 8{\cdot}85 \times 10^{-12} \frac{\text{coulomb}^2}{\text{newton - m}^2}$$

$$h = 6{\cdot}6 \times 10^{-34} \text{ joule - second}$$

$$\therefore \quad \phi = \frac{9 \times 10^{-31} \times (1{\cdot}6 \times 10^{-19})^4}{8 \times (8{\cdot}85 \times 10^{-12})^2 \times (6{\cdot}6 \times 10^{-34})^2 \times 1{\cdot}6 \times 10^{-19}}$$

$$\phi = \mathbf{13{\cdot}51 \text{ eV}}.$$

Example 12.7. *The ionisation potential for hydrogen atom is* 13·51 eV. *Calculate the Planck's constant. Given that*

$$e = 1{\cdot}6 \times 10^{-19} \text{ coulomb}, \quad m = 9 \times 10^{-31} \text{ kg}$$

$$\varepsilon_0 = 8{\cdot}85 \times 10^{-12} \frac{\text{coulomb}^2}{\text{newton–m}^2}$$

In *rationaliseed MKS units* or *SI units,*

$$\phi = \frac{m\,e^4}{8\,\varepsilon_0^2\,h^2} \text{ joules}$$

$$\phi = 13{\cdot}51 \text{ eV} = 13{\cdot}51 \times 1{\cdot}6 \times 10^{-19} \text{ joule}$$

$$m = 9 \times 10^{-31} \text{ kg}, \quad e = 1{\cdot}6 \times 10^{-19} \text{ coulomb}$$

$$\varepsilon_0 = 8{\cdot}85 \times 10^{-12} \frac{\text{coulomb}^2}{\text{newton - m}^2}$$

$$h = \left[\frac{m\,e^4}{8\varepsilon_0^2 \phi}\right]^{\frac{1}{2}}$$

$$= \left[\frac{9 \times 10^{-31} \times (1{\cdot}6 \times 10^{-19})^4}{8 \times (8{\cdot}85 \times 10^{-12})^2 \times 13{\cdot}51 \times 1{\cdot}6 \times 10^{-19}}\right]^{\frac{1}{2}}$$

$$= \mathbf{6{\cdot}6 \times 10^{-34} \text{ joule - second.}}$$

12.15. Zeeman Effect

A spectral line emitted by the excited atoms is split up into a doublet or a triplet when the emitting atoms are placed in a magnetic field. This effect of the splitting of a spectral line under the action of a magnetic field is known as *normal Zeeman effect.*

To produce Zeeman effect, the source of light such as a sodium lamp or a mercury arc or gas discharge in a Geissler tube, is placed between the poles of a powerful electromagnet [Fig. 12.13 (*a*)]. The light coming from the source is examined by means of a spectroscope of high resolving power. In order to view the light parallel to the magnetic field, a hole is drilled in one of the pole-pieces along the axis of the magnet.

When no magnetic field is applied, the spectroscope is focussed on one of the lines in the spectrum of the source of light. When a magnetic field is applied, it is observed that :

(1) When the light is viewed in a direction perpendicular to the magnetic field, three component lines are observed. One of the lines is in the same position as the original line and the other two lines are on the two sides of the original line. The outer two lines, when observed by means of a nicol prism as an analyser, are polarized at right angles to the undisplaced line. This effect is known as *transverse Zeeman effect.*

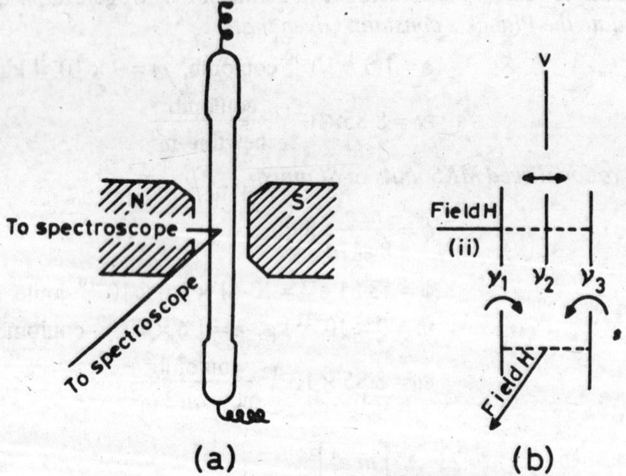

Fig. 12.13. Normal Zeeman effect (i) No field; (ii) Perpendicular to the field, and (iii) Parallel to the field.

(2) When the light is viewed in a direction parallel to the direction of the field, there is no line in the position of the original line, only two outer lines are

present. These lines are found to be circulary polarised in opposite directions. The effect is known as longitudinal *Zeeman effect.*

The normal Zeeman effect is explained on Lorentz electron theory. Consider an electron moving in a circular orbit of radius r with a velocity v (Fig. 12.14). The centripetal force,

$$F = \frac{mv^2}{r}.$$

If an external magnetic field is applied, an additional force acts which is directed perpendicular to the direction of motion of the electron. This force is also perpendicular to the direction of the magnetic field and is along the radius. When this force acts inwards along the radius, the velocity of the electron increases. When this force acts outwards along the radius, the velocity of the electron decreases. Suppose, this force due to the magnetic field=F_1 and let the velocity of the electron be increased to v_1 by the application of the magnetic field. Then F_1=Hev_1. Suppose, this force is directed towards the centre, the total force along the radius

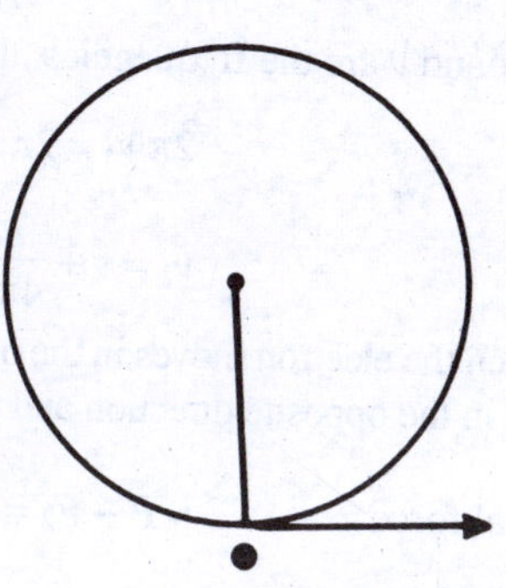

Fig. 12.14.

$$= F + F_1 = \frac{mv^2}{r} + Hev_1$$

Total force $= \dfrac{mv_1^2}{r}$

$$\therefore \quad \frac{mv_1^2}{r} = \frac{mv^2}{r} + Hev_1 \quad \ldots (1)$$

But, $v = r\omega$ and

$v_1 = r\omega_1$

where ω and ω_1 are the respective angular velocities.

From equation (1)

$$\frac{mr^2\,\omega_1^2}{r} = \frac{mr^2\,\omega^2}{r} + Her\,\omega_1 \quad \ldots (2)$$

$$nr\,\omega_1^2 - mr\,\omega^2 = Her\,\omega_1$$

$$\omega_1^2 - \omega^2 = \frac{eH\,\omega_1}{m}$$

$$(\omega_1 + \omega)\,(\omega_1 - \omega) = \frac{eH\,\omega_1}{m}$$

$$\omega_1 - \omega = \frac{eH\,\omega_1}{m\,(\omega_1 + \omega)}$$

$(\omega_1 + \omega)$ is approximately equal to $2\omega_1$

$$\therefore \qquad \omega_1 - \omega = \frac{eH\,\omega_1}{2m\,\omega_1} = \frac{eH}{2m}$$

$$\therefore \qquad \omega_1 = \omega + \frac{eH}{2m}$$

If ν_1 and ν are the frequencies, then, $\omega_1 = 2\pi\,\nu_1,\ \omega = 2\pi\nu$

$$\therefore \qquad 2\pi\,\nu_1 = 2\pi\nu + \frac{eH}{2m}$$

$$\nu_1 = \nu + \frac{eH}{4\pi\,m} \qquad \ldots (3)$$

When the electron moves in the opposite direction, the magnetic field produces a force in the opposite direction and the velocity decreases to v_2. In that case

$$\text{Total force} = F - F_2 = \frac{mv^2}{r} - Hev_2$$

$$\therefore \qquad \frac{mv_2^2}{r} = \frac{mv^2}{r} - Hev_2$$

But, $\quad v_2 = r\omega_2 \quad$ and $\quad v = r\omega$

$$\therefore \qquad \frac{mr^2\omega_2^2}{r} = \frac{mr_0^2\,\omega^2}{r} - Her\,\omega_2$$

$$mr\,\omega_2^2 - mr\,\omega^2 = -Her\,\omega_2$$

$$\omega_2^2 - \omega^2 = -\frac{eH\,\omega_2}{m}$$

$$(\omega_2 - \omega) = \frac{eH\,\omega_2}{m\,(\omega_2 + \omega)}$$

$$\omega_2 - \omega = -\frac{eH}{2m} \qquad (\because \omega_2 + \omega = 2\omega_2 \text{ approx.})$$

$$2\pi\nu_2 - 2\pi\nu = -\frac{eH}{2m}$$

$$\therefore \qquad \nu_2 = \nu - \frac{eH}{4\pi m} \qquad \ldots (4)$$

From (3) and (4) we get

$$\Delta\nu \text{ (in each case)} = \frac{eH}{4\pi m} \qquad \ldots (5)$$

This shows that the two lines are displaced equally on the two sides of the original line.

$$\nu_1 - \nu_2 = \frac{eH}{2\pi m} \quad \ldots (6)$$

In general $$\Delta\nu = \pm\frac{eH}{4\pi m} \quad \ldots (7)$$

Alternative treatment. Consider an electron moving in a circular orbit of radius r with a velocity v.

The centripetal force $F = \frac{mv^2}{r} = mr\omega^2$. If an external field H is applied, the change in the force,

$$dF = \pm Hev = \pm Her\omega \quad \ldots (1)$$

$$F = mr\omega^2$$

Differentiating, $$dF = 2mr\,\omega\,d\omega \quad \ldots (2)$$

Equating (1) and (2)

$$\pm Her\omega = 2mr\omega\,d\omega$$

$$\frac{e}{m} = \pm\frac{2d\omega}{H} \quad \ldots (3)$$

Also $$\omega = 2\pi\nu = \frac{2\pi c}{\lambda}$$

Differentiating $$d\omega = \frac{-2\pi c}{\lambda^2}\,d\lambda.$$

Substituting the value of $d\omega$ in equation (3)

$$\frac{e}{m} = \pm\frac{4\pi\,c\,d\lambda}{\lambda^2 H} \quad \ldots (4)$$

$$d\lambda = \pm\frac{eH\lambda^2}{4\pi\,mc} \quad \ldots (5)$$

$\therefore$ $$\lambda_1 = \lambda + \frac{eH\lambda^2}{4\pi\,mc}$$

$$\lambda_2 = \lambda - \frac{eH\lambda^2}{4\pi\,mc}$$

$$\lambda_1 - \lambda_2 = \frac{eH\lambda^2}{2\pi\,mc} \quad \ldots (6)$$

$$\nu = \frac{c}{\lambda}$$

Differentiating, $$d\nu = -\frac{c}{\lambda^2}\,d\lambda$$

Substituting the value of $d\lambda$ from equation (5)

$$d\nu = \pm\frac{eH}{4\pi m} \quad \ldots (7)$$

$$\nu_1 = \nu + \frac{e}{4\pi m}$$

$$\nu_2 = \nu - \frac{eH}{4\pi m}$$

$$\nu_1 - \nu_2 = \frac{eH}{4\pi m} \qquad \ldots (8)$$

The quantity $\frac{eH}{4\pi m}$ is known as *normal Zeeman separation.* Knowing the values $\Delta\nu$ of and H, e/m can be calculated. The value of e/m calculated from the measurements of Zeeman effect $= 1{\cdot}757 \times 10^7$ emu/g and it is in agreement with the value of e/m of the electron obtained from Thomson's experiment.

This experiment established that the electron in the atom is responsible for the emission of spectral lines.

Example 12.7. *Calculate the wavelength separation between the two component lines which are observed in the normal Zeeman effect. The magnetic field used* = 4000 gauss, $e/m = 1{\cdot}76 \times 10^7$ emu/g *and* $\lambda = 6000$ Å.

Here $\lambda = 600\ \text{Å} = 6000 \times 10^{-8}$ cm

$$\frac{e}{m} = 1{\cdot}76 \times 10^7 \text{ emu/g, and } H = 4000 \text{ gauss}$$

$$\therefore \text{Wavelength separation } \lambda_1 - \lambda_2 = \frac{eH\lambda_2}{2\pi m\, c}$$

$$= \frac{e}{m} \times \frac{H\lambda^2}{2\pi c}$$

$$= \frac{1{\cdot}76 \times 10^7 \times 4000 \times (6000 \times 10^{-3})^2}{2 \times 3{\cdot}14 \times 3 \times 10^{10}}$$

$$= 1{\cdot}335 \times 10^{-10} \text{ cm} = \mathbf{0{\cdot}01335\ \text{Å}}.$$

12.16. Structure of the Nucleus

The scattering of α-particles by Rutherford showed that the atomic nucleus has a positive charge and its radius is 10^{-12} to 10^{-13} cm. Therefore, the volume of the nucleus $= \frac{4}{3}\pi r^3 = \frac{4}{3}\pi\, 10^{-36}$cc. As the mass of the hydrogen atom is about $1{\cdot}6 \times 10^{-24}$ gram, therefore, the density of the lightest nucleus

$$= \frac{\text{Mass}}{\text{Volume}} = \frac{1{\cdot}6 \times 10^{-24}}{\frac{4}{3}\pi \times 10^{-36}} \text{ g/cc.}$$

This shows that the density of the nucleus is of the order of 10^{12} g/cc. In actual practice it is impossible to get a substance of this density. All along, attempt has been made to find the contents of the nucleus. The main function of nuclear physics is to correlate the experimental facts with the theoretical considerations.

The *satellite of sirius* consists of a substance 60,000 times denser than water 1 cc. of this substance will weigh 60 kg on the earth. In ordinary conditions such enormous density is absolutely unthinkable because the space between atoms in solid bodies is too small to allow for any compression. In this case, it has been assumed that it is a *different matter* and the atoms have parted with their electrons circling around the nuclei. When the atoms lose electrons, the atomic diameter is reduced several thousand fold ; but the mass remains the same. Due to the terrific pressures in the stellar space, these diminished *atomic nuclei* come thousands of times closer to each other than nuclei of normal atoms. This results in a substance of density 60,000 g/cc on sirius' satellite. Moreover, the density of a substance at *Van Maanen* star has been found to be 400,000 g/cc. This shows that 1 cc of this substance would weigh the same as *seven men* on the surface of the earth.

12.17. Proton-Electron Hypothesis

The fact that radio-active elements emit α and β-particles led to the idea that the nucleus is built up of elementary constituents. Prout suggested in 1816 that the atomic weights of atoms are whole numbers and are integral multiples of the atomic weight of the hydrogen atom. This hypothesis was discarded when it was found that some elements have fractional atomic weights *e.g.*, chlorine 35·46 and copper 63·54.

The discovery of isotopes in the twentieth century supported the whole number rule of Prout's hypothesis. Later, Aston formulated whole number rule and suggested that the atomic weights of the elements are close to the whole number. The experiments on isotopes and analysis of the positive rays from various substances showed that the lightest possible positively charged particle found, has the same mass as the hydrogen atom and its charge was equal to the electronic charge in magnitude. This particle which was the nucleus of the hydrogen atom has a mass very close to one atomic mass unit. This particle was given the name *proton* and is the fundamental constituent of all atoms.

According to *proton electron hypothesis*, for the nucleus having an atomic weight nearer the integer A, it was necessary that the nucleus contained A protons. As the atomic number of the atom is Z, the positive charge in the nucleus can be only equal to Z and not equal to A. To overcome this difficulty it was assumed that the nucleus contained A protons and $(A - Z)$ electrons. Thus the net positive charge on the nucleus = Z, which neutralizes the negative charge on the extranuclear electrons round the nucleus and makes the atom electrically neutral.

A is known as the mass number and Z the atomic number. The emission of a β-particle was explained on this hypothesis because the nucleus contains electrons. Therefore, the ejection of an electron from the nucleus is a β-particle. The emission of an α-particle by the radio-active nucleus was explained by assuming that an α-particle could be formed by the combination of four protons and two electrons. But this hypothesis could not explain modern phenomena and has been *discarded.*

12.18. Failure of Proton-Electron Hypothesis (why an Electron cannot be present in the Nucleus?)

1. The angular momentum property of the nucleus has led to the failure of this hypothesis. When lines of the spectral series were examined with a spectroscope of very high resolving power, it was found that each of these components is split into a number of lines lying close together. This further splitting is known as hyperfine structure. The hyperfine structure could not be explained to be due to extranuclear electrons. Pauli in 1924 accounted the hyperfine structure to be due to the angular momentum or spin of the nucleus. The magnetic moment is associated with the angular momentum of the nucleus. The experimental determination of the angular momentum has shown that angular momentum I depends on the mass number A of the nucleus. (1) If A is even, I is an integer or zero and (2) if A is odd I has an odd half integral value. It means, that if A is even, I=0 1, 2, 3, ..., etc., and if A is odd $I = \frac{1}{2}, \frac{3}{2}, \frac{5}{2},$ etc. This has led to the failure of proton-electron hypothesis. In the case of nitrogen, A =14 and Z = 7. It means it should contain 14 protons and 7 electrons. The contribution of 14 protons and 7 electrons *i.e.*, total 21 particles is equal to odd half integral multiples. But the angular momentum of nitrogen has been found experimentally to be I = 1. Thus the experimental result is contrary to the hypothesis. The isotopes of mercury Z = 80 and A = 199 or 201 should have an angular momentum equal to zero or an integral value. But the value of angular momentum found experimentally is odd half integers.

(2) The measurements of the nuclear magnetic moment has led to the fact that nucleus cannot contain an electron. The magnetic moment of an electron

$\mu_B = \frac{eh}{4\pi\, mc}$, where m is the mass of the electron.

μ_B is known as *Bohr magneton* and its value = $0{\cdot}92 \times 10^{-20}$ ergs/gauss. The measured nuclear magnetic moments are of the order of 10^{-23} ergs/gauss. The nuclear magneton has to value

$$\mu_N = \frac{eh}{4\pi\, Mc} = 0{\cdot}505 \times 10^{-23} \text{ ergs/gauss}$$

where M is the mass of the proton. As the measured value of nuclear magnetic moment is far less than the Bohr magneton, an *electron cannot exist inside the nucleus.*

(3) The wave mechanics is also against the existence of free electrons in the nucleus. According to the uncertainly principle, $\Delta\ x\ \Delta p \sim h$. The free electron confined to the nucleus would have a kinetic energy of the order of 60 MeV, and a velocity 0·999 c. But the electrons of β -particles emitted by radio-active nuclei have never been found to have energies more than 4 MeV. The wave mechanics confirms the existence of protons in the nucleus.

Thus, there cannot be electrons in the nucleus. Rutherford in 1920 suggested that an electron and a proton might be so closely combined as to form a neutral particle. This *hypothetical particle* was given the name *neutron.*

12.19. Proton-Neutron Hypothesis

The discovery of neutron by Chadwick and the failure of *electron-proton hypothesis* led Heisenberg to suggest the *Proton-Neutron hypothesis* in 1932. According to this hypothesis, the total number of particles in the nucleus is equal to the mass number A. The number of protons in the nucleus is equal to the atomic number Z and the number of neutrons is equal to $(A–Z)$. The total charge of the nucleus is due to the protons only. The proton-neutron hypothesis is in agreement with wave mechanics regarding angular momentum and magnetic moments. This is also consistent with the phenomenon of radio-activity. The absence of electron in the nucleus show that the β -particle is created in the act of emission. The β-particle is emitted when a neutron changes into a proton, an electron and neutrino in the nucleus. This has been found to be true both experimentally and theoretically. The emission of an α -particle is due to the combination of two protons and two neutrons. Neutrons and protons are also known as *nucleons.*

12.20. Electron Volt

This is the unit of work and energy in nuclear physics. It is defined as the work done in taking an electron through a difference of potential of one volt. The work done is the energy given to the electron. One volt is defined as the potential difference between two points when one joule of work is done is taking a charge of l coulomb from one point to the other.

Since the charge of the electron = $1{\cdot}6 \times 10^{-19}$ coulomb,

1 electron volt =$1{\cdot}6 \times 10^{-19}$ joule of work or energy.

An electron accelerated through a potential difference of 10^4 volts will have 10^4 eV energy and 10^4 eV = $10^4 \times 1{\cdot}6 \times 10^{-19}$ joule.

12.21. De Broglie Model of the Atom

The dual behaviour of photons inspired De Broglie to suggest a model of the atom. The experiments on interference, diffraction and polarisation established the wave nature of light. The experiments on photo-electric effect established the corpuscular nature of light. Einstein's equation $E = mc^2$, wherre E is the energy, m is the mass and c is the velocity of light, gives the relation of conversion of mass into energy. De Broglie noted that the corpuscular properties were more obvious for very energetic light *i.e.*, light of shorter wavelength. Thus, the light particle known as a *photon* has corpuscular as well as wave nature.

Thinking on the same lines, De Broglie suggested that particles like electron have also a wave nature.

For a particle, Einstein's equation is

$$E = mc^2 = \frac{m_0c^2}{\sqrt{1-\left(\frac{v^2}{c^2}\right)}}$$

Here m_0 is the rest mass.

The energy of a wave of frequency ν according to Planck is

$$E = h\nu \qquad \ldots (2)$$

From (1) and (2) $h\nu = mc^2 = \dfrac{m_0c^2}{\sqrt{1-\left(\frac{v^2}{c^2}\right)}}$.

But $\nu\lambda = c$

$\therefore \quad \dfrac{hc}{\lambda} = mc^2$

or $\quad \lambda = \dfrac{h}{mc} \qquad \ldots (3)$

For an electron of velocity v

$$\lambda = \frac{h}{mv} = \frac{h\sqrt{1-\left(\frac{v^2}{c^2}\right)}}{m_0v} \qquad \ldots (4)$$

For electrons, the wavelength is longest, while for protons it is smaller than the electrons. If the values of h, m and v are substituted in equation (4), λ is found to be very small.

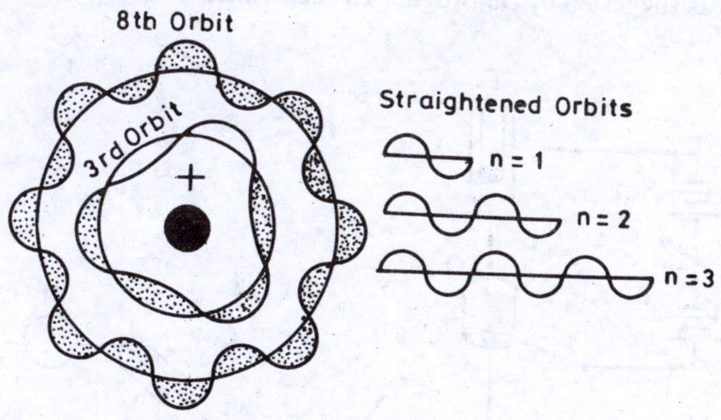

Fig. 12.15. De Broglie model of Atom.

Thus the De Broglie model of an atom has electrons in various orbits and the electrons behave as matter waves of wavelength $\lambda = \frac{h}{mv}$ · The electron exists as a standing wave in each orbit. The energy levels and 'orbits' of the Bohr model are retained. Moreover, this model explains the Bohr's postulate. The electron can be only in those orbits whose circumferences can contain the complete waves of the electron or the length of the orbit is a whole number multiple of the wavelength (Fig. 12.15).

The electron cannot be in an orbit whose length is not a whole number multiple of the wavelength. The necessary condition for the electron in a particular orbit can be calculated mathematically. The length of the orbit of radius $r = 2\pi r$ and $2\pi r = n\lambda$, where n is a whole number.

But, $$\lambda = \frac{h}{mv} \quad \text{[from equation (4)]}$$

$\therefore$ $$2\pi r = \frac{nh}{mv}$$

or $$mvr = \frac{nh}{2\pi} \quad \ldots (5)$$

Equation (5) is in agreement with the Bohr's quantisation hypothesis. Thus the De Broglie model seems to be more exact and the electrons are matter waves in various orbits round the nucleus. Davisson and Germer experiment on the diffraction of electrons demonstrated the fact that the material particles like electrons exhibit a wavelength given by the De Broglie equation, $\lambda = h/mv$.

12.22. Davisson-Germer Experiment

Davisson and Germer performed an experiment to study the wave mature of electrons as suggested by De Broglie. The experimental arrangement is shown

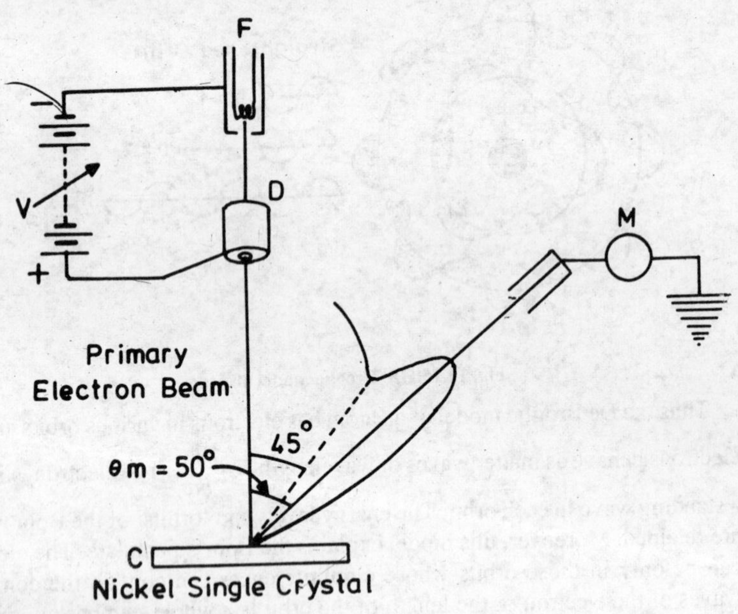

Fig. 12.16.

in Fig. 12.16. Electrons from a filament *F* are rendered into a fine beam by applying a positive potential to the cylinder *D*. A fine narrow beam of electrons is incident on the nickel crystal *C*. The electrons are reflected and the intensity of the reflected electrons is measured at various angles with the help of a current meter *M*.

Different positive potentials can be applied to the cylinder *D* and the velocity of the electrons can be varied. The curve represents a polar graph indicating the intensity of the reflected electrons at various angles. It is found that the intensity is maximum at 50° for a critical energy of 54 eV. By altering the potential applied to the cylinder *D*, the energy of the electrons can be changed. It is round that the nature of the intensity graph is similar and the spur (maximum intensity) occurs always at 50°.

The results obtained are similar to X-ray diffraction experiments. This experiment shows that the electrons behave like matter-waves as suggested by De-Broglie.

12.23. G.P. Thomson Experiment

The experimental arrangement is shown in Fig. 12.17.

The apparatus is evacuated with the help of evacuation pumps. The electrons from the filament F are accelerated through a potential difference of 50,000 volts by applying a positive potential to the cylinder D. The energy of each electron is 50,000 eV. This beam of electrons is incident on a thin gold foil G and after passing through the foil, it is incident on a photographic plate P. The pattern obtained in the photographic plate consists of concentric circular rings of varying intensity.

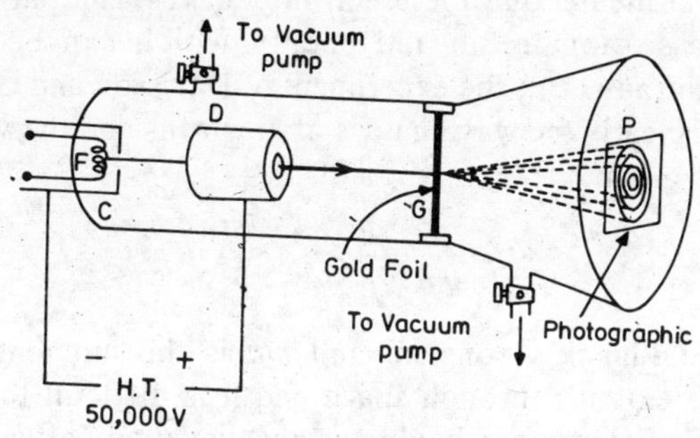

Fig. 12.17

If the electrons behaved only as corpuscles, the electrons passing through the foil should have been scattered through a wide angle. But the pattern obtained is similar to the Laue-pattern obtained with X-rays, which is possible only in the case of waves. Hence electrons also behave as waves.

Example 12.8. *Calculate the De-Broglie wavelength for a beam of electrons whose enegry is* 100 eV. *Take* $h = 6{\cdot}6 \times 10^{-34}$ joule–second

and $m = 9{\cdot}1 \times 10^{-31}$ kg.

Energy $E = \frac{1}{2} mv^2 = 100 \text{ eV} = 100 \times 1{\cdot}6 \times 10^{-19}$ joule

$$v^2 = \frac{2E}{m} \quad \text{or} \quad v = \sqrt{\frac{2E}{m}}$$

Momentum, $mv = m\sqrt{\frac{2E}{m}} = \sqrt{2mE}$

$$\lambda = \frac{h}{mv} = \frac{h}{\sqrt{2mE}} \text{ metres.}$$

Here $h = 6{\cdot}6 \times 10^{-34}$ joule–second

$m = 9{\cdot}1 \times 10^{-31}$ kg

$$\therefore \qquad \lambda = \frac{6{\cdot}6 \times 10^{-34}}{[2 \times 9{\cdot}1 \times 10^{-31} \times 100 \times 1{\cdot}6 \times 10^{-19}]^{1/2}}$$

$$= 1{\cdot}23 \times 10^{-10} \text{ metre}$$

$$= \mathbf{1{\cdot}23\ \text{Å}}$$

12.24. Heisenberg's Uncertainty Principle

In the experiments for the determination of specific charge of an electron, it was shown that electron gets deflected by the application of electric and magnetic fields. It means that the electron behaves as a particle and it has definite, mass, momentum and energy which can be measured with desired accuracy. But the experiment of Davisson and Germer, and Thomson showed clearly the wave nature of electrons and the wavelength of the electron is given by

$$\lambda = \frac{h}{mv} = \frac{h}{p}.$$

Here h is the Planck's constant and mv is the momentum of the electron. A wave extends through space and it is difficult to locate the exact position of an electron behaving as a wave, at any given instant of time.

This dual behaviour of electron as a particle and as a wave presents difficulty in locating the exact position and momentum at the same time. This difficulty is overcome according to *Heisenberg's uncertainty principle.*

The principle states that the exact position and momentum of a particle (say electron) cannot be determined simultaneously with a desired accuracy. Taking Δx as the error in determining its position and Δp the error in determining its momentum at the same instant, these quantities are related as follows :

$$\Delta x \, \Delta p \sim \frac{1}{2} \times \frac{h}{2\pi} \qquad \ldots (1)$$

The product of the two errors is approximately of the order of Planck's constant.

Multiplying and dividing the left hand side of equation (1) by v, the velocity of the particle,

$$\frac{\Delta x}{v} \Delta pv \sim \frac{1}{2} \times \frac{h}{2\pi}$$

or $$\Delta t \, \Delta E \sim \frac{1}{2} \times \frac{h}{2\pi} \qquad \ldots(2)$$

Here ΔE represents the uncertainty in the measurement of the energy of a particle and Δt the uncertainty in the measurement of time.

In equation (1), the product Δx, Δp is of the order of $h/2\pi$. If Δx is small, Δp will be large and *vice versa*. It means that if one quantity is measured accurately, the other quantity becomes less accurate. A similar argument holds good for $\Delta E \, \Delta t$ also.

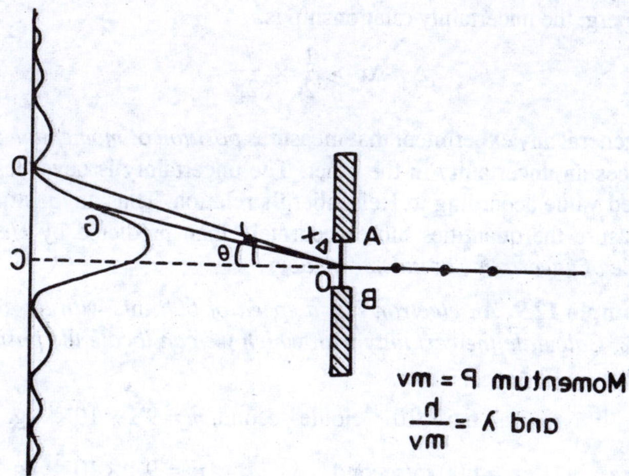

Fig. 12.18.

Consider a beam of electrons travelling in the direction shown in Fig. 12.18. The slit *AB* of width Δx is perpendicular to the path of the electrons. Before entering the slit, the electron has a definite momentum $p = mv$. After passing through the slit the electron gets diffracted and acquires a momentum along *OG*. The angular deflection θ depends upon the component of the momentum parallel to the slit. *i.e.* $\Delta p = p \sin\theta = p\,\theta$ for small angular deflection. The angle θ_0 corresponding to the direction of first minimum of the diffraction pattern is

$$\theta_0 = \frac{\lambda}{\Delta x}, \text{ for small value of } \theta_0$$

or $$\Delta x = \frac{\lambda}{\theta_o}$$

$$\therefore \qquad \Delta p\, \Delta x \sim p\lambda \times \frac{\theta}{\theta_0}$$

But $\qquad p\lambda = h$

$$\therefore \qquad \Delta p\, \Delta x \sim h \frac{\theta}{\theta_0}$$

Taking, $\qquad \theta = \theta_0$ approximately,

$$\therefore \qquad \Delta p\, \Delta x \sim h$$

The probable deflection θ of the electron is less than θ_o and according to Heisenberg, the uncertainty relationship is,

$$\Delta p\, \Delta x \geqslant \frac{1}{2} \times \frac{h}{2\pi}$$

In general any experiment that measures *position* or *momentum* accurately, introduces an uncertainty in the other. The uncertainty is never less than the predicted value according to Heisenberg's relation. Thus any instrument cannot measure the quantities more accurately than predicted by *Heisenberg's principle of uncertainty* or *indeterminacy.*

Example 12.9. *An electron has a speed of 600* m/s *with an accuracy of 0·005 %. Calculate the certainty with which we can locate the position of the electron.*

$$h = 6{\cdot}6 \times 10^{-34} \text{ joule-second, } m = 9{\cdot}1 \times 10^{-31} \text{ kg.}$$

Here $\qquad v = 600$ m/second, $\qquad m = 9{\cdot}1 \times 10^{-31}$ kg.

$$h = 6{\cdot}6 \times 10^{-34} \text{ joule-second}$$

Momentum of electron

$$= mv = 9{\cdot}1 \times 10^{-31} \times 600 \text{ kg-m/s}$$

$$\Delta p = \left(\frac{0{\cdot}005}{100}\right) mv$$

$$= (5 \times 10^{-5} \times 9{\cdot}1 \times 10^{-31} \times 600) \text{ kg-m/s}$$

$$\Delta x\, \Delta p \geqslant \frac{1}{2}\, \frac{h}{2\pi} > \frac{h}{4\pi}$$

$$\Delta x \geqslant \frac{h}{4\pi \times \Delta p}$$

$$\Delta x \geqslant \frac{6{\cdot}6 \times 10^{-34}}{4\pi \times 5 \times 10^{-5} \times 9{\cdot}1 \times 10^{-31} \times 600}$$

$$\Delta x > 0{\cdot}001923 \text{ m}$$

$$\Delta x > \mathbf{1{\cdot}923 \text{ mm}}$$

.If the momentum of the electron is determined to the given accuracy, then the position of the electron cannot be measured to an accuracy, less than 2 mm. It means the concept of an electron as a tiny particle does not hold good.

Example 12.10. *A bullet of mass* 0·025 kg *is moving with a speed of* 400 m/s. *The speed is measured accurate to 0·02%. Calculate the certainty with which we can locate the position of the bullet. Take* $h = 6{\cdot}6 \times 10^{-34}$ joule second.

Here $v = 400$ m/s, $m = 0{\cdot}025$ kg.

Momentum of the bullet,

$$p = mv$$

$$= 0{\cdot}025 \times 400 = 10 \text{ kg-m/s}$$

$$\Delta p = \left(\frac{0{\cdot}02}{100}\right) \times p$$

$$\Delta p = (2 \times 10^{-4})\,(10) = 2 \times 10^{-3} \text{ kg-m/s}$$

$$\Delta x\, \Delta p \geq \frac{h}{4\pi}$$

$$\Delta x \geq \frac{h}{4\pi \times \Delta p}$$

$$\Delta x \geq \frac{6{\cdot}6 \times 10^{-34} \times 7}{4 \times 22 \times 2 \times 10^{-3}}$$

$$\Delta x \geq \mathbf{2{\cdot}625 \times 10^{-32} \text{ m}}.$$

Here Δx is a very small quantity and cannot be measured. Therefore, in heavy bodies the particle concept prevails.

12.25. Luminiferous Ether

Newton referred to the fixed stars as the frame of reference to explain his laws. A search for a new frame of reference which is more fixed than the stars was carried on for a number of years. According to Maxwell, light waves are electromagnetic waves. Light waves are transverse waves and sound waves are longitudinal waves. Polarization confirmed that light waves are transverse waves.

A material medium is a necessity for the propagation of a wave. For light waves a material medium called 'luminiferous ether' was supposed to fill the free space.

Transverse waves require shearing forces and these forces can occur in solids only. It means that ether must be a rigid solid filling the whole space. The velocity of wave propagation depends on the elasticity of the medium. As the velocity of light is very high, the elasticity of ether must be very high. Thus the whole free space must be filled up with this medium called ether which is difficult to conceive.

If light is propagated through the ether medium, then in optical experiments, one would expect a change in velocity (drift) depending upon the direction in which the light is propagated with respect to the motion of the apparatus. Considering earth to be moving in its orbit around the sun with a velocity of 3×10^4 m/s and the velocity of light 3×10^8 m/s the ratio between the two is 10^{-4}. Earth can be considered to be moving through the stationary ether. This led to the experiments of Michelson and Morley in 1881 – 1887 to detect the luminiferous ether.

12.26. Michelson Morley Experiments

Light from a monochromatic source S is rendered parallel with the lens L. It is divided into two portions by the half silvered plate A (Fig. 12.19). One portion of the beam travels to the mirror M_1 and is reflected back.

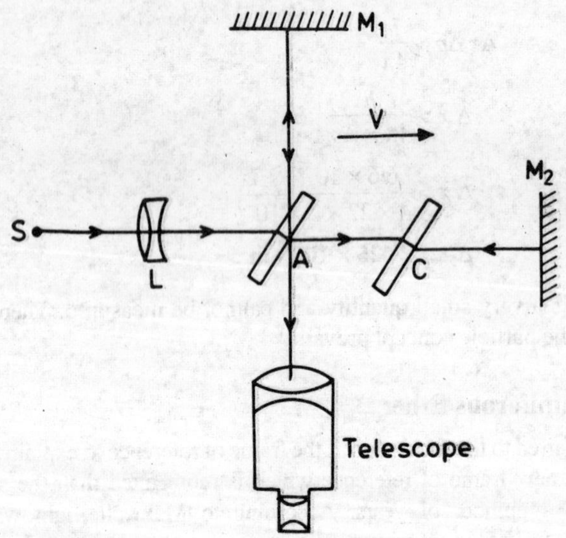

Fig. 12.19.

The other portion of the beam travels towards M_2 and is reflected back. The two reflected beams interfere and the interference fringes are viewed with the help of a telescope.

The apparatus is arranged to move along the direction of the earth's orbit round the sun. The speed of movement of the apparatus is equal to v *i.e.*, speed of the earth in its orbit. The whole apparatus is floated on mercury and can be adjusted such that it is always moving along the direction of the earth's orbit round the sun. Ether is assumed to be stationary.

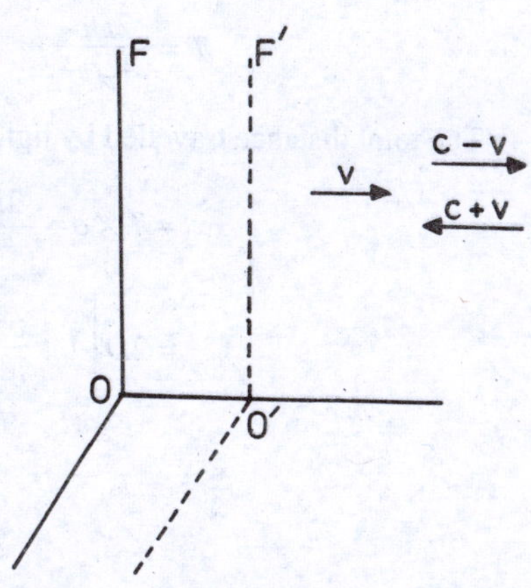

Fig. 12.20.

According to the Galilean frame of reference, F is a fixed frame corresponding to the ether medium (Fig.12.20). F' is the frame of reference moving with a velocity v in the direction of the movement of the apparatus *i.e.* in the direction of the movement of the earth in its orbit round the sun. The velocity of light in the direction of the movement of the frame $F' = c - v$ and in the opposite direction it is $= c + v$. Let T_1 be the time taken by light to travel from A to M_2' and T_2 the time taken from M_2' to A' (Fig.12.21).

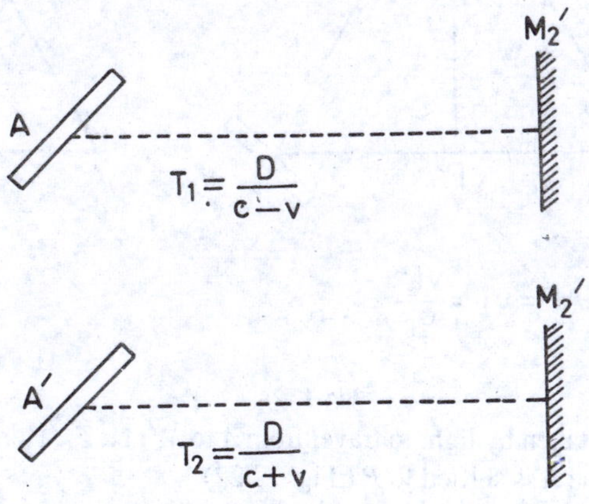

Fig. 12.21.

Here $D = AM_1 = AM_2$

The total time taken by light (to and fro)

$$T = T_1 + T_2 = \frac{D}{c - v} + \frac{D}{c + v}$$

$$T = \frac{2Dc}{c^2 - v^2}.$$

The total distance travelled by light

$$x_1 = T \times c = \frac{2Dc^2}{c^2 - v^2}$$

$$= 2D\left[1 + \frac{v^2}{c^2}\right] \qquad \ldots (1)$$

(neglecting the higher powers)

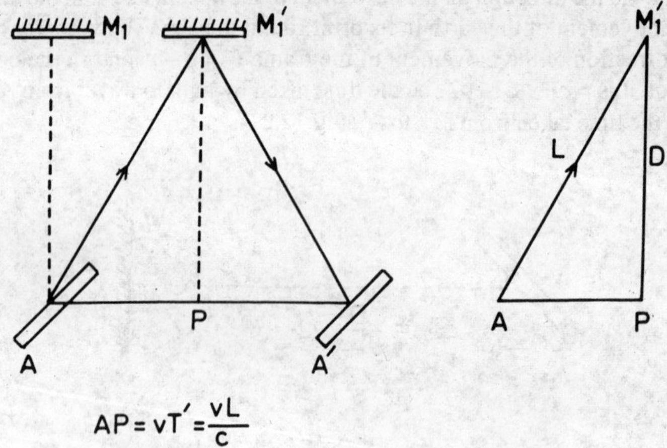

Fig. 12.22.

Let the time taken by light to travel from A to M'_1 be T'. Here M_1 is shifted to M'_1 in the time A is shifted to P (Fig. 12.22).

The distance $\qquad AP = vT'$

But $\qquad T' = \frac{L}{c}$

$$AP = \frac{vL}{c}$$

$$L^2 = D^2 + \left[\frac{vL}{c}\right]^2$$

$$L^2\left[1 - \frac{v^2}{c^2}\right] = D^2$$

$$L^2 = \frac{D^2}{\left[1 + \frac{v^2}{c^2}\right]}$$

$$L = D\left[1 + \frac{v^2}{2c^2}\right]$$

$\therefore$ Total distance travelled by light in time T in going from A to A'_1 and back to A'

$$= 2L = 2D\left[1 + \frac{v^2}{2c^2}\right]$$

$$x_2 = 2D\left[1 + \frac{v^2}{2c^2}\right] \qquad \ldots (2)$$

From equations (1) and (2), the path-difference,

$$x_1 - x_2 = 2D\left[1 + \frac{v^2}{c^2}\right] - 2D\left[1 + \frac{v^2}{2c^2}\right]$$

$$= \frac{Dv^2}{c^2} \qquad \ldots (3)$$

If the apparatus is turned through 90°, the path-difference will be $= \frac{Dv^2}{c^2}\cdot$ The displacement in the interference fringes is $\frac{2Dv^2}{c^2}\cdot$

The displacement expected is about 0·04 of a fringe width.

But in this experiment no displacement of the fringes was observed. This shows that this experiment is a negative experiment. The experiment of Michelson and Morley was repeated with certain modifications by a number of workers but no drift was observed. Even with monochromatic light from a modern laser, the drift has not been detected.

This negative result suggests that the *velocity of light is invariable and remains constant in all directions*. Moreover the effects of ether are undetectable. This has been verified by Doppler's principle in light which shows that Doppler's principle in light is symmetric *i.e.*, the speed of light is independent of the motion of the source or the observer.

Thus all attempts to make ether as a fixed frame of reference failed.

12.27. Einstein's Theory of Relativity

Einstein in 1905 propounded the special theory of relativity and in 1915 proposed the general theory of relativity. The special theory deals with the problems in which one frame of reference moves with a *constant linear velocity* relative to another frame of reference. The general theory of relativity deals with problems in which one frame of reference is accelerated with respect to another frame of reference. Einstein assumed that all observers will notice that their motion in space will make no difference in the velocity of light with respect to them. He assumed that a fixed frame of reference cannot be located. All laws must be so stated that they are applicable in any frame of reference.

According to Einstein's special theory of relativity :

(1) The observations on any particular frame of reference are not preferred to those on the other frames of reference. It means that the laws of physics are applicable equally well for an observer in the frame of reference A moving with a constant linear velocity with respect to an observer in another frame of reference B and *vice versa*.

(2) The velocity of light in free space is constant and is independent of the motion of the observer in any frame of reference.

Einstein proved the following facts based on his theory of relativity. Let v be the velocity of the spaceship with respect to a given frame of reference. The observations are made by an observer in that reference frame.

1. All clocks on the spaceship will go slow by a factor.

$$\sqrt{1-\frac{v^2}{c^2}}$$

2. All objects on the spaceships will have contracted in length by a factor.

$$\sqrt{1-\frac{v^2}{c^2}}$$

3. The mass of the spaceship increases by a factor.

$$\left[\frac{1}{\sqrt{1-\frac{v^2}{c^2}}}\right]$$

4. Mass and energy are interconvertible.

$$E = mc^2$$

5. The speed of a material object can never exceed the velocity of light.
6. If two objects A and B are moving with velocities u and v with respect to each other along the x-asis, the relative velocity of A with respect to B is given by

$$V_x = \frac{u - v}{1 - \frac{uv}{c^2}}$$

Here u and v are comparable with the value of c.

Example 12.11. *Two spaceships A and B are moving in opposite directions each with a speed of* 0·90 c. *Find the relative velocity of B with respect to A.*

Here $u = 0{\cdot}9\ c$

$$v = -0{\cdot}9\ c$$

$$V_x = \frac{u - v}{1 - \frac{uv}{c^2}} = \frac{0{\cdot}9c + 0{\cdot}9c}{1 + \frac{0{\cdot}9c \times 0{\cdot}9c}{c^2}}$$

$$= \mathbf{0{\cdot}994\ c.}$$

Example 12.12. *Two photons A and B are moving in opposite directions each with a speed of c. Calculate the relative velocity of the photon A with respect to B.*

Here $u = c,\ v = -c$

$$V_x = \frac{u - v}{1 - \frac{uv}{c^2}} = \frac{c + c}{1 + \frac{c \times c}{c^2}} = \mathbf{c.}$$

Example 12.13. *An electron is moving with a speed of* 0·85 c *in a direction opposite to that of a moving photon. Calculate the relative velocity of the electron with respect to the photon.*

Here, $u = 0{\cdot}85\ c,\ v = -c$

$$V_x = \frac{u - v}{1 - \frac{uv}{c^2}}$$

$$V_x = \frac{0{\cdot}85c + c}{1 + \frac{0{\cdot}85c \times c}{c^2}} = \mathbf{c}$$

Note. The above examples show that the velocity of any object cannot exceed the velocity of light.

12.28. Lorentz Transformations

Consider two frames of reference A and B (Fig.12.23). A is fixed and B is moving with a constant speed v along the direction of the x-axis. Initially both the frames have the same origin of coordinates. After a time t the frame of reference B has moved a distance $OO' = vt$. For the point P in space, the coordinates are (x, y, z) with reference to the frame A and (x', y', z') with reference to the frame B.

According to Galilean frame of reference

$$x' = x - vt$$

$$y' = y$$

$$z' = z$$

$$t' = t \qquad \ldots (1)$$

Differentiating the first equation in (1)

$$\frac{dx'}{dt} = \frac{dx}{dt} - v \qquad \ldots (2)$$

$$c' = c - v \qquad \ldots (3)$$

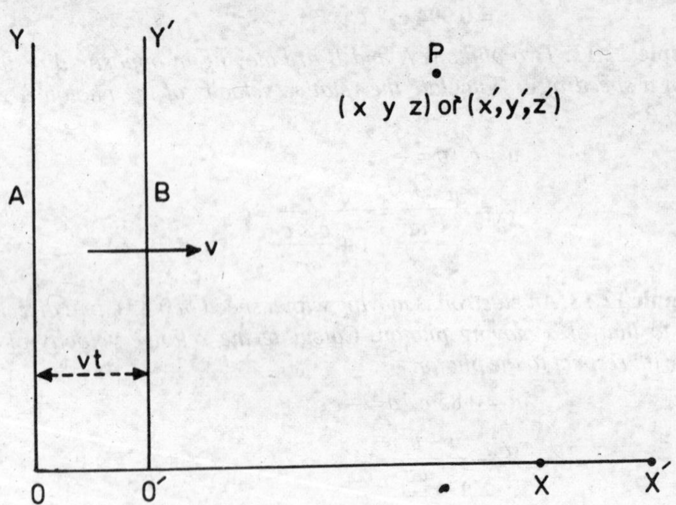

Fig. 12.23.

It means that if a person is moving in a spaceship, the speed of the passing light will be $(c - v)$. But all attempts to show that the velocity of light changes with the motion of the observer have failed. The velocity of light remains constant in free space, according to the postulates of the special theory of relativity. The equation $x' = x - vt$ is in accordance with the ordinary laws of mechanics. The new transformation for the x-coordinate must be similar to this equation when the value of v is extremely small as compared to the velocity of light, c. The simplest possible form of this equation can be

$$x' = k(x - vt) \quad \ldots (4)$$

Here k depends only on the value of v and does not depend upon the values of x and t. Equations (4) is linear and x' has only one value for a given value of x.

According to the first postulate of the special theory of relativity, observations made in the frame of reference B must be identical to those made in A, except for a change in the sign of v and having the same value for the constant of proportionality k.

$$x = k(x' + vt') \quad \ldots (5)$$

As the relative motion of A and B is confined to only the x direction,

$$y' = y \quad \ldots (6)$$

$$z' = z \quad \ldots (7)$$

But this equality does not hold good for t and t'.

Equations (4) and (5) will not hold, if $t' = t$.

The value of x' from equation (4) is substituted in equation (5). This gives

$$x = k[k(x - vt) + vt']$$

$$x = k^2(x - vt) + kvt' \quad \ldots (8)$$

$$kvt' = x - k^2(x - vt)$$

$$t' = \frac{x - k^2 x + k^2 vt}{kv}$$

$$t' = \frac{k^2 vt}{kv} + \frac{x(1 - k^2)}{kv}$$

$$t' = kt + \left(\frac{1 - k^2}{kv}\right)x \quad \ldots (9)$$

To find the value of k, consider two reference frames A and B. The spaceship in reference frame A measures the time t and the spaceship in reference frame B measures the time t', for a flash of light. The x-coordinates for both the ships will be

$$x = ct \quad \ldots (10)$$

$$x' = ct' \quad \ldots (11)$$

It is so because the value of c must remain constant in both the frames of reference according to the postulate of relativity.

Substituting the values of x' and t' in equation (11)

$$k(x - vt) = c\left[kt + \left(\frac{1 - k^2}{kv}\right)x\right]$$

$$kx - kvt = ckt + cx\left(\frac{1-k^2}{kv}\right)$$

$$kx - cx\left(\frac{1-k^2}{kv}\right) = ckt + kvt$$

$$x\left[k - c\left(\frac{1-k^2}{kv}\right)\right] = ctk\left[1 + \frac{v}{c}\right]$$

$$x = \frac{ctk\left[1 + \dfrac{v}{c}\right]}{k\left[1 - c\left(\dfrac{1-k^2}{k^2 v}\right)\right]}$$

$$x = ct\left[\frac{1 + \dfrac{v}{c}}{1 - \left(\dfrac{c}{v}\right)\left(\dfrac{1}{k^2} - 1\right)}\right] \quad \ldots (12)$$

This value of x must be equal to the value of x in equation (10).

$$\therefore \quad ct = ct\left[\frac{1 + \dfrac{v}{c}}{1 - \left(\dfrac{c}{v}\right)\left(\dfrac{1}{k^2} - 1\right)}\right] \quad \ldots (13)$$

$$1 - \left(\frac{c}{v}\right)\left(\frac{1}{k^2} - 1\right) = 1 + \frac{v}{c}$$

$$-\left(\frac{c}{v}\right)\left(\frac{1}{k^2} - 1\right) = \frac{v}{c}$$

$$1 - \frac{1}{k^2} = \frac{v^2}{c^2}$$

$$\frac{1}{k^2} = 1 - \frac{v^2}{c^2}$$

$$k^2 = \frac{1}{1 - \dfrac{v^2}{c}}$$

$$k = \frac{1}{\sqrt{1 - \frac{v^2}{c^2}}} \quad \ldots (14)$$

The value of k is in accordance with the experimental result. When the value of v is extremely small in comparison to c, he value of k is practically equal to 1.

When the value of k is substituted in equations (4), (6), (7) and (9) we get

$$x' = k(x - vt)$$

$$x' = \frac{x - vt}{\sqrt{1 - \frac{v^2}{c^2}}} \quad \ldots (15)$$

$$y' = y \quad \ldots (16)$$

$$z' = z \quad \ldots (17)$$

and

$$t' = kt + \left(\frac{1 - k^2}{kv}\right)x$$

$$t' = kt + \frac{x}{kv} - k\left(\frac{x}{v}\right)$$

$$t' = kt + \frac{x}{v}\left[\frac{1}{k} - k\right]$$

$$t' = kt + k\frac{x}{v}\left(\frac{1}{k^2} - 1\right)$$

$$\frac{t + \frac{x}{v}\left(1 - \frac{v^2}{c^2} - 1\right)}{\sqrt{1 - \frac{v^2}{c^2}}}$$

$$t' = \frac{t - \frac{vx}{c^2}}{\sqrt{1 - \frac{v^2}{c^2}}} \quad \ldots (18)$$

Equations (15), (16), (17) and (18) are called Lorentz transformations :

These equations give the conversion of the measurements of time and space made in the stationary frame A to their counterparts in the moving frame B.

The inverse Lorentz transformation of measurements of time and space made in the frame B to their counterparts in frame A can be written as follows :

$$x = \frac{x' + vt'}{\sqrt{1 - \frac{v^2}{c^2}}} \quad \ldots (19)$$

$$y = y' \quad \ldots (20)$$

$$z = z' \quad \ldots (21)$$

and

$$t = \frac{t' + \frac{vx'}{c^2}}{\sqrt{1 - \frac{v^2}{c^2}}} \quad \ldots (22)$$

12.29. Lorentz-Fitzerald Contraction

Measurements of space and time are not absolute but depend on the relative motion of the observer and the observed objects.

Consider a rod of length L_0 parallel to the x-axis and having the coordinates x_1 and x_2 in the reference frame A. An observer in the reference frame A measures the length of the rod as $L_0 = x_2 - x_1$.

Consider a second reference frame B moving with a velocity v along the x-axis, with respect to the reference frame A. An observer in the reference frame B measures the end coordinates of the rod as x_1' and x_2'. The length L as observed by the observer in B is

$$L = x_2' - x_1'$$

The relation between x_1 and x_1', and also between x_2 and x_2' according to inverse Lorentz transformation will be,

$$x_1 = \frac{x_1' + vt'}{\sqrt{1 - \frac{v^2}{c^2}}}$$

$$x_2 = \frac{x_2' + vt'}{\sqrt{1 - \frac{v^2}{c^2}}}$$

$$L_0 = x_2 - x_1$$

$$L_0 = \frac{x_2' + vt'}{\sqrt{1 - \frac{v^2}{c^3}}} - \frac{x_1' + vt'}{\sqrt{1 - \frac{v^2}{c^2}}}$$

$$L_0 = \frac{x_2' - x_1}{\sqrt{1 - \frac{v^2}{c^2}}}$$

$$L_0 = \frac{L}{\sqrt{1 - \frac{v^2}{c^2}}}$$

$$L = L_0\sqrt{1 - \frac{v^2}{c^2}} \qquad \ldots(1)$$

This shows that the length of a stationary object with respect to an observer in motion is shorter than the length measured by the observer at rest. Similarly, when the object is in motion with respect to a stationary observer, again the object is shortened in length. This relativistic contraction works both ways, *i.e.* whether the object is in motion or the observer is in motion. This is called Lorentz-Fitzerald contraction.

Lorentz-Fitzerald contraction is appreciable only when the velocity v is comparable to the velocity of light c.

Suppose $\qquad v = 0{\cdot}6\,c$

$$L = L_0\sqrt{1 - \frac{v^2}{c^2}}$$

$$L = L_0\sqrt{1 - 0{\cdot}36}.$$

$$L = 0{\cdot}8\,L_0$$

The contraction in length $= L_0 - L = 0{\cdot}2\,L_0$

i.e., contraction in length is 20%.

When the velocity of the body negligibly small as compared to c, the contraction in length is negligible and

$$L = L_0$$

Suppose $\qquad v = 0{\cdot}01c$

$$L = L_0\sqrt{1 - (10^{-4})}$$

$$L = 0{\cdot}9999\,L_0$$

i.e., $\qquad L \approx L_0$

Due to this reason, the shortening in length cannot be visually observed for objects moving on the surface of earth. However, Lorentz-Fitzerald contraction has been verified by experimental measurement.

12.30. Time Dilation

According to the special theory of relativity, time intervals are also affected by relative motion between two frames of reference under consideration. Suppose the reference frame A is stationary and reference frame B is moving with a velocity v along the x-axis.

Suppose at any instant, the two reference frames coincide at $t = t' = 0$. The observer in B notes the time at any instant in his clock as t_1' whereas an observer in A notes the time as t_1.

According to inverse Lorentz transformation,

$$t_1' + \frac{\left(\frac{vx'}{c^2}\right)}{\sqrt{1 - \frac{v^2}{c^2}}} \qquad \ldots (1)$$

Let t_2' and t_2 be the times measured by observers in B and A respectively at the same instant.

Here

$$t_2' + \frac{\left(\frac{vx'}{c^2}\right)}{\sqrt{1 - \frac{v^2}{c^2}}} \qquad \ldots (2)$$

Let t_0 be the interval of time as measured by an observer in B and t the interval of time measured by an observer in A.

$$t_0 = t_2' - t_1' \qquad \ldots (3)$$

$$t = t_2 - t_1 \qquad \ldots (4)$$

$$t = \frac{t_2' + \frac{vx'}{c^2}}{\sqrt{1 - \frac{v^2}{c^2}}} - \frac{t_1' + \frac{vx'}{c^2}}{\sqrt{1 - \frac{v}{c^2}}}$$

$$t = \frac{t_2' - t_1'}{\sqrt{1 - \frac{v^2}{c^2}}}$$

$$t = \frac{t_0}{\sqrt{1 - \frac{v^2}{c^2}}} \qquad \ldots (5)$$

It means, an interval of time observed in a moving frame of reference will be less than the same interval of time observed in a stationary frame of reference. It shows that the clocks in the moving spaceships will appear to go slower than the clocks on the surface of the earth. It is called time dilation.

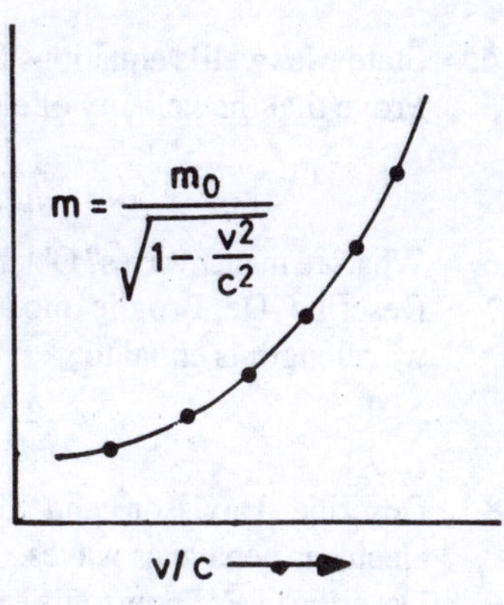

Fig. 12.24.

12.31. Relativity of Mass

According to relativity, the measurements of an object or an event will be different for different observers who are in relative motion. If the rest mass of the body is m_0, then the mass of the body in motion is given by,

$$m = \frac{m_0}{\sqrt{1 - \frac{v^2}{c^2}}}$$

Here v is the velocity of the body in motion and c is the velocity of light.

It shows that $m > m_0$. Thus an object will appear to have more mass while in motion than when it is at rest.

The relative mass of the electron at various speeds is shown in the graph (Fig.12.24). The relative mass of the electron increases enormously when its velocity approaches the velocity of light.

If $v = c$,

$$m = \frac{m_0}{\sqrt{1 - \frac{v^2}{c^2}}}$$

$$m = \frac{m_0}{0} = \infty$$

i.e., an electron moving with the speed of light will have an infinite mass which is not possible. Hence no material particle can have a velocity equal to the velocity of light.

EXERCISES

1. Distinguish between wave velocity and group velocity and obtain expressions for their values.
2. Show that in the case of a non-dispersive medium, phase velocity and group velocity are equal.
3. How are electromagnetic waves produced and detected ?

4. State Maxwell's equations for electromagnetic waves.
5. Prove that the velocity of electromagnetic wave in free space is equal to
$$\frac{1}{\sqrt{\mu_o \varepsilon_o}}.$$
6. What are matter waves? Obtain an expression for wavelength of matter waves.
7. Describe De Broglie model of an atom and show that De Broglie wavelength is equal to
$$\frac{h}{mv}$$
8. Describe Davisson and Germer experiment to show that moving electrons behave as waves.
9. Describe G.P. Thomson's experiment to demonstrate the existence of De Broglie waves.
10. Discuss briefly Hiesenburg's uncertainty principle.
11. Discuss special theory of relativity and show that
$$m = \frac{m_0}{\sqrt{1 - \frac{v^2}{c^2}}}$$
12. Discuss Lorentz-Fitzerald contraction.
13. Show that, according to special theory of relativity
$$t = \frac{t_0}{\sqrt{1 - \frac{v^2}{c^2}}}.$$
14. What is De Broglie's concept of matter waves? Derive an expression for the wavelength of De Broglie wave associated with an electron in Angstrom units, when its energy is given in electron volts. Give the experimental verification of matter waves. *(Delhi, 1976)*
15. Establish the relation between wave velocity and group velocity in a dispersive medium. *(IAS, 1976)*
16. Give a critical account of the Bohr theory of the atom and the role of earlier empirical spectroscopic investigations in providing support to the theory. What are the limitations of this theory? *(IAS, 1975)*
17. Obtain the relation between group and phase velocity in a dispersive medium. *[Delhi (Hons.) 1992]*
18. Find an expression for the velocity of the waves formed on the surface of a liquid under the combined action of gravity and surface tension. Obtain an expression for the minimum velocity of these waves. *(Delhi. 1991)*
19. Distinguish between the gravity waves and ripples. (Delhi, 1991)

20. If μ and μ_g are the refractive index and group refractive index of a material at a certain wavelength λ, prove that

$$\mu_g = \mu - \lambda\left(\frac{d\mu}{d\lambda}\right)$$ *[Delhi (Hons.) 1993]*

21. Write short notes on :

(i) Wave velocity and group velocity
(ii) Maxwell's equations
(iii) Velocity of electromagnetic waves
(iv) Poynting vector
(v) De Broglie waves
(vi) Matter waves
(vii) Electron Diffraction experiments
(viii) Heisenburg's uncertainty principle
(ix) Special theory of relativity
(x) Relativity of mass.

Index